PERSPECTIVES ON
THE COORDINATION OF MOVEMENT

ADVANCES IN PSYCHOLOGY

61

Editors:

G. E. STELMACH

P. A. VROON

NORTH-HOLLAND
AMSTERDAM · NEW YORK · OXFORD · TOKYO

PERSPECTIVES ON THE COORDINATION OF MOVEMENT

Edited by

Stephen A. WALLACE
Department of Kinesiology
University of Colorado
Boulder, Colorado, U.S.A.

1989

NORTH-HOLLAND
AMSTERDAM · NEW YORK · OXFORD · TOKYO

ELSEVIER SCIENCE PUBLISHERS B.V.
P.O. Box 211, 1000 AE Amsterdam, The Netherlands

Distributors for the United States and Canada:
ELSEVIER SCIENCE PUBLISHING COMPANY, INC.
655 Avenue of the Americas
New York, N.Y. 10010, U.S.A.

ISBN: 0 444 88053 4

Transferred to digital printing 2007

To Penny and Makaila

I do not see any way to avoid the problem of coordination and still understand the physical basis of life.

H. H. Pattee

PREFACE

As I am typing this preface, I am amazed how easily certain words and phrases seemingly flow out from my finger tips. To type the word *the* takes absolutely no effort at all even though it requires alternating key presses from first the index finger of my left hand, then the index finger of my right hand, and finally the middle finger of my left hand. Typing this word, which is easy for me to do now after years of practice, requires temporal order, the use of three physiologically distinct components (my fingers) on two different hands, and knowledge of the spatial configuration of the keyboard. Any mix-up on my part of any of these three elements of the task would result in an error. Clearly, some type of coordination of my fingers is needed if I am to type as I was taught in high school.

Equally interesting is my ability to reach and grasp my coffee cup. Vision is clearly helpful to me, but I can come close to finding my cup even with my eyes closed. Of couse, I wouldn't dare try this if the cup were filled to the brim with hot coffee—yet I would be more inclined to try if I knew in advance that the cup were only half filled! I notice that my fingers and thumb open and close down on the cup at the right time no matter how fast I reach. It also doesn't seem to matter where I begin my movement, whether from the keyboard of my personal computer, from the top of my head, which I have just scratched, or after a button press on my remote control terminating the distractions coming from my television set—I successfully grasp my cup regardless of initial conditions. Additional investigation of this feat reveals an apparent coordination among several joints spanning my shoulder, elbow, wrist, and fingers. This everyday task is easy for me to do, but I shudder to think how many computer program statements it would take to instruct a robot of similar complexity to do the same.

Rarely are we required to do only one thing or control only one component at a time in performing motor skills. Indeed, every task I can think of, no matter how trivial or complex, appears to require coordination among many nerve impulses, muscles, and joints. This insight is not new. Bernstein and von Holst were the pioneers in the field of biological coordination who defined and investigated important concepts such as absolute and relative coordination, and synergies. Only recently have researchers undertaken systematic programs to address the issues raised by von Holst and Bernstein. Why? I believe part of the problem has been technical and economical. To study coordination requires that several components of movement be monitored simultaneously over the duration of the activity. Sophisticated, real-time, motion analysis systems capable of monitoring many joint movements, for example, have only recently been developed; and high-speed, high-capacity computers for operating these systems and storing the resulting vast amounts of data have only recently become economically feasible for the scientist. Of course, technological breakthroughs are not the only reason for this relatively sudden outbreak of coordination research. New theoretical developments, some of which are discussed in this volume, also account for the heightened interest. Thus, with more affordable systems, advanced technology, and new theoretical fuel, research in the coordination of movement is on the upswing.

But is there a right way to study coordination? What experimental paradigms are appropriate? Are there laws and principles that the biological system uses to coordinate movement? Do all biological systems—human and otherwise—share these same principles? Is coordination inherited or acquired? Is it a central nervous system, muscular, or mechanical problem? Indeed, what is coordination and how can it be quantified?

This volume represents my attempt to help answer some of these questions by bringing together a collection of conceptual approaches to and empirical investigations of the coordination of movement. It will be evident to the reader that no one discipline monopolizes the study of coordination and that the complete uncovering of its mysteries will no doubt require an interdisciplinary approach. For this reason, I hope this volume will be of interest to students and scientists in many fields, including biology, psychology, engineering and robotics, physical education, physical therapy, kinesiology, and physiology. The volume is intended as a graduate-level text, but advanced undergraduates should find the chapters readable and challenging. Although the volume is technical in places, the authors were asked not to take their jargon for granted and to define terms that may be foreign to those in other fields. For the most part, I believe they have.

The volume is composed of thirteen invited chapters and has been divided into four sections. The first section is entitled "Conceptual Approaches to the Study of Coordination" and contains six chapters from authors who differ in their research strategies. The second section, "Developmental Issues," contains chapters on the motor behavior of infants and young children. The third section concentrates on the "Coordination of Adult Motor Behavior" and includes experimental studies of prehensile, multilimb, and locomotor movements. The last section, "Coordination and Movement Disorder," focuses on the impairment of limb and speech movement.

This volume was never meant to be all-inclusive or completely representative of the different research strategies and experimental approaches that researchers are using to investigate coordination. It is my hope that the volume will, at a minimum, contribute to our appreciation of complex problems of coordination and represent a continuing effort by many serious researchers throughout the world to help solve some of its mysteries.

I have many to thank for their help and support of my effort to bring this volume together. Doug Weeks, my graduate student, and Penny McCullagh helped in correspondence duties and editing of the chapters. I thank Barbara Cooper in the Institute of Cognitive Science, University of Colorado, Bryon Coe and Barbara Miller in the Department of Kinesiology, University of Colorado, and Betty Harvey at the Center for Complex Systems, Florida Atlantic University for typing chapters and correspondence to the authors. Dr. Martha C. Polson, assistant director of the Institute of Cognitive Science, generously made the resources of the Institute available to me. I don't know where I'd be without the help of Janet Grassia, who served as technical editor of the volume. Her enthusiasm kept me going throughout the project. Of course, without support and guidance from North-Holland Publishers, this volume would not have been possible. Finally, my hat goes off to all the contributing authors for their willingness to take the time to prepare the invited chapters. I hope they will forgive me for being a bit overbearing in my editorial duties.

I was on sabbatical at the Center for Complex Systems, Florida Atlantic University while this volume was prepared, and I received financial support from the University of Colorado-Boulder, my home institution, and from grant N00014-88-J-1191 Office of Naval

Research, Perception Science, and grant MH 42900-02 National Institutes of Mental Health, Neurosciences Research Branch, awarded to Professor J. A. S. Kelso. My special thanks to him for a very exciting year at the Center.

Stephen A. Wallace

CONTENTS

CONTRIBUTORS

J. H. ABBS
University of Wisconsin Medical School and University of Wisconsin

S. G. ADAMS
University of Wisconsin-Madison

S. ATHÈNES
CNRS Cognitive Neuroscience Unit

N. P. CONNOR
University of Wisconsin Medical School and University of Wisconsin

J. C. FENTRESS
Dalhousie University

C. von HOFSTEN
Umeå University

J. J. JEKA
Florida Atlantic University

J. A. S. KELSO
Florida Atlantic University

R. D. KENT
University of Wisconsin-Madison

P. N. KUGLER
University of Illinois at Urbana-Champaign

P. V. McDONALD
University of Illinois at Urbana Champaign

F. MÜLLER
University of Wisconsin-Madison

K. M. NEWELL
University of Illinois at Urbana-Champaign

J. G. PHILLIPS
University of Wisconsin-Madison

R. C. SCHMIDT
University of Connecticut and Haskins Laboratories

D. E. SHERWOOD
University of Colorado, Boulder

J. F. SOECHTING
University of Minnesota

E. THELEN
Indiana University

M. T. TURVEY
University of Connecticut and Haskins Laboratories

R. E. A. van EMMERIK
University of Illinois at Urbana-Champaign

A. M. WING
MRC Applied Psychology Unit

D. A. WINTER
University of Waterloo

SECTION 1

CONCEPTUAL APPROACHES TO THE STUDY OF COORDINATION

Perspectives on the Coordination of Movement
S.A. Wallace (Editor)

THE DYNAMIC PATTERN APPROACH TO COORDINATED BEHAVIOR: A TUTORIAL REVIEW[§]

John J. JEKA and J. A. S. KELSO*

Center for Complex Systems and Department of Psychology Florida Atlantic University

ABSTRACT

We elaborate, in tutorial fashion, a theoretical framework that originated from observations of phase transitions in human movement coordination. Based upon theories of self-organization and pattern formation in dissipative dynamical systems (in particular, Haken's [1983] synergetics), this theoretical but operational language is aimed at understanding the behavioral patterns produced by biological systems. The key concepts are the identification of *collective variables* (or *order parameters*) for behavioral patterns and the determination of their *dynamics* obtained through study of the *stability* (and *loss of stability*) of behavioral patterns. Methods for calculating stability measures are defined and discussed (e.g., fluctuations, relaxation times, time scale relations). Such measures, when obtained in experiment, yield results that agree with theoretical predictions. *Behavioral information* is shown to contribute to the pattern dynamics, attracting the system toward the (e.g., environmentally specified, intended, learned) behavioral pattern. Such behavioral information is

---

*Address correspondence to: J. A. S. Kelso, Center for Complex Systems, Bldg. MT-9, Florida Atlantic University, Boca Raton, FL 33431, U.S.A.

[§]This research is supported by a NIMH (Neurosciences Research Branch) grant MH42900-01 and contract N00014-88-J-1191 from the U.S. Office of Naval Research.

defined in the same space as the collective variables that characterize the patterns and thus is meaningful and specific to biological functions or tasks. Although *dynamic pattern theory* (e.g., Kelso & Schöner, 1987; Schöner & Kelso, 1988a) was formulated in the context of movement coordination, other experimental systems (e.g., speech), other types of behavioral patterns (e.g., locomotory gaits, action-perception patterns), and other levels of description (e.g., neuronal activity) are accessible to this level-independent approach.

PROLOGUE

This chapter is dedicated to the genius of the behavioral physiologist Erich von Holst, whose unique contributions to the understanding of coordinated behavior anticipated current advances in the behavioral and brain sciences. As a result of his extensive comparative studies of locomotion (2 miles of tracings!), von Holst (1939/1973, pp. 119-120) synthesized the following rules, paraphrased for present purposes:

1. Only a certain proportion of the extremely wide range of behavioral forms is actually realized. The ones observed are distinguished from others by their greater stability.

2. This stability is expressed in the fact that with smooth or gradual alteration of internal or external conditions, periodic forms maintain themselves until a critical limiting condition is reached. Transference to another equilibrium relationship occurs—usually abruptly—which is then maintained over a particular range of conditions.

3. The stability that characterizes the periodic forms as a whole does not apply to individual temporal subdivisions, in which disequilibrium states are more likely to occur. These disequilibria are exactly balanced within the temporal unit of the entire period.

4. There is a general tendency towards transference to equilibrium states of ever-increasing stability. The degree of stability increases with the simplicity of the frequency relation-

ships. Increasing degree of complexity is accompanied by decreasing stability.

As revealed in the present chapter, von Holst's rules can now be cast in a theoretical language that has evolved over the last decade (the mathematical concepts and tools of nonlinear dynamical systems), leading to testable predictions for specific experimental model systems and deeper insights into the nature of coordinated behavior.

The human brain possesses 10^{14} neurons and neuronal connections, is influenced by hundreds of active chemicals, and displays highly complex patterns of electrical activity. New concepts and tools are needed if the inherent complexity of the most complex system of all—the brain and its relation to behavior—is to be understood. Presently, there is a huge void between what a single neuron does (which we know a lot about) and what many of them do when they cooperate. Why is it crucial to discover the principles of coordination among large numbers of interacting components? The answer is that this cooperative behavior lies at the root of understanding ourselves and the world we live in—how we touch, see, hear, plan, and act. Such fundamental behavioral functions depend on temporally coherent functional units distributed throughout different regions of the brain and are not elucidated by standard methods.

When we use the word *how* in this chapter, we mean the discovery or identification of laws or principles of coordination at a chosen level of observation (e.g., kinematic, muscular, neuronal). Given that the nervous system is high-dimensional, as is the environment within which nervous systems have evolved, laws of coordination are expected to be instantiated at numerous scales of description. Further, it is possible that the long sought for link between neuronal activities (microscopic events) and behavior (macroscopic events) resides in collective effects (pattern formation) at the microscopic level that create macroscopic order (and disorder). Thus, we view the problem of coordination as continuous with efforts to understand pattern formation in complex systems with many interacting components, in particular, Haken's (1975; 1983) *Synergetics*, a theory of self-organization in nonequilibrium systems. In synergetics, methods have been found to compress system complexity, for example, in various physical, chemical, and biochemical systems that contain many degrees of freedom, to only one or a few degrees of freedom, the so-called order parameters, whose dynamics (equations of motion) are low-dimensional (Haken's slaving principle). The beauty of the resulting dynamics, which are in general nonlinear, is that they give rise to complex behavioral patterns, including multistability, multiple patterns, flexibility, and even deterministic chaos. Thus, the two seemingly diametrically opposed views in science of surface simplicity arising from deep complexity and surface complexity arising from deep simplicity (Yates, 1987) are both, in fact, part and parcel of nature's design for complex systems (Kelso, 1988).

Of course, the amount of information necessary to describe the individual states of neurons and muscles is very large, and ways must be found to select the relevant quantities to compress the amount of information (see also Haken, 1987). In the case of large-scale neuronal systems like the brain, this is difficult to do if we treat the brain as a general purpose machine capable of producing arbitrary outputs to arbitrary inputs. An alternative strategy, exemplified here, is to treat the brain more as a "special purpose device" (Runeson, 1977) that temporarily self-organizes for particular tasks (e.g., Kelso & Scholz, 1985; Schöner & Kelso, 1988a; Sejnowski, Koch, & Churchland, 1988). Many neurons, muscles, and joints must cooperate in the performance of behavioral functions. Evolving patterns of activity among these components may best be understood with respect to their functional significance for the organism. Thus, we argue here, it is *when the nervous system is involved in performing certain behavioral tasks that one sees it "living" in the low-dimensional space of order parameters.* This is where the laws of coordinated behavior lie.

Elsewhere, building on the concepts and tools of synergetics, we have elaborated an operational approach to biological coordination that embraces both theory and experiment (Kelso & Scholz, 1985; Kelso & Schöner, 1987, 1988; Kelso, Schöner, Scholz, & Haken, 1987; Schöner & Kelso, 1988a, 1988b, 1988c). Rather than describe this "dynamic pattern" theory again, here we adopt a more tutorial attitude intended to (a) communicate the essentials of the approach, which involves a synergy between theory, computation, and experiment; and (b) demonstrate the broader significance of the approach for understanding coordination in different experimental systems and at different levels of description. Moreover, we show that dynamical laws are fundamental, in the sense of the need for their prior identification, if certain *essentially* biological and psychological functions such as learning, adaptation, and intentional behavioral change are to be better understood. Because the language we use may be new to students of motor coordination, we present the essentials in a series of questions and answers, a kind of dialogue. This language, we stress, is not at all a question of neologisms but rather emphasizes the operational character of the approach, which requires that all theoretical constructs must explicitly relate to experiment. In this way, we think, scientists who study coordination, the collective behavior of many interacting components, at different levels of description may communicate in an unambiguous fashion. Relatedly, as we shall show, the language of nonlinear dynamics provides a way of linking levels of description and many different phenomena.

Q. *In any given experimental system, how do you identify the relevant degrees of freedom?*

A. In biology, we don't know what the relevant degrees of freedom are, a priori. In this sense, engineering or robotics approaches are not especially helpful. The engineer designs the system and thus can explicitly define the degrees of freedom in terms of the type of action possible for a given rigid segment. Depending upon the motion required and the number of orthogonal planes in which the motion is executed, the degrees of freedom for a joint may vary from one to three. It is very tempting to view the body as a collection of mechanical linkages in which one rigid part is connected to another with various restrictions on the motions possible.

As the name implies, an aim of dynamic pattern theory is to identify the degrees of freedom corresponding to patterns, where the word *pattern* is viewed always in terms of a particular function or task. These patterns are not fixed by the conventions of mechanics; rather, they are flexibly assembled in order to satisfy certain boundary or task conditions. In speech, for example, there is good evidence for certain constriction points (e.g., the closing of the lips, preserving a tongue-palate relationship) that are crucial if a given sound is to be communicated (e.g., a /b/, a /p/, an /f/, or a /z/). The *relevant collective* variable then, around which the many components (e.g., jaw, lips, tongue, velum, and pharynx) are self-organized, is a task- or sound-specific constriction point (Abbs, Gracco, & Cole, 1984; Kelso, Tuller, & Fowler, 1982; Kelso, Tuller, Vakaitos-Bateson, & Fowler, 1984; see also Saltzman & Kelso, 1987).

A key to the precise definition of degrees of freedom corresponding to patterns is to find phase transitions, that is, situations in which the system's behavior changes qualitatively. As one varies a task dimension (in psychology we might say "manipulates an independent variable," although, as we shall see, that language is not appropriate for a variety of reasons), many measurable quantities may change smoothly or stay the same. Qualitative change, however, allows one to clearly distinguish one pattern from another and enables one to specify which dimension of the pattern is relevant. In addition, differential effects of the transition make it possible to study the relative stability of different patterns. (For another approach, which, however, does not study transitions as a tool for understanding coordinated movement, see Kugler & Turvey, 1987.) From the dynamic pattern view, the discovery of a phase transition enables one to identify the *order parameter*, or collective variable, corresponding to the pattern itself, and the *control parameter* or *parameters* that lead the system through these patterns. Control parameters, in the dynamic pattern approach, are *unspecific* to the resulting patterns; they carry no information whatsoever about the pattern that emerges. Under continuous changes in a control parameter, patterns may emerge spontaneously. In fact, this is a signature feature of self-organization. That is, patterns arise solely as a function of the dynamics of the system. There is no specific ordering influence from the outside and no homunculus sitting inside.

It is always crucial to establish theoretical notions in a concrete, experimental situation. Thus, the discovery of phase transitions in studies of human bimanual coordination formed the cornerstone of dynamic pattern theory and more generally of the synergetic approach to biology (Haken, 1987). The observations were as follows: Kelso (1981; 1984) had subjects rhythmically move their index fingers or hands under two initial conditions, one in which limb segments move in the same direction and electromyographic (EMG) activity of pertinent muscles fires synchronously (homologous muscles contracting *in-phase*) and an anti-phase condition in which homologous muscles contract in an alternating fashion. Through the use of a pacing metronome, frequency of oscillation was systematically increased. Figure 1.1 shows a time series when the hands were prepared initially in the anti-phase mode. Obviously, at a certain critical frequency, switching occurs spontaneously from the anti-phase to the in-phase mode. This switch is re-

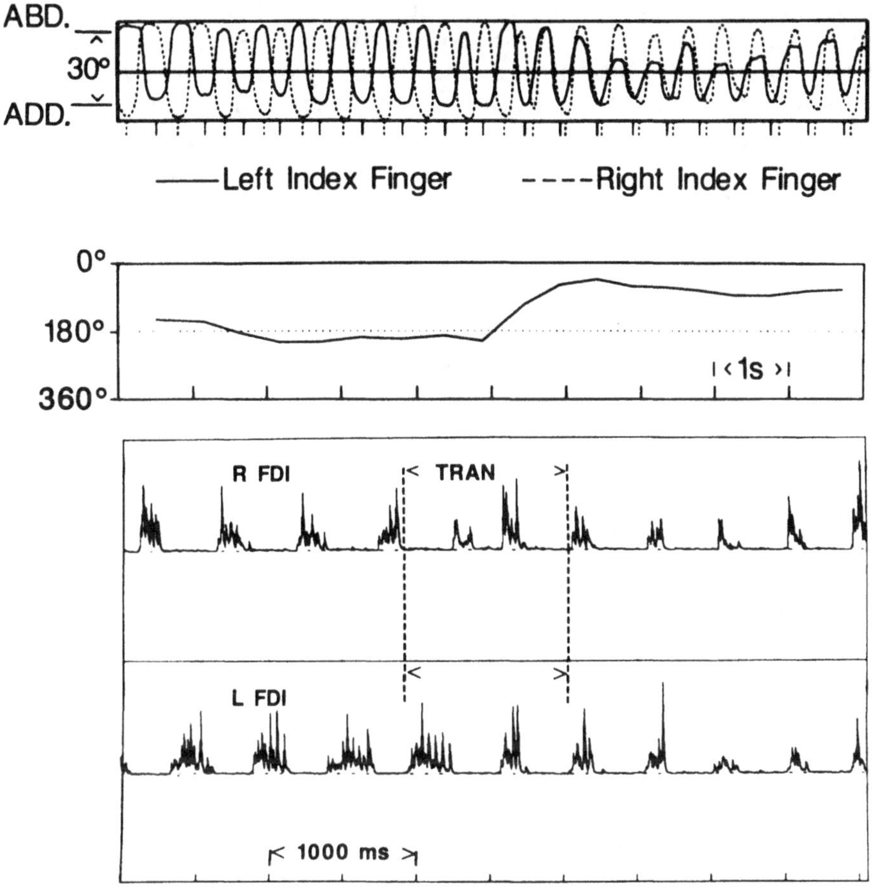

Figure 1.1. Top: time series of left and right finger position; middle: point estimate of relative phase; bottom: EMG recordings of first dorsal interossei muscle for the left and right fingers (integrated and rectified).

flected in the point estimate of relative phase in the top graph in Figure 1.1 as well as on a different level of description (EMG) in the bottom graph of Figure 1.1. No switching in the reverse direction occurs when the subject starts in the in-phase mode. Thus, although there are two stable patterns for low frequency values, only one pattern remains as frequency is scaled beyond a critical region. This transition behavior can be monitored by calculation of the relative phase between the two fingers. A point estimate of relative phase is the latency of one finger with respect to the other finger's cycle, as determined from peak-to-peak displacement. A continuous estimate of relative phase (i.e., at the sampling rate of 200 Hz) can be obtained from the phase plane trajectories of both fingers (the velocities may be obtained by a central difference numerical differentiation procedure). When the finger oscillations are normalized to the unit circle, the phases of the individual fingers can be obtained from the arctangent ($\dot{x}/x$) if x is normalized finger position and $\dot{x}$ the velocity (see Kelso & Tuller, 1987). Relative phase is just the difference between these individual phases. Often the relative phase fluctuates before the transition and stabilizes thereafter (e.g., Kelso, 1984; Kelso & Scholz, 1985; see following discussion of fluctuations).

What is the relevant degree of freedom in this case—or in the language of dynamic patterns, what is the order parameter or collective variable? From our discussion of phase transitions, the relative phase, ϕ, is a suitable candidate because (a) ϕ characterizes all observed coordinative patterns; (b) ϕ changes abruptly at the transition and is only weakly dependent upon parameters outside the transition; and (c) ϕ has very simple dynamics in which the ordered, phase-locked patterns correspond to attractors (we will define the term *attractor* shortly). Because the prescribed frequency of oscillation, manipulated during the experiment, is followed closely (i.e., is not affected by the relative phase) and because frequency drives the system through the collective states, frequency may be considered the control parameter.

In summary, we want to stress again that the phase transition methodology allows one to identify relevant degrees of freedom. Phase transitions represent singular boundaries that separate, as it were, different realms of existence, in our case, movement patterns. Nonequilibrium phase transitions are a universal feature of all complex systems; oscillating fingers are simply a window into establishing this fact for biological coordination.

Q. *Once the relevant degrees of freedom or order parameters are found empirically, how do you model?*

A. A first step is to provide a mathematically accurate description of the main qualitative features of one's data. The key idea is to map observed patterns onto attractors of a dynamical model. The meaning of *dynamical* here has nothing to do with forces or masses in the conventions of mechanics; rather, it refers to the temporal evolution of (in our case) a collective variable, that is, how this variable changes or stays the same as time flows (formally, the flow of a vector field). Given that we have identified a relevant degree of freedom, x, characteristic of the pattern, underlying the dynamic pattern view is the assumption that $x = x(t)$, where t is time, obeys a dynamical law:

$$\dot{x} = f(x_t, \textit{ parameters, noise}) \tag{1}$$

For a large class of functions f, special solutions of Equation (1) exist called *attractors*. By definition, an attractor is asymptotically stable; that is, all neighboring solutions of Equation (1) converge in time to the attractor solution. Nonequilibrium systems generally obey dissipative dynamics, the word *dissipative* meaning that many independent trajectories of the system with different initial conditions eventually converge on a certain limit set, the attractor. The simplest attractor type is a stable fixed point, that is, a constant solution of Equation (1) to which all neighboring trajectories converge. Another important attractor type in biology is the *limit cycle*, a stable periodic solution of Equation (1). Many more complicated attractor types exist (see also Newell et al., this volume), and their identification has proved significant in many branches of science (see Campbell, 1987; Kelso, Mandell, & Shlesinger, 1988) and medicine (see Koslow, Mandell, & Shlesinger, 1987). Attractors play a key role in the modeling process because the behavior of the collective variable in time (the dynamics of the collective variable) may be mapped onto attractors, the layout of which may be altered as a control parameter is changed continuously.

For the hand experiments of Kelso and colleagues, Haken, Kelso, and Bunz (1985) were able to determine the dynamics of relative phase, ϕ, from a few basic postulates. First, the observed stationary states of ϕ at 0° and ±180° are modeled as point attractors. This is a minimality strategy in which only the observed attractor type appears in the model. *It cannot be overemphasized that these are point attractors in the space of collective variables*, that is, of the system's relevant degrees of freedom. Second, the model must reproduce the observed bifurcation; that is, two patterns are available below a critical frequency, a condition called *bistability*, whereas only one is stable above the critical point (*monostability*). Third, due to the angular character of ϕ, the dynamics have to be 2π periodic. Another way of saying this is that because ϕ occurs only under cosine or sine functions, the properties of the physical system must not change when ϕ is replaced by $\phi + 2\pi$. Fourth, both hands are assumed to play a symmetric role; that is, the behavior of the system does not depend on the way we label the right hand and the left hand. This means that the model is symmetric under the transformation ϕ to $-\phi$. This assumption fits the data well; evidence for hand preferences in the bimanual experiments is weak at best.

In synergetics, the equations for order parameters are often of the form

$$\dot{\phi} = -\frac{\partial V}{\partial \phi} \tag{2}$$

where V is the so-called potential function. Following this strategy, the simplest and most general model obeying the four postulates is a potential, V, that is a superposition of two cosine functions:

$$V = -a \cos(\phi) - b \cos(2\phi) \tag{3}$$

which is an explicit model of the dynamics of relative phase with two parameters, a and b.

The behavior of the system described by Equations (2) and (3) can be readily visualized if ϕ is identified with the coordinate of a particle that moves in an overdamped fashion in the potential, V. When the total superposition (Equation 3) is taken and the ratio b/a is changed, a whole series of potential fields can be traversed (see Schöner, Haken, & Kelso, 1986, for how the parameters a and b are calculated from real data). Now prepare the system, as in the state shown by the black ball in Figure 1.2 ($\phi = \pm\pi$). Decrease the ratio b/a, which corresponds to increasing the experimental frequency. At a critical value of the parameters, the black ball falls to the lower minimum at $\phi = 0$. This corresponds to the transition from the anti-phase (antisymmetric: $\phi = \pm\pi$) state to the in-phase (symmetric: $\phi = 0$) state. When frequency is further increased (b/a tends to 0), the hand movements remain in the symmetric pattern. Note also that if the system is prepared in the symmetric pattern and b/a is decreased, no transition to $\phi = \pm\pi$ occurs. Also, following a transition to $\phi = 0$, if b/a is increased again (corresponding to a decrease in experimental frequency), the system remains in that state. This hysteresis phenomenon is well known in many physical and biological systems and was also a feature of the hand experiments.

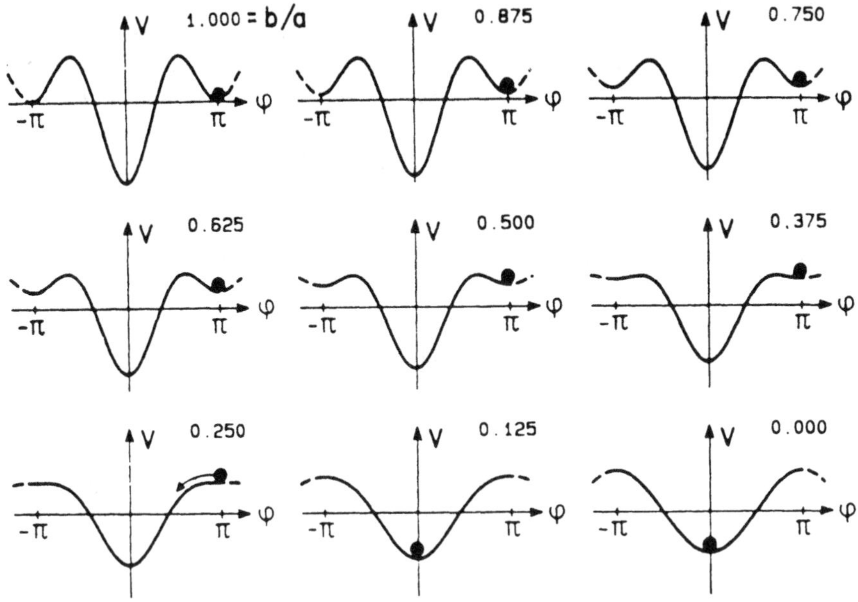

Figure 1.2. The potential (3) as the ratio b/a is changed (numbers refer to b/a). The system is initially prepared anti-phase (b/a = 1.0). As b/a decreases, the little ball illustrates the system's transition to in-phase (b/a = 0.0) cycling behavior. *Note:* from "A Theoretical Model of Phase Transitions in Human Hand Movements" by H. Haken, J. A. S. Kelso, and H. Bunz, 1985, *Biological Cybernetics, 39*, p. 150. Copyright 1985 by Springer-Verlag. Reprinted by permission.

Q. *Your theoretical strategy enables you to provide a compact description of the hand experiment results, but how do you really know that stability and change of biological movement patterns correspond to nonequilibrium phase transitions?*

A. The answer is that we don't know in advance and that we have to find out. Certainly, not all changes correspond to phase transitions. For example, as frequency was increased in the bimanual experiments, there was often a parallel, smooth decrease in movement amplitude in both fingers (see Kay, Kelso, Saltzman, & Schöner, 1987). However, such changes are best viewed as parametric. On the other hand (pardon the pun), in systems close to transition points, certain specific phenomena are predicted to occur jointly. One of these predictions involves fluctuations around the mean state of the collective variable. These fluctuations arise from the dynamics of numerous subsystems to which the system is coupled. The collective effect of these underlying processes acts as a perturbation in the form of noise. As the system approaches a critical point, one may observe an accompanying increase in these fluctuations—so-called *enhancement of fluctuations*—reflecting the growing inability of the system to maintain a particular pattern. In the critical region itself, the system briefly displays transient behavior, in which no definitive pattern is apparent. The system then evolves to a new or different pattern, apparent from the new value of the collective variable. The switch to a new pattern is accompanied by a marked decrease in fluctuations, signifying that the transition to a new stationary state is complete (cf. Prologue).

At values of the control parameter where fluctuations are minimal, the pattern is considered to be more stable than in control parameter regimes where higher fluctuations are observed. Stability, therefore, is not just an intuitive descriptive label; rather, it is a well-defined concept that is central to dynamic pattern theory (for further discussion see Kelso et al., 1987; Schöner & Kelso, 1988a). Stability serves a dual purpose in linking theory and experiment: Not only does it characterize the states in which the system resides, but also *loss of stability* in the order parameter is hypothesized to be the chief mechanism that effects a change of pattern. More explicitly, fluctuations may be considered stochastic forces, acting as continuously applied perturbations that drive the system away from its present state. In the bimanual experiments, these fluctuations were measured as the standard deviation of the collective variable, relative phase. In other nonequilibrium systems, other observables such as the output of laser light or the molecular concentration in chemical reactions undergo large fluctuations (see Haken, 1983, for many examples).

Kelso and Scholz (1985; see also Kelso, Scholz, & Schöner, 1986) analyzed the mean relative phase and its standard deviation in each of the two coordinative patterns as frequency (a control parameter) was increased. When the motions were prepared initially in the symmetric pattern, the mean relative phase and standard deviation remained relatively constant. However, clear enhancement of fluctuations both before and during the transition were observed when the movements were prepared anti-phase. Furthermore, after the switch from the anti-phase to the in-phase pattern, fluctuations decreased dramatically to levels comparable to the symmetric, in-phase condition. In Figure 1.2, it is quite easy to see what enhancement of fluctuations means. Any small fluctuation, when the attractors are well defined (top left), will be quickly damped; that is, the effect of fluctuations in this parameter regime is small. However, the same fluctuations, when parameters flatten the minimum at $\phi = \pi$, will be seen to be greatly amplified.

A second prediction of dynamic pattern theory and synergetics in general that further characterizes the differential stability of the attractors at different values of a control parameter concerns *critical slowing down*. Simply stated, when a system is close to a transition point, the system reacts more slowly to external perturbations than it does when it is far removed from the critical point. If a small perturbation is applied to the system, driving it away from its stationary state, the time it takes for the system to return to that state, the local relaxation time (τ_{rel}), is a measure of the stability of the attractor. The smaller τ_{rel}, the more stable the attractor. Obviously, as the phase transition regime is approached, enhanced fluctuations should be reflected in a parallel increase in the duration of τ_{rel} (actually, they grow as a square root function of the relaxation times; cf. Scholz, Kelso, & Schöner, 1987; Schöner et al., 1986). As well, once the critical point is crossed, the sharp decrease in fluctuations that follows the emergence of a new pattern should be accompanied by a decrease in local relaxation time. In terms of the potential of Figure 1.2, critical slowing down is reflected in a flattening of the pretransition mimimum (as b/a is decreased and the transition regime is approached). Thus, when the little ball is pushed away from the minimum by a perturbation, it takes longer to return to the minimum in this flattened state than it does when the shape of the potential is steeper (i.e., at lower values of the control parameter where restoring forces are large, or after the transition is complete and the system resides in a new, well-articulated basin of attraction. The attractor basin is defined as the set of all initial points from which trajectories converge to a given attractor.) At even higher values of the control parameter, the initial minimum disappears altogether, and the little ball falls to the sole remaining basin of attraction.

The critical slowing down prediction of synergetics was tested in the bimanual paradigm (Scholz & Kelso, in press-b; Scholz et al., 1987). A torque pulse (50 ms duration) perturbed one of the index fingers as they moved at different frequencies. This pulse acted as a perturbation that moved the bimanual pattern temporarily away from its prepared state and enabled calculation of the time to return to the initially prepared pattern. The results were extremely consistent with predictions: (a) Except for the lowest frequencies, the relaxation time in the anti-phase mode was consistently higher than in the in-phase mode; and (b) as the system approached the transition, the relaxation time increased in the anti-phase mode but remained constant or decreased in the in-phase mode. Thus, through the use of an entirely different experimental observable, support for a nonequilibrium phase transition in biological coordination was provided. In sum, observations of critical slowing down and enhancement of fluctuations are typical features of *self-organization* in synergetic systems. New states evolve without specific influence from the outside; that is, the control parameter does not anticipate or indicate the new state but rather creates the necessary conditions for the system to acquire it.

When collective variables for patterns are identified, relaxation times and fluctuation measures are well defined and open to observation. As we have noted, the key step is to link such measures to the concept of stability. Time scales play a crucial role in this regard, that is, in the interpretation of observed patterns as attractor states of a dynamical system. Up to now, we have introduced local relaxation time, τ_{rel}, as a measure of the time it takes the system to relax to an attractor once it is nearby. But two other time scales in addition to τ_{rel} are important: (a) observation time (τ_{obs}), the typical time scale on which the experimenter observes the system in a given preparation and over which statistical averages are performed; and (b) equilibration time (τ_{equ}), the time it takes the system to reach a stationary probability distribution, or stationary state, from a typical initial distribution. In the bimanual case, τ_{equ} is the time it takes the system to travel from one basin of attraction, for example, at $\phi = \pm\pi$, to the other basin of attraction at $\phi = 0$. In order to interpret observed states as (local) attractor states, the following time scales relation must be fulfilled:

$$\tau_{rel} \ll \tau_{obs} \ll \tau_{equ} \tag{4}$$

Here, the time it takes the system to relax back to its stationary state is shorter than the time over which the system is observed (e.g., for a given value of the control parameter) and far less than the time necessary to reach its most stable state. In terms of Figure 1.2, this relation means that the rolling ball may be pushed away from its present minimum, but the push is not enough to force it over the potential hill. That is, the system relaxes to its (local) stationary state on the observed time scale. However, as a state loses stability, its local relaxation time increases until, at the phase transition, the time scales relation (4) is violated, and switching occurs. The nature of the transition (e.g., whether critical fluctuations will be observed) depends on another time scale, the time scale of parameter change, τ_{par}. This time scale reflects the fact that in biological systems, the control parameter that brings about the instability is itself changing in time. For example, if

$$\tau_{rel} \ll \tau_{par} \ll \tau_{equ} \tag{5}$$

then the system remains at a control parameter value much longer than the time taken to return to its present state after being perturbed, allowing the system to relax to a locally stable state. In such cases, typical features of critical phenomena (enhancement of fluctuations and critical slowing down) are predicted. Thus, in our experimental system, if time scales relation (5) holds, it is possible to maintain a particular pattern even as fluctuations increase, and a transition is observed only as the old state becomes unstable. If, however,

$$\tau_{rel} \ll \tau_{equ} \ll \tau_{par} \tag{6}$$

meaning that the control parameter is fixed at a value that is quite long relative to the time it takes the system to find a global state of stability, we may see no enhancement of fluctuations because the system seeks out the lowest potential minimum before the old state actually becomes unstable. One can readily see the general implications of these time scale relations for the design of experiments. In the bimanual work, because τ_{par} and τ_{obs} are of the same order in the experiment, we expect time scales relation (5) to hold up to the transition. This requires us to differentiate two parameter regimes: the noncritical regime in which the system is stationary in the sense of relation (5), and the critical regime in which the system exhibits transient behavior. In these regimes, the full stochastic dynamics of Equation (3) can be solved numerically (Schöner et al., 1986) with pretransitional information about the standard deviation of relative phase and relaxation time of the anti-phase mode. The stochastic model accounts very well for the transient behavior without adjusting any parameters. Moreover, we should emphasize that all time scales are measurable. In the case of τ_{obs} and τ_{par}, measurement is often quite obvious; for τ_{equ}, several measurement techniques exist. One direct measure is called *mean first passage time*, which is the average length of time before the system first changes state. In the bimanual experiments, mean first passage time was determined directly using experimental data (Scholz et al., 1987), enabling a direct test of relation (5) and its breakdown at the phase transition. Switching indeed ocurred as the time scale relation was violated. Thus, time scale relations that govern the switching dynamics among collective states are directly observable and become even more important when other essentially biological features such as learning (Schöner & Kelso, 1988f) and development (Thelen, Kelso, & Fogel, 1987) are considered.

The stochastic version of Equation (3) (Schöner et al., 1986) contains another novel feature that pins down the nonequilibrium phase transition interpretation of stability and change in patterns of coordination. This feature is the duration of the transient from the anti-phase state to the in-phase state—which we call *switching time*. Due to the stochastic aspect of the dynamic model, switching does not occur as soon as the critical frequency is reached. Instead, during the transition, the probability density of relative phase, initially concentrated at $\phi = \pm 180°$, flows to $\phi \approx 0°$ and accumulates there until the "new" peak at $\phi \approx 0°$ is dominant and stationary. The model predicts the duration of this process in terms of both its mean and distribution. Consistent with our operational approach, switching times were also extracted from experimental data (Scholz & Kelso, in press-b; Scholz et al., 1987). In most cases, they were easy to calculate as the time between the relative phase value immediately before the transition and the value assumed immediately following the transition. The striking agreement between theoretical prediction (Schöner et al., 1986) and empirical data is particularly interesting because it shows that the switching process itself is quite closely captured by the stochastic dynamics of Equation (3) with noise added. The language of phase transitions is thus adequate for understanding the present phenomenon and opens the way to explore others (see Kelso & Schöner, 1987, for further examples).

Q. *Are there other ways to identify attractors of the collective variable dynamics?*

A. A technique that has proven useful in our experimental data is a sampling technique known as a return map. Here we use the information provided by relative phase values (point estimate) to determine whether a relation exists between one relative phase value (ϕ_n) and the next (ϕ_{n+1}). Essentially, this method tests for a deterministic structure in the dynamics by grouping discrete relative phase values into pairs (ϕ_n, ϕ_{n+1}) and plotting each as a point over time. If the relative phase values are 2π-periodic, then each pair plots the same point, a pattern suggesting the presence of a point attractor in this particular state space. Similarly, a limit cycle would appear as the same points revisited in a particular order. At the opposite extreme, a purely random process would appear as a scatter of points reflecting the lack of any history in the succession of values (see, e.g., Shaw, 1984).

Before giving an example, however, we emphasize that discreteness may be imposed in a number of ways. If we consider a 3-D trajectory in the space of (x, $\dot{x}$, t), we may sample, or "strobe," the trajectory at a specified time within each cycle of its path. Essentially, this sampling defines a plane perpendicular to the (x, $\dot{x}$) plane. Each time the trajectory passes through this plane, a point is plotted (ϕ_n, ϕ_{n+1}), the successive values of which enable us to construct a return map. In our example, which uses peak-to-peak relative phase, time itself is not the sampling variable. By choosing the peak displacement of each cycle, we define a sampling surface perpendicular to the ($\dot{x}$, t) space whose shape is a deformed plane. In other words, the point at which we sample will vary with changes in peak displacement values. Otherwise, if the peak displacement value is exactly the same in each cycle, then the sampling plane is uniform. Notice also that if the oscillatory components are perfect sinusoids, then peak values define uniform perpendicular planes in both the (x, $\dot{x}$) and ($\dot{x}$, t) space because peaks define constant time and displacement values.

Shaw (1981) draws attention to three factors on which the success of this method depends. First, the important attractor properties are topological; that is, almost any set of coordinates that can be manufactured will suffice to describe the attractor. Second, the system is nearly deterministic for short times; that is, the point-to-point structure of the attractor is well defined. Third, the dimensionality of the attractor is in fact small enough to be tractable (see Kay, Saltzman, & Kelso, 1988, for application of dimensionality calculations to motor behavior; also Kay, in press).

Figure 1.3 is a series of relative phase return maps for the cycling behavior of the right arm and right leg of a human seated in an experimental apparatus that allows all four human limbs to cycle up and down in the sagittal plane (Jeka & Kelso, 1988; Kelso & Jeka, in preparation). By convention, relative phase values close to 180° correspond to the arm moving up as the leg moves down; relative phase values around 0° correspond to both limbs moving up and down in the same direction. Cycling frequency was scaled as in previous bimanual experiments. Relative phase values for cycle *n* are plotted on the horizontal axis versus relative phase for cycle *n*+1 on the vertical axis. Each of the four maps represents two of eight successive frequency plateaus (from 1.25 to 3.00 Hz in steps of 0.25 Hz). The first two frequency plateaus appear in the upper left map, with following plateaus graphed in a clockwise direction. In each map, the phase values of the first plateau pair are plotted with the white squares, and the second plateau appears as black triangles.

The convention we have adopted to plot phase values in these maps is that points appearing in any of the four corners of this coordinate plane denote approximately equivalent states that vary around 0° (or 360°). They differ only in that when the arm reaches its peak displacement slightly ahead of the leg, the relative phase values are slightly less than 360°, whereas if the arm is slightly behind the leg, relative phase values are slightly greater than 0°. Points lying on either end of an imaginary 45° line from the origin are the special case in which $\phi_n = \phi_{n+1}$, that is, when the lead-lag relationship remains the same in successive cycles. Points at either end of the opposing diagonal occur when the lead-lag relationship between limbs changes in successive cycles. All points clustered around an anti-phase (180°) value, however, will appear grouped around the center of the coordinate plane.

Beginning with the upper left map, one sees a cluster of points grouped around a 180° anti-phase relationship, corresponding to the initially prepared pattern. The arrows reflect the fact that signs of instability are already emerging by the second frequency plateau, as relative phase briefly wanders toward an in-phase pattern, before returning to an anti-phase relationship, by the end of the plateau. The next map shows even more transient behavior, as points travel in a clockwise direction toward the origin and then around three corner points corresponding to an in-phase pattern

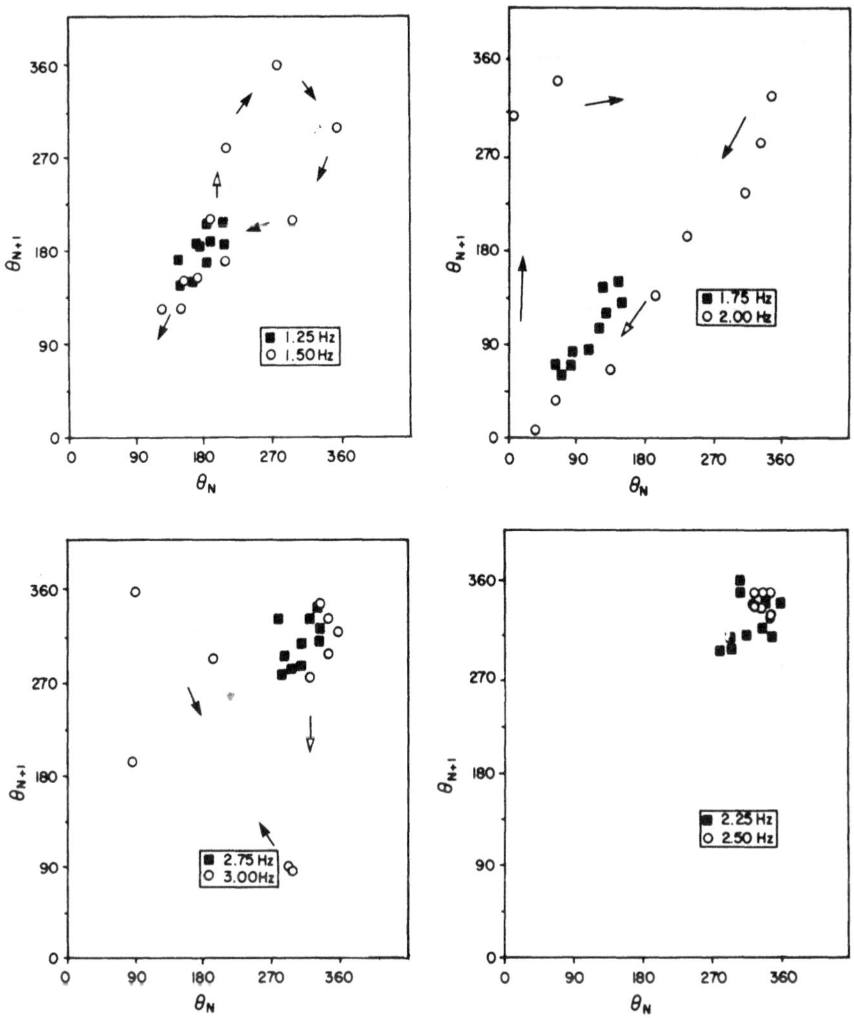

Figure 1.3. Return maps: the relative phase of cycle *n* (abscissa) versus cycle *n*+1 (ordinate). Each map contains two successive frequency plateaus beginning clockwise from top left. Arrows depict the change between successive relative phase values. White arrows show the initial change from a temporary stationary value of relative phase.

with changing lead-lag relationships between the limbs. The map in the bottom right corner indicates that the transition is now complete, as points remain clustered close to 360° (in-phase) for two complete plateaus as well as the first plateau of the last map (bottom left). The final map also demonstrates the transition created by the limbs operating at different frequencies, a transition that occurs when the leg can no longer follow the required frequency as set by a metronome. One now observes relative phase values jumping from one side of the 45° line to the other, possibly suggesting a bifurcation to limit cycle behavior in which phase values visit a number of repetitive values (states) in a specific order. We must emphasize, however, that this pattern does not prove the existence of an attractor but merely suggests that some underlying deterministic process may be driving the interaction between the limbs. Such a suggestion is legitimate only if it is supported with many more samples and a quantititative analysis (Kelso & Jeka, in preparation).

The important point illustrated by the return map, which has been discussed previously, is that a system with numerous components may be represented by a small number of parameters. We are essentially mapping a higher dimensional space (minimally, the position and velocity of each limb with possible coupling degrees of freedom) onto a low-dimensional map from which one can observe transitions in the collective variable, relative phase, and the emergence of new, stable phasing relationships. Such maps allow one to explore generalities that may apply to many systems that would otherwise be difficult to compare. For example, the well-known logistic map (e.g., May, 1976), $f(x_{n+1}) = \lambda x_n (1 - x_n)$, possesses a single parameter, λ, which has certain universal properties—known as Feigenbaum constants (Feigenbaum, 1983)—that have been quantitatively measured in a number of experimental systems. Such one-dimensional maps are capable of modeling the complexity of behavior typically seen in high-dimensional systems and illustrate again how important it is to identify the relevant degrees of freedom in complex systems.

We must stress that graphical characterizations indicative of biological attractors are not without historical precedent, even though the significance of such characterizations may not have been realized when they were first introduced. For example, Erich von Holst (1939/1973), whose work on the oscillatory rhythms of fish fins was one of the earliest efforts to characterize stability in a biological system, formulated an empirical technique known as the time and speed tables. Researchers have used this technique more recently to characterize cardiac rhythms in cats (Reid, 1969) and cellular bursting rhythms in the nervous systems of lobsters (Ayers &

Selverston, 1979; see also Stein, 1976). Two rhythms from von Holst's own work are shown in the top half of Figure 1.4. Even though both signals show periodicity, there are also nonuniform features to the lower signal. The question is whether these changes connote an influence between the two rhythms or whether they are merely "noisy" fluctuations in otherwise unrelated physiological signals.

To distinguish such possibilities, von Holst (1939/1973) first recorded the "relative phase" of the two rhythms on a cycle-to-cycle basis. He then measured the half-cycle duration for both the upstroke (fin movements to the right) and the downstroke (fin movements to the left) of each cycle. The result of these measurements appears in the bottom half of Figure 1.4, which shows the time table for the rhythms directly above it, now with cycle time normalized around a mean value. Relative phase is plotted on the abscissa versus the individual cycle time on the ordinate. The *x* curve corresponds to upstroke cycle times, and the *o* curve represents downstroke cycle times. We show only the time table here because the speed table is conceptually identical but uses average velocity within a half cycle instead of the half-cycle duration.

The time table reveals that the duration of a particular half cycle is dependent upon the momentary phase relationship between the two rhythms. This interaction is revealed by the behavior of points surrounding the intersection of the horizontal mean half-cycle duration line and the *x* curve, where the *x* curve has a maximum negative slope. Any point one chooses just below the line on the *x* curve represents a half-cycle duration that is slightly shorter than its mean value. As a consequence, the *next* half-cycle is brought closer to a relative phase relationship with the uniform rhythm, resulting in a duration that is now greater than the mean value. Again, the next upstroke is shifted further away from the vertical midline, this shift resulting in a smaller half-cycle duration, and so on. Thus, the intersection of the *x* curve and the mean value line is an "attractive" point in the sense that all combinations of cycle duration and relative phase tend toward this point (in the present language, it is a stable fixed point). The strength of that influence depends on how close the *x* point lies to the attractive point. A similar look at the point at which the *x* curve's slope is *increasing* through the mean cycle duration line reveals that it repels all nearby values (i.e., it is an unstable fixed point). It is noteworthy that the exact values of these intersection points correspond to entrained states in which

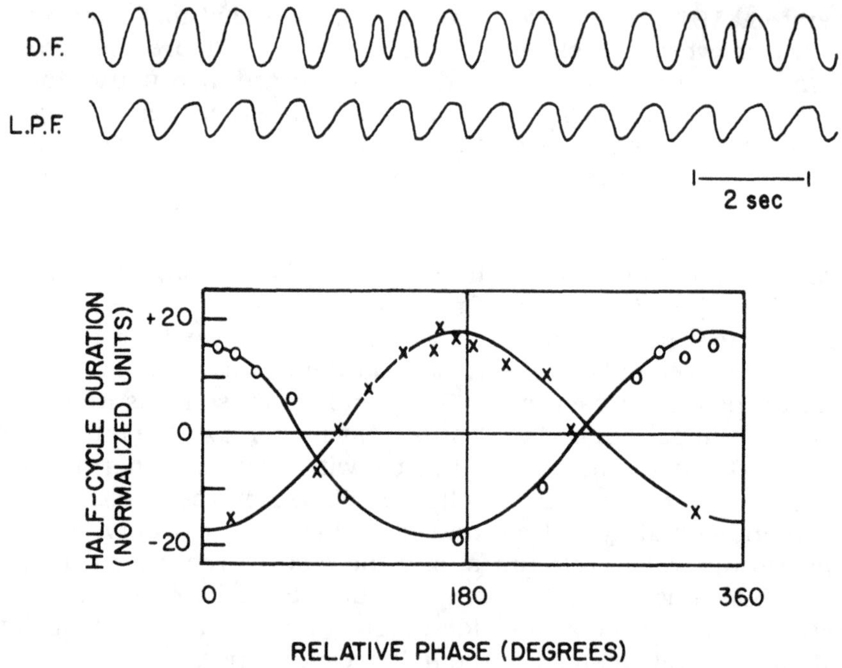

Figure 1.4. Time table for the rhythms recorded from the left pectoral and dorsal fin movements of Labrus. For each cycle, the relative phase of the two rhythms appears on the abscissa versus the percentage deviation from the mean value (at 0) of the half-cycle duration on the ordinate. The *x* curve depicts fin movements to the right; the *o* curve, fin movements to the left. *Note:* from "Relative Coordination as a Phenomenon and as a Method of Analysis of Central Nervous Function" by E. von Holst, 1973, in R. Martin (Ed. and Trans.), *The Collected Papers of Erich von Holst: Vol. 1. The Behavioral Physiology of Animals and Man* (p. 59), Coral Gables, FL: University of Miami Press. Copyright 1973 by the University of Miami Press. Adapted by permission.

cycle duration and phase are locked. It is only when one moves away from these points that the nature of their influence emerges. The same analysis holds for the *o* curve, corresponding to downstroke cycles.

Although von Holst (1939/1973) was most interested in characterizing the coordination of irregular rhythms compared to the extreme cases of entrained and random patterns, he actually provided a technique that characterizes stability in a qualitative yet precise fashion. In essence, he graphically identified fixed-point attractors; one that is stable and attracts all neighboring states and one that is unstable, on which the system remains only if it sits precisely upon this point (note again in the space of an adequate collective, temporal variable). Any slight deviation from the unstable fixed point leads to further divergence. To better visualize the existence of two such fixed points within the same system, consider the case of a freely swinging pendulum (Abraham & Shaw, 1982). The stable fixed point corresponds to the bottom of its circular arc (i.e., 6 o'clock) because no matter where the pendulum begins its swing, the combination of frictional and gravitational forces ensures its final position at this fixed point. Theoretically, however, there is one exception to this case, in which these same forces are perfectly balanced so that the pendulum remains perched at the very top of its swing (i.e., 12 o'clock). Intuitively, it is easy to see why this point is unstable: Even the slightest puff of wind will perturb the pendulum away from this fixed point toward its more stable partner. It is interesting to note that a simple, damped pendulum forced by a cosine function exhibits chaotic behavior; that is, a small uncertainty in the initial state rapidly makes it impossible to forecast future states (for a review, see Grebogi, Ott, & Yorke, 1987). Thus, even in seemingly simple systems, crisscrossing the border between stable and unstable regions of state space may result in enormous behavioral complexity.

Q. *In this picture, once you have identified the patterns and their dynamical laws, how do you derive these laws? I'm asking how different levels of description may be related.*

A. A key feature of the discussion thus far has been to characterize coordinated patterns in terms of the dynamics of macroscopic collective variables. But the strategy allows one to address the subsystems themselves and how these are coupled to produce coordinated patterns. This feature distinguishes the present operational approach from others in the human motor control literature. Yet it is entirely consistent with typical scientific procedure, namely, find the macroscopic laws first and then derive them from a more microscopic level of description. In this regard, we cannot help noting that the words *macro* and *micro* are always used in a relative sense, as is the case in science generally. For instance, the relevant micro levels of the particle physicist, the condensed matter physicist, the chemist, the biologist and so forth are all very different.

In our experimental model system, the individual oscillatory components are the next level down from the patterns they produce when they are coordinated. One may define the dynamics of individual components using each limb's position (x) and velocity ($\dot{x}$) as collective variables, but collective now in terms of the *next* lower level of description (e.g., agonist-antagonist muscle combinations). The stable and reproducible oscillatory behavior of each hand is modeled as an attractor in the phase plane (x, $\dot{x}$), in this case, a limit cycle. This step again is based on an accurate description of the experimental data. In a study related to the bimanual finger work (Kay et al., 1987), the cycling movements of individual hands were observed. Subjects rotated their hands about the wrist at six metronome-paced frequencies from 1 to 6 Hz in 1-Hz steps. Although subjects were given no explicit instructions about movement amplitude, the results demonstrated that amplitude remained roughly the same within a given frequency but decreased as frequency increased. It was found that with certain minimal assumptions, a nonlinear oscillator model captured these experimental features successfully. In mapping the observed oscillatory state onto a limit cycle, the notion of stability is once again a key feature of the theory. This can be checked by measuring the relaxation time of the individual components in a way similar to that described earlier. Along with the observed kinematic relations, relaxation time measures allowed us to determine all the individual oscillator parameters. A further assumption, that the oscillation is autonomous and not explicitly time-dependent, was tested in the perturbation paradigm using phase-resetting techniques (see Kay et al., 1988). The dimensionality of the attractor (Grassberger & Procaccia, 1983) was also calculated and corresponded to a limit cycle attractor with noise ($D \sim 1.2$; see Kay et al., 1988).

How can the components with their dynamics give rise to the phase-locked coordinative modes? Obviously, their dynamics have to be coupled. Haken et al. (1985) have determined coupling structures that can account for the observed phase-lockings. The simplest model that achieves this is a van-der-Pol-like coupling of the form:

$$\ddot{x}_1 + f(x_1, \dot{x}_1) = (\dot{x}_1 - \dot{x}_2)\,[A + B(x_1 - x_2)^2] \tag{7}$$

$$\ddot{x}_2 + f(x_2, \dot{x}_2) = (\dot{x}_2 - \dot{x}_1)\,[A + B(x_2 - x_1)^2] \tag{8}$$

where f is the aforementioned oscillator function and A and B are coupling constants. The experimental observation that the kinematic relations (e.g., amplitude-frequency relation) are not significantly different between the coordinative modes and the single hand movements shows that the coupling constants A, B are small compared to the corresponding coefficients of the oscillator function (see Haken et al., 1985; Kay et al., 1987). In spite of this, the coupling structure described by Equations (7) and (8) gives rise to the two phase-locked states. Indeed, Haken et al. (1985) were able to derive the equation for relative phase (3) from Equations (7) and (8) using the slowly varying amplitude and rotating wave approximations. These results not only provide further support for the dynamical model on the collective variable level but also rigorously establish the relation of the two levels of description. It is important to note that the coupling functions are quite unspecific to the emergent patterns of coordination because several functional forms may result in the same pattern of phase-locking (Haken et al., 1985). Furthermore, coordinative changes may also emerge in different ways. For example, Kelso and Scholz (1985) showed computer simulations in which the phase transition was effected by various combinations of coupling strength and noise level. Also, simply keeping the coupling function constant and changing only the eigenfrequencies of the component oscillators bring about changes in coordinative pattern. Thus, the collective properties of the system are attributable more to the coordinated system as a whole than to the actual coupling terms. This fact denotes a consistency across the physiological and mathematical domains; in both, different mechanisms may give rise to the same pattern (Kelso & Scholz, 1985; Schöner & Kelso, 1988a).

Q. *It might be helpful now if you would summarize the main features of dynamic pattern theory.*

A. This theory (Kelso & Schöner, 1987, 1988; Schöner & Kelso, 1988a, 1988b, 1988c, 1988d) builds upon the concepts of synergetics (Haken, 1975, 1983), a theory of self-organization and pattern formation in nonequilibrium systems. Synergetics provides a theoretical but operational language. The main idea is to view patterns of coordination, or more generally, behavioral patterns, at one's chosen level of observation in terms of their nonlinear dynamics. We can summarize the theory as follows:

1. Patterns of coordination at a given level of description are characterized by low dimensional collective variables or order parameters whose dynamics are function-specific.

2. Observable (i.e., reproducible, stationary over a certain time scale) patterns of coordination are mapped onto attractors of the order parameter dynamics.

3. Biological boundary conditions (e.g., environmental, task, and energetic constraints) act as parameters on the collective dynamics in the sense that they can modify the behavioral patterns but are themselves not dependent on these patterns. A parameter that moves the system through different collective states is a control parameter in the sense of synergetics. These parameters may be quite unspecific to the resulting behavioral patterns.

4. Fluctuations determine not only the stability of the observed pattern but also different time scales (global and local relaxation times). Time scale relations govern the switching dynamics among multiple coordinative patterns and account for observed multistability, that is, coexistence of several patterns under the same conditions.

5. Loss of stability leads to switching of pattern, and switching is governed by stochastic order parameter dynamics.

6. If coordinative patterns are thus characterized at different levels of description, these levels may be related without introducing additional concepts.

Points 1 to 3 provide a framework through which a dynamic pattern description may be obtained. These steps must be taken for any particular experimental system at any level of description in order to give the concepts a concrete meaning. Once a consistent description in the sense of Points 1 to 3 is reached, Points 4 and 5 enable one to test crucial predictions of the theory. It is important to emphasize that the linkage among levels of observation (Point 6) is possible only if a dynamic pattern analysis is available on both levels in question.

Q. *Now that you've provided the building blocks, as it were, of your approach, what further extensions are possible?*

A. Although dynamic pattern theory was originally formulated in the context of patterns of movement coordination, other behavioral patterns, functions, and experimental systems (a few examples follow) are open to analysis. There are really two cornerstones to the whole approach that make these extensions possible. The first is the necessity of identifying, in any given problem, the order parameter dynamics in the absence of any specific parametric influence. We refer to these dynamics as *intrinsic dynamics*, which simply means that the patterns arise as a result of *nonspecific* changes in control parameters (see Point 3). Nothing about "hard wiring" in these intrinsic dynamics is implied. However, the main point is that one has to discover the intrinsic dynamics to know *what* is modifiable, by the environment, by intentions, or by learning, for example. Once one knows the patterns and their dynamics, one can begin to talk rigorously of what it is that is changed (or indeed changeable) by *specific* parametric influences.

The specific parametric effects due to environmental requirements, intentional or purposeful needs, tasks to be learned, and so forth allow us to introduce the second, not yet discussed, cornerstone of the theory, namely, the concept of *behavioral information*. Such information may be expressed in the form of required behavioral patterns, that is, as part of the dynamics that attracts a behavioral pattern (defined by the intrinsic dynamics) toward the required pattern (Schöner & Kelso, 1988b). In this explicit sense, *information is meaningful and specific to the biological system only to the extent that it contributes to the order parameter dynamics attracting the system to the required (e.g., perceived, learned, memorized, intended) behavioral pattern.* An important consequence of this definition is that information is defined in the same space as the collective variables that characterize the pattern. In fact, this operational definition of information has no meaning whatsoever outside its influence on a set of collective variables and their (intrinsic) dynamics. The crux of this formulation is that information is not arbitrary with respect to the dynamics it modifies.

This statement is not merely a claim or a philosophical commitment (Kelso, Holt, Kugler, & Turvey, 1980; Kugler, Kelso, & Turvey, 1980; 1982) but rather can be shown to work (in the sense of an explicit mapping between experiment and theory) in a number of cases, including the formulation of perception-action patterns (Kelso, Delcolle, & Schöner, in press), the modification of action patterns by the environment (experiments by Tuller & Kelso, 1985, in press; see Schöner & Kelso, 1988c, 1988d for theory), and learned and memorized behavioral patterns (Yamanishi, Kawato, & Suzuki, 1980; Schöner & Kelso, 1988b, 1988c, 1988d). Many other experimental systems are open to a similar theoretical analysis: locomotory gait patterns and gait changes (Schöner, Jiang, & Kelso, 1988), the coordination of rhythmic movement between two human subjects (Schmidt, Carello, & Turvey, 1989), the mode-lockings studied by Kelso and DeGuzman (1988), and interlimb coordination of discrete movements (Kelso, Southard, & Goodman, 1979).

Let us fix the concept of behavioral information in a particular context that we have not discussed so far, namely, the ability of biological systems to change patterns of behavior flexibly in a purposeful or intentional fashion. Our previous empirical and theoretical work on spontaneous switching of coordination patterns leads to two general predictions. First, the intrinsic dynamics should influence the process of intentional change among available behavioral patterns. That is, the time it takes to switch from one pattern to the other depends on the stability of the patterns themselves. For example, if it is true that the anti-phase coordination pattern is less stable than the in-phase pattern, the system should switch faster in that direction than vice versa. The second prediction is that an intention, defined now as an intended behavioral pattern (and thus included as part of the behavioral dynamics) can change the dynamical chacteristics, such as the stability, of the behavioral patterns. Thus, intentional information can be viewed as a perturbation of the intrinsic dynamics, attracting the system to the intended behavioral pattern. Note that the conceptual distinction between the intrinsic dynamics and the contribution of intention is meaningful only if the intrinsic dynamics are identified by experiments that do not involve intentional behavioral change. It is this essential feature of the approach that makes predictions about the process of behavioral change possible in the first place.

An easy way to see the consequences of these predictions is through the potential pictures shown in Figure 1.5. On the top left we show the standard intrinsic dynamics (Equation 3) of Haken et al. (1985) with two minima of the collective variable at $\phi = 0°$ and $\phi = 180°$. On the top right we show the potentials corresponding to an intentional perturbation,

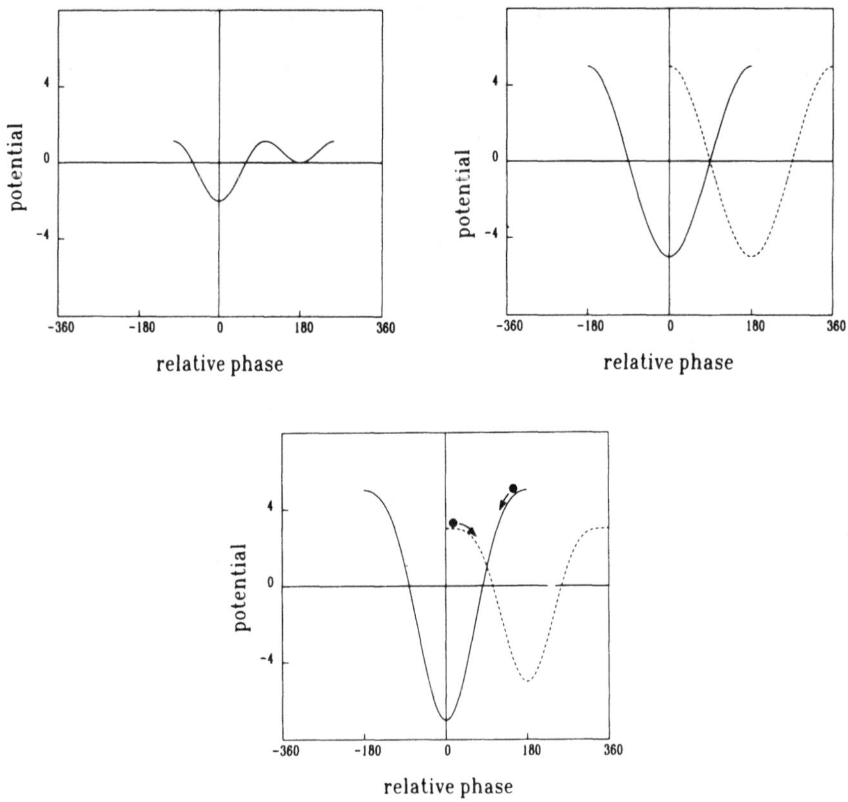

Figure 1.5. Modification of the intrinsic dynamics by behavioral information, in this case, an intended behavioral pattern. Upper left depicts the intrinsic dynamics according to the potential (3). Upper right shows the potential (9) specifying the intentional perturbation. Lower graph is the result of summating the top two to arrive at the full dynamics. The little ball travels much faster along the steeper slope of $\phi = 0$ than along the slope of $\phi = 180$, consistent with empirical switching time data of Kelso et al. (1988).

$$V_{\psi}(\phi) = - c_{intent} \cos(\phi - \psi) \qquad (9)$$

where the solid line represents an intentional pattern that is in-phase ($\psi = 0$) and the dotted line an intentional pattern that is anti-phase ($\psi = \pi$). A single parameter, c_{intent}, determines the strength of the intention to produce one pattern or the other. Note that in this figure the intrinsic dynamics are absent, so the two patterns are equally stable.

The full dynamics, that is, $V(\phi) + V(\phi)_{intent}$, is shown bottom center. Now it is easy to visualize the consequences of the theory applied to intentional behavior (see Schöner & Kelso, 1988d, 1988e, for details). First, it is obvious that an intention can change the intrinsic dynamics; that is, it can destabilize one pattern and stabilize the other. Second, because the attractor at $\phi = 0°$ is more stable than at $\phi = 180°$, the system will run faster to the in-phase pattern than to the anti-phase pattern. We have illustrated this situation with black balls in Figure 1.5. A recent experiment nicely demonstrated the influence of the intrinsic dynamics on intentional switching (Kelso, Scholz, & Schöner, 1988; Scholz & Kelso, in press-a). The task was initially to cycle the fingers either in-phase or anti-phase. Subjects were paced for 10 cycles by a metronome, which was then turned off. Instructions were to continue cycling at the initial frequency until an auditory tone signaled a switch to the opposite mode of coordination. The results, measured as mean switching time, show that switching from in-phase to anti-phase is about twice as slow as in the opposite direction, further evidence that the anti-phase mode is intrinsically less stable than the in-phase state.

Two points, one methodological and the other theoretical, emerge from these results. The first is that switching time, the tool employed here, reveals dynamic constraints on intentional switching that may be put to more general use. For instance, one may use the method "backwards" to identify the relative stabilities of behavioral patterns in those patterns for which it is difficult, if not impossible, to find phase transitions. The second point is that two languages that are often considered irreconcilable, namely, the language of intentionality and the language of dynamics, are actually captured in one unified picture. In the bimanual case, an intention acts in the same space of collective variables as that in which the intrinsic patterns are measured. Intentional information defines an attractor in that space and is meaningful to the extent that it attracts the system towards an intended behavioral pattern. At the same time, intentions are restricted by the intrinsic dynamics, in that the ability to perform a particular pattern is influenced by the relative stability of the available patterns. In short, intentions parameterize the dynamics but are in turn constrained by the dynamics.

Q. *You have told me mostly about rhythmic movement of two components. What about other systems, other levels of description?*

A. Earlier, we mentioned that speech research may benefit from the strategy we have formulated, particularly because this discipline has been searching for its own relevant variables for some time. The acoustic instantiation accompanying the production of a word varies with many factors, including stress pattern, the rate at which the word is produced, and the nature of the surrounding speech segments. However, this variation is not reflected in our perception of a spoken word. We seem able to decipher the linguistic content of an utterance despite the context-dependent nature of its production. Previously it was thought that a one-to-one relationship might exist between the electromyographic activity of individual muscles and the corresponding speech segments produced (Liberman, Cooper, Shankweiler, & Studdert-Kennedy, 1967). However, such activity was found to be highly context-dependent (MacNeilage & DeClerk, 1969). This finding led researchers to study the articulatory gestures associated with the speech segments (Liberman et al., 1967; Lindblom, 1963; Kozhevnikov & Chistovich, 1965) to determine whether the robust nature of speech perception lies in the articulatory movements themselves. These movements, however, displayed the same lack of invariance found in electromyographic activity. Presently, the conceptual basis for invariance is being challenged because this invariance means no change in the face of numerous metrical changes such as stress and speaking rate. But the noisiness of biological systems makes it difficult to characterize invariance as anything more than a statistical effect (see also Abbs & Connor, this volume).

Just as it has proven useful in the bimanual experiments, relative timing may be conceptually better suited to understanding of the speech system. This most recent tack (e.g., Kent & Netsell, 1971; Lofqvist & Yoshioka, 1981) has shown that relationships between articulators vary less across metric change than do absolute measures. This tack has also enabled investigators (Kelso, Saltzman, & Tuller, 1986) to introduce *stability* as a means of characterizing patterns in articulatory gestures. We stress, however, that this is not merely an exchange of terms. The advantage that accompanies stability is that fluctuations are no longer seen as a confound but as essential to characterizing the change from one pattern (phonemic utterance) to another.

An example of this insight applied to dynamic pattern theory is a study by Tuller and Kelso (in press), who investigated the relative timing between peak glottal opening and minimum lip aperture in two utterance conditions. Subjects were instructed to repetitively produce either a consonant-vowel (/pi/) or a vowel-consonant (/ip/) utterance to the beat of a visual metronome whose frequency was scaled similarly to that in previous bimanual experiments in our lab. The results displayed a consistent relative phase value for /pi/ across all frequencies. However, although /ip/ displayed a distinctly different value for relative phase than /pi/ did at lower frequencies, a transition to the relative phase corresponding to /pi/ ensued at higher frequencies. Thus, we have empirical evidence that the order parameter dynamics of relative phase is not unique to the bimanual system, but is useful in characterizing pattern stability and change in a completely different context. Furthermore, in a followup experiment using tape recordings of the actual trials from the production experiment, listeners were found to perceive the shifts in relative phase as corresponding shifts from /ip/ to /pi/ precisely at the point at which this transition occurred in the articulatory data. This remarkable fit between the production measures and perception of syllable form suggests that the relative phase dynamics may provide a means for understanding the coupling between speech production and perception.

A further step may be taken to link dynamic pattern theory to a completely different level of observation. Although the idea has not been explicitly or systematically tested to date, examples in the neurobiological literature have suggested that certain experimental aspects of neuronal behavior may be realizable within a dynamic pattern framework (Kelso et al., 1987; Schöner & Kelso, 1988a). In fact, typical phenomena of temporal ordering are widespread in the neuronal pattern generation literature, for example, sychronization, frequency locking, and phase locking (Rand, Cohen, & Holmes, 1988). These phenomena are typically depicted through measures of relative phase and latency among components, neuronal burst frequency, and frequency differences among neurons (e.g., Croll, Davis, & Kovac, 1985). Such findings strongly suggest that collective variables for temporal order at the neuronal level can be defined. Furthermore, observation at this level has uncovered numerous candidates for control parameters such as serotonin, whose concentration changes the elicitation threshold for rhythmic feeding patterns in Helisoma (see, e.g., Selverston & Moulins, 1987) or electrical stimulation, which has proven to modify locomotory patterns in decerebrate cats (Shik, Severin, & Orlovskii, 1966) and fleeing behavior in hens (von Holst & von Saint Paul, 1973). Finally, we note that stability measures, although not explicity formalized in the neurobiological literature, are evident. For example, fluctuation measures (Wyman, 1965) have been used in the analysis of flight patterns in the blowfly at the individual neuronal level, although collective variables were not identified and thus stability was not concretely ascertained. Nevertheless, once collective variables for neuronal patterns have been identified, the relaxation time and fluctuation measures are well defined and can be calculated through perturbation techniques and fluctuational analysis. These theoretical concepts and empirical techniques are the key to understanding pattern stability and change at the neuronal level.

Q. *Your discussion of pattern switching leads me to inquire about pattern selection. How, in this picture, does one particular pattern emerge from those available?*

A. Processes of pattern formation and selection occur throughout nature. Typical nonequilibrium systems will have many possible configurations of pattern available, not all of which are stable. A popular image is of multiple attractors with many basins of temporary attraction (Gleick, 1987). One of the challenges in nonlinear science is to understand the dynamics of sequencing among multiple patterns, namely, which patterns are explored and eventually selected. Again, such questions are relevant at multiple scales of observation and to multiple functions, from learning to the immune system (see contributions in, e.g., Koslow et al., 1987). Here are a few ways to think about pattern selection in the paradigm of nonlinear science. The list is far from inclusive, of course, and is restricted to issues germane to this book on motor coordination.

1. In our example of intentional switching among behavioral patterns, we saw that an "intention" can select a pattern, but which pattern can be most easily selected is determined by the pattern's stability. Selection, in other words, is influenced by the (quantitatively measurable) relative stability of the patterns available to the system. Our focus has been to identify *constraints* on the pattern selection rather than the dynamics of the selection process itself. For example, other factors (e.g., selection on the basis of cost or other performance criteria) are not accounted for in the behavioral dynamics. Such accounting would require degrees of freedom

in addition to those included in the intrinsic dynamics (cf. Schöner & Kelso, 1988e).

2. In the language of dynamic pattern theory, boundary conditions play a parametric role in pattern selection. Such parametric influences can be unspecific or specific. Generally, in nonequilibrium systems, boundary conditions may cause the system to "pin" itself to the most stable modal configuration (for examples see Campbell, 1987; Haken, 1983). The shape of the environment, for example, in Benard convection favors the formation of certain patterns; that is, whether the enclosure is circular or rectangular will result in quite different patterns under the same values of the temperature gradient. Instabilities can be viewed as playing a role in pattern selection in the sense that these instabilities seek out the most stable pattern. Fluctuations are crucial because they probe the environment of the collective state, "selecting" (or leading to the emergence of) a new pattern.

3. In a number of experimental situations, for example, the perception-action patterns studied by Tuller and Kelso (in press) and the mode-locking studies of Kelso and DeGuzman (1988), the pattern that emerges (or is selected) is a direct consequence of *cooperative* and *competitive* interactions between the intrinsic dynamics and external influences. In the Tuller and Kelso data, for example, when an environmentally required relative phase coincides with one of the basic intrinsic patterns, $\phi = 0$ or $\phi = \pi$, the minimum of the potential is exactly at the required relative phase, and its shape is well articulated (i.e., variation in ϕ is small). There is, in other words, a *cooperation* between extrinsic (environmentally defined) and intrinsic dynamics. In contrast, if the environmentally required relative phase does not correspond to one of the intrinsic patterns, a competition ensues, pulling the minimum away from the required relative phase (i.e., variation in ϕ is large). In short, the balance between intrinsic and extrinsic dynamics can be seen as a cooperative or competitive one. These processes thus determine (or select) which pattern is observed.

As a general point, nonlinear dynamical systems—even deterministic equations of motion—are enormously sensitive to initial conditions and parameters in certain regions of parameter space. Very complex behavior can emerge in systems governed by simple rules (e.g., the logistic equation [see May, 1976] or the circle map [see, e.g., Kelso & DeGuzman, 1988]). Until we understand these systems better, we should probably resist ascribing a priori the process of pattern selection to an agency residing inside the system.

CONCLUSION

We have presented, in a dialogue, some of the key features of the dynamic pattern strategy, which is aimed at understanding the coordination of behavior, its stability and change. The language of dynamic patterns stresses a close relation between theory and experiment and is level-independent in the sense that the central ideas and observables are applicable on several, potentially multiple, levels of observation. Key concepts are the characterization of behavioral patterns by collective variables, the determination of the dynamics of these patterns, and the study of their stability (and lack thereof). As Gould (1988) has recently remarked, "To know the reasons for infrequent change, one must understand the ordinary rules of stability" (p. 23). But the stability of what? The paradox, from our perspective, is that to understand stability, one must understand how it is lost because, as we have shown, loss of stability is central to behavioral change.

Identifying collective variables (order parameters in the language of synergetics) proves to be crucial to defining those aspects of behavior that are modifiable. Information with meaning is expressed in dynamic pattern language in terms of the same set of collective variables by which the behavioral patterns are characterized. This is a step toward the reduction, if not the elimination, of a traditional demarcation between mental (linguistic) and physical (dynamical) modes of description (cf. Pattee, 1976). When such information is included as part of the dynamics, the match between mathematically formulated theoretical predictions and empirical results is quite adequate.

The strategy of dynamic pattern theory offers a recipe, but the scientist must provide his or her own ingredients. It still demands insight and knowledge, particularly from the experimentalist, who must know the system sufficiently well to define the parameters that promote nonlinear behavior. It is a nontrivial step to find a qualitative change of behavior and to identify the conditions under which such change will emerge. Phase transitions are nevertheless crucial to understanding the dynamics that underlie both stability and change of coordinated behavior. They provide the physical foundation upon which to build a deeper understanding of those essential biological and psychological functions that we all care about.

ACKNOWLEDGMENTS

We thank P. G. Zanone for comments on earlier drafts of this chapter and G. Schöner for sharing his knowledge and insights on nonlinear dynamical systems.

REFERENCES

Abbs, J. H., Gracco, V. L., & Cole, K. J. (1984). Control of multimovement coordination: Sensorimotor mechanisms in speech motor programming. *Journal of Motor Behavior, 16*, 195-232.

Abraham, R., & Shaw, C. (1982). *Dynamics: The geometry of behavior: 1. Periodic behavior*. Santa Cruz, CA: Aerial Press.

Ayers, J. L., & Selverston, A. I. (1979). Monosynaptic entrainment of an endogenous pacemaker network: A cellular mechanism for von Holst's magnet effect. *Journal of Comparative Physiology A, 129*, 5-17.

Campbell, D. K. (1987). Nonlinear science: From paradigm to practicality. *Los Alamos Science, 15*, 218-262.

Croll, R. P., Davis, W. J., & Kovac, M. P. (1985). Neural mechanisms of motor program switching in the mollusc Pleurobranchaea. I. Central motor programs underlying ingestion, egestion and the neural rhythms. *Journal of Neuroscience, 5*, 48-55.

Feigenbaum, M. J. (1983). Universal behavior in nonlinear systems. *Physica D, 7*, 16-39. (Reprinted from *Los Alamos Science*, 1980, *1*, 4-27)

Gleick, J. (1987). *Chaos: Making a new science*. New York: Viking Penguin.

Gould, S. J. (1988). A web of tales. *Natural History, 97*, 16-25.

Grassberger, P., & Procaccia, I. (1983). Characterization of strange attractors. *Physics Review Letters, 50*, 346-349.

Grebogi, C., Ott, E., & Yorke, J. A. (1987). Chaos, strange attractors, and fractal basin boundaries in nonlinear dynamics. *Science, 238*, 632-638.

Haken, H. (1975). Cooperative phenomena in systems far from thermal equilibrium and in non-physical systems. *Review of Modern Physics, 47*, 67-121.

Haken, H. (1983). *Synergetics, an introduction: Non-equilibrium phase transitions and self-organization in physics, chemistry and biology* (3rd ed.). Berlin: Springer.

Haken, H. (1987). Information compression in biological systems. *Biological Cybernetics, 56*, 11-17.

Haken, H., Kelso, J. A. S., & Bunz, H. (1985). A theoretical model of phase transitions in human hand movements. *Biological Cybernetics, 39*, 139-156.

Jeka, J. J., & Kelso, J. A. S. (1988). Dynamic patterns in multi-limb coordination. In J. A. S. Kelso, A. J. Mandell, & M. F. Shlesinger (Eds.), *Dynamic patterns in complex systems* (p. 403). Singapore: World Scientific.

Kay, B. A. (in press). The dimensionality of movement trajectories and the degrees of freedom problem: A tutorial. *Human Movement Science.*

Kay, B. A., Kelso, J. A. S., Saltzman, E. L., & Schöner, G. (1987). The space-time behavior of single and bimanual movements: Data and model. *Journal of Experimental Psychology: Human Perception and Performance, 13*, 178-192.

Kay, B. A., Saltzman, E. L., & Kelso, J. A. S. (1988). *Steady-state and perturbed rhythmical movements: Dynamical modeling using a variety of analytical tools.* Manuscript submitted for publication.

Kelso, J. A. S. (1981). On the oscillatory basis of movement. *Bulletin of the Psychonomic Society, 18*, 63.

Kelso, J. A. S. (1984). Phase transitions and critical behavior in human bimanual coordination. *American Journal of Physiology: Regulatory, Integrative and Comparative Physiology, 15,* R1000-R10004.

Kelso, J. A. S. (1988). Introductory remarks: Dynamic patterns. In J. A. S. Kelso, A. J. Mandell, & M. F. Shlesinger (Eds.), *Dynamic patterns in complex systems* (pp. 1-5). Singapore: World Scientific.

Kelso, J. A. S., & DeGuzman, G. C. (1988). Order in time: How cooperation between the hands informs the design of the brain. In H. Haken (Ed.), *Neural and synergetic computers* (pp. 180-196). Berlin: Springer.

Kelso, J. A. S., Delcolle, J. D., & Schöner, G. (in press). Action-perception as a pattern formation process. In M. Jeannerod (Ed.), *Attention and Performance XIII.* Hillsdale, NJ: Erlbaum.

Kelso, J. A. S., Holt, K. G., Kugler, P. N., & Turvey, M. T. (1980). On the concept of coordinative structures as dissipative structures: II. Empirical lines of convergence. In G. E. Stelmach & J. Requin (Eds.), *Tutorials in motor behavior* (pp. 49-70). Amsterdam: North-Holland.

Kelso, J. A. S., & Jeka, J. J. (in preparation). *Dynamic patterns in multi-limb coordination.*

Kelso, J. A. S., Mandell, A. J., & Shlesinger, M. F. (1988). *Dynamic patterns in complex systems.* Singapore: World Scientific.

Kelso, J. A. S., Saltzman, E., & Tuller, B. (1986). The dynamical perspective in speech production: Data and theory. *Journal of Phonetics, 14,* 29-60.

Kelso, J. A. S., & Scholz, J. P. (1985). Cooperative phenomena in biological motion. In H. Haken (Ed.), *Complex systems: Operational approaches in neurobiology, physical systems and computers* (pp. 124-149). Berlin: Springer.

Kelso, J. A. S., Scholz, J. P., & Schöner, G. (1986). Non-equilibrium phase transitions in coordinated biological motion: Critical fluctuations. *Physics Letters, A118,* 279-284.

Kelso, J. A. S., Scholz, J. P., & Schöner, G. (1988). Dynamics governs switching among patterns of coordination in biological movement. *Physics Letters A, 134*, 8-12.

Kelso, J. A. S., & Schöner, G. (1987). Toward a physical (synergetic) theory of biological coordination. In R. Graham & A. Wunderlin (Eds.), *Lasers and Synergetics: Springer Proceedings in Physics*, (Vol. 19, pp. 224-237). Berlin: Springer.

Kelso, J. A. S., & Schöner, G. (1988). Self-organization of coordinative movement patterns. *Human Movement Science, 7*, 27-46.

Kelso, J. A. S., Schöner, G., Scholz, J. P., & Haken, H. (1987). Phase-locked modes, phase transitions and component oscillators in biological motion. *Physica Scripta, 35*, 79-87.

Kelso, J. A. S., Southard, D. L., & Goodman, D. (1979). On the nature of human interlimb coordination. *Science, 203*, 1029-1031.

Kelso, J. A. S., & Tuller, B. (1987). Intrinsic time in speech production: Theory, methodology and preliminary observations. In E. Keller & M. Gopnik (Eds.), *Motor and sensory processes of language* (pp. 203-221). Hillsdale, NJ: Erlbaum.

Kelso, J. A. S., Tuller, B., & Fowler, C. (1982). The functional specificity of articulatory control and coordination. *Journal of the Acoustical Society of America, 72* (Suppl. 1), s103.

Kelso, J. A. S., Tuller, B., Vakaitos-Bateson, E., & Fowler, C. (1984). Functionally specific articulatory cooperation following jaw perturbations during speech: Evidence for coordinative structures. *Journal of Experimental Psychology: Human Perception & Performance, 10*, 812-832.

Kent, R. D., & Netsell, R. (1971). Effects of stress contrasts on certain articulatory parameters. *Phonetica, 24*, 23-44.

Koslow, S. H., Mandell, A., & Shlesinger, M. F. (Eds.). (1987). *Annals of the New York Academy of Sciences: Vol. 504. Perspectives in biological dynamics and theoretical medicine*. New York: New York Academy of Sciences.

Kozhevnikov, V. A., & Chistovich, L. A. (1965). *Speech: Articulation and perception* (U.S. Department of Commerce, Clearinghouse for Federal Scientific and Technical Information, Trans., JPRS 30543). Washington, DC: Joint Publication Research Service.

Kugler, P. N., Kelso, J. A. S., & Turvey, M. T. (1980). On the concept of coordinative structures as dissipative structures: I. Theoretical lines of convergence. In G. E. Stelmach & J. E. Requin (Eds.), *Tutorials in motor behavior* (pp. 1-47). New York: North-Holland.

Kugler, P. N., Kelso, J. A. S., & Turvey, M. T. (1982). On the control and coordination of naturally developing systems. In J. A. S. Kelso & J. E. Clark (Eds.), *The development of movement control and coordination* (pp. 5-78). New York: Wiley.

Kugler, P. N., & Turvey, M. T. (1987). *Information, natural law and the self-assembly of rhythmic movement.* Hillsdale, NJ: Erlbaum.

Liberman, A. M., Cooper, F. S., Shankweiler, D. P., & Studdert-Kennedy, M. (1967). Perception of the speech code. *Psychological Review, 74*, 431-461.

Lindblom, B. (1963). Spectrographic study of vowel reduction. *Journal of the Acoustic Society of America, 35*, 1773-1781.

Lofqvist, A., & Yoshioka, H. (1981). Interarticulator programming in obstruent production. *Phonetica, 38*, 21-34.

MacNeilage, P. F., & DeClerk, J. (1969). On the motor control of coarticulation in CVC monosyllables. *Journal of the Acoustical Society of America, 78*, 1548-1553.

May, R. (1976). Simple mathematical models with very complicated dynamics. *Nature, 261*, 459-467.

Pattee, H. H. (1976). Physical theories of biological coordination. In M. Grene & E. Mendelsohn, (Eds.), *Boston studies in the philosophy of science: Vol. 27. Topics in the philosophy of biology* (pp. 153-173). Boston: Reidel.

Rand, R. H., Cohen, A., & Holmes, P. J. (1988). Systems of coupled oscillators as models of central pattern generators. In A. Cohen, S. Rossignol, & S. Grillner (Eds.), *Neural control of rhythmic movements in vertebrates* (pp. 333-367). New York: Wiley.

Reid, J. V. O. (1969). The cardiac pacemaker: Effects of regularly spaced nervous input. *American Heart Journal, 78*, 58-64.

Runeson, R. (1977). On the possibility of smart perceptual mechanisms. *Scandinavian Journal of Psychology, 18*, 172-179.

Saltzman, E., & Kelso, J. A. S. (1987). Skilled actions: A task dynamic approach. *Psychological Review, 94*, 84-106.

Schmidt, R. C., Carello, C., & Turvey, M. T. (1989). *Phase transitions and critical fluctuations in visually coupled oscillators.* Manuscript submitted for publication.

Scholz, J. P., & Kelso, J. A. S. (in press-a). Intentional switching between patterns of bimanual corrdination is dependent on the intrinsic dynamics of the patterns. *Journal of Motor Behavior.*

Scholz, J. P., & Kelso, J. A. S. (in press-b). A quantitative approach to understanding the formation and change of coordinated movement patterns. *Journal of Motor Behavior.*

Scholz, J. P., Kelso, J. A. S., & Schöner, G. (1987). Non-equilibrium phase transitions in coordinated biological motion: Critical slowing down and switching time. *Physics Letters A, 123*, 390-394.

Schöner, G. Haken, H., & Kelso, J. A. S. (1986). A stochastic theory of phase transitions in human hand movement. *Biological Cybernetics, 53*, 442-452.

Schöner, G., Jiang, W., & Kelso, J. A. S. (1988). *A dynamic theory of quadruped gaits and gait transitions.* Manuscript submitted for publication.

Schöner, G., & Kelso, J. A. S. (1988a). Dynamic pattern generation in behavioral and neural systems. *Science, 239*, 1513-1520.

Schöner, G., & Kelso, J. A. S. (1988b). A synergetic theory of environmentally-specified and learned patterns of movement coordination: I. Relative phase dynamics. *Biological Cybernetics, 58*, 71-80.

Schöner, G., & Kelso, J. A. S. (1988c). A synergetic theory of environmentally-specified and learned patterns of movement coordination: II. Component oscillator dynamics. *Biological Cybernetics, 58*, 81-89.

Schöner, G., & Kelso, J. A. S. (1988d). Dynamic patterns in biological coordination: Theoretical strategy and new results. In J. A. S Kelso, A. J. Mandell, & M. F. Shlesinger (Eds.), *Dynamic patterns in complex systems* (pp. 77-102). Singapore: World Scientific.

Schöner, G., & Kelso, J. A. S. (1988e). A dynamic theory of behavioral change. *Journal of Theoretical Biology, 135*, 501-524.

Schöner, G., & Kelso, J. A. S. (1988f). A theory of learning and recall in biological coordination. In J. A. S Kelso, A. J. Mandell, & M. F. Shlesinger (Eds.), *Dynamic patterns in complex systems* (p. 409). Singapore: World Scientific.

Sejnowski, T. J., Koch, C., & Churchland, P. S. (1988). Computational neuroscience. *Science, 241*, 1299-1306.

Selverston, A. I., & Moulins, M. (1987). *The crustacean stomatogastric system.* Berlin: Springer.

Shaw, R. S. (1981). Strange attractors, chaotic behavior and information flow. *Zeitschrift Naturforsch, 36a*, 800-112.

Shaw, R. S. (1984). *The dripping faucet as a model chaotic system.* Santa Cruz, CA: Aerial.

Shik, M. L., Severin, F. V., & Orlovskii, G. N. (1966). Control of walking and running by means of electrical stimulation. *Biophysics, 11*, 1011.

Stein, P. S. G. (1976). Mechanisms of interlimb phase control. In P. M. Herman, S. Grillner, P. S. G. Stein, & D. G. Stuart (Eds.), *Neu-*

ral control of locomotion (pp. 465-487). New York: Plenum Press.

Thelen, E., Kelso, J. A. S., & Fogel, A. (1987). Self-organizing systems and infant motor development. *Developmental Review, 7,* 39-65.

Tuller, B., & Kelso, J. A. S. (1985, November). *Coordination in normal and split-brain patients.* Paper presented at the meeting of the Psychonomic Society, Boston, MA.

Tuller, B., & Kelso, J. A. S. (in press). Environmentally-specified patterns of movement coordination in normal and split-brain subjects. *Experimental Brain Research.*

von Holst, E. (1973). Relative coordination as a phenomenon and as a method of analysis of central nervous function. In R. Martin (Ed. and Trans.), *The collected papers of Erich von Holst: Vol. 1. The behavioral physiology of animals and man* (pp. 33-135). Coral Gables, FL: University of Miami Press. (Original work published 1939)

von Holst, E., & von Saint Paul, U. (1973). On the functional organization of drives. In R. Martin (Ed. and Trans.), *The collected papers of Erich von Holst: Vol. 1. The behavioral physiology of animals and man* (pp. 220-258). Coral Gables, FL: University of Miami.

Wyman, R. (1965). Probabilistic characterization of simultaneous nerve impulse sequences controlling dipteron flight. *Biophysics Journal, 5,* 447-471.

Yamanishi, T., Kawato, M., & Suzuki, R. (1980). Two coupled oscillators as model for the coordinated finger tapping by both hands. *Biological Cybernetics, 37,* 219-225.

Yates, E. F. (1987). *Self-organizing systems: The emergence of order.* New York: Plenum Press.

Perspectives on the Coordination of Movement
S.A. Wallace (Editor)

ELEMENTS OF COORDINATED ARM MOVEMENTS IN THREE-DIMENSIONAL SPACE[§]

J. F. SOECHTING*

Laboratory of Neurophysiology
University of Minnesota

ABSTRACT

In this chapter some of the basic problems in the study of coordinated limb motion are defined, an outline of how they might be solved by the central nervous system is presented, and pertinent experimental results are summarized.

One problem is that the central nervous system has at its disposal extra degrees of freedom, and therefore limb motion is in principle indeterminate. Because the observed behavior is unique, one can conclude that this uniqueness results because suitable constraints have been imposed. Such constraints, if they can be identified, can provide insights concerning the way movements are planned and controlled.

A second issue concerns the need for sensorimotor transformations and the existence of frames of reference in which information about limb movement is represented. The premise of this chapter is that coordinated limb movements involve a succession of three such transformations: between extrinsic and intrinsic kinematic coordinates, between kinematics and dynamics, and between dynamics and muscle activation

*Address correspondence to: J. F. Soechting, Department of Physiology, 6-255 Millard Hall, University of Minnesota, Minneapolis, MN 55455, U.S.A.

§The author thanks the U.S. Public Health Service (NS-15018) and the National Science Foundation (BNS-8418539) for their continued support.

patterns. The rationale for this hypothesis is outlined, and pertinent experimental data are presented.

INTRODUCTION

Even though it was in the 1930s that Bernstein (1967) began to outline the problems the central nervous system must deal with in producing coordinated movements, it is only in the last decade that investigators have begun to study these problems systematically. In part, the reason for this tardiness is technical; it is only recently that the means have been developed for transducing motion in three-dimensional space without unduly encumbering the subject (Woltring, 1974). Perhaps another reason is conceptual. Researchers, in the reductionist tradition, felt that they could gain an understanding of motor behavior by focusing on simple behaviors, namely, movements restricted to one joint. They reasoned that the insights gained from such studies could be extrapolated to apply to more complicated behaviors involving motion at more than one joint. It was implicit in this approach that the problem of movement coordination is one of timing: A compound, multijoint movement would be nothing more than the sum of a number of single-joint movements whose temporal evolutions were appropriately related to one another.

Today we know that this is not the case and that the study of multijoint movements brings to the fore the coordination of movement in the spatial domain as well as in the temporal domain. The work in the past decade has led to a clearer understanding of what the problems are. How they are solved by the central nervous system is not as clear, although some tentative suggestions have emerged in this regard as well. In this chapter, I will attempt to define some of the basic problems in the study of coordinated limb motion, to outline how they might be solved, and to summarize pertinent experimental results.

DEGREES OF FREEDOM

Simply stated, the number of degrees of freedom of a system corresponds to the minimum number of parameters that are necessary to fully describe its state. The central problem, as stated by Bernstein (1967), is that the central nervous system has at its disposal redundant degrees of freedom; that is to say, a task can in principle be achieved in a variety of ways. A few examples will suffice to illustrate this point.

Consider the task of grasping an object (Figure 2.1). Three parameters are sufficient to specify its location in space. For example, they might be the distances x, y, and z in the anterior, lateral, and vertical directions. For the sake of simplicity, assume that the subject's trunk is fixed and that motion is restricted to the arm. Even if one ignores the hand and the wrist, a variety of postures of the arm are adequate to grasp the object. Figure 2.1 shows one example, but any rotation of the arm about the axis passing from the shoulder to the wrist (the dashed line in Figure 2.1) will not change the location of the wrist in space. Such rotations are possible because the arm has four degrees of freedom (three at the shoulder and one at the elbow), and this extra degree of freedom leads to the indeterminacy of the movement.

So far, we have viewed the object as if it were a point in space, three parameters being adequate to specify its location. Three more parameters will be necessary to define the orientation of a physical object such as a mug. (For example, the mug might be upright with the handle at the right or left, or it might be tilted at some angle. Two parameters are needed to specify the angle of tilt and one to describe the orientation of the handle.) The wrist also has three degrees of freedom (flexion, abduction, and supination), and thus the indeterminacy remains: the arm (shoulder, elbow, and wrist) having seven degrees of freedom, although only six parameters are needed to specify the location and orientation of an object in space.

Now consider the following task: Beginning from a prescribed posture of the limb, move toward and grasp the object. This task is doubly indeterminate. For example, the wrist could move in a straight line from the starting point to the end point, but in principle any curvilinear path passing through these two points is equally possible. Furthermore, at each point along the path, the arm could take on a variety of orientations.

The number of possible ways in which the task could be executed will obviously increase if some of the constraints are removed; that is, if motion of the trunk is also permitted. However, in this chapter

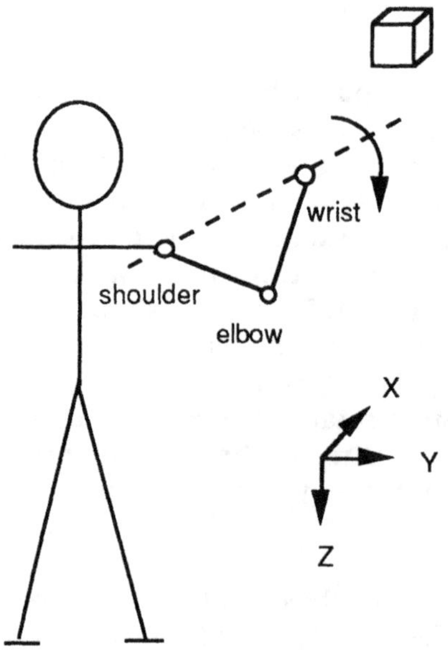

Figure 2.1. Kinematic redundancy. The diagram indicates one orientation of the arm appropriate to grasp the object. However, any other orientation produced by a rotation about the axis from shoulder to wrist will not change wrist position.

the problem will be simplified, and only arm motion with the trunk fixed will be considered.

The indeterminacies described so far are kinematic in that they relate to the motion or posture of different segments of a limb. Such motion or posture ultimately results from the forces produced by activating muscles that span the shoulder and elbow joints. At this level also there is indeterminacy: The number of degrees of freedom represented by the muscles exceeds the number of degrees of freedom necessary to specify the dynamical state of the limb, that is, the torques at the shoulder and elbow. A simple example, shown in Figure 2.2, illustrates this point. Here the task is to hold a weight represented by the force F_1. This weight will exert a torque at the shoulder and elbow joints tending to extend the arm and is resisted by

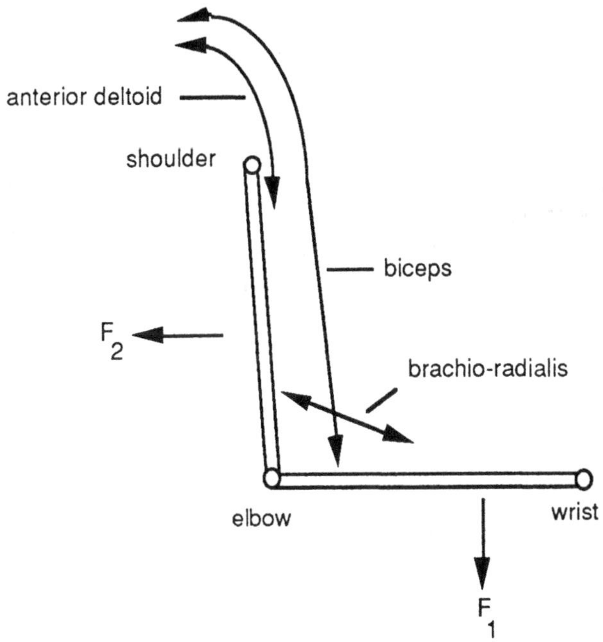

Figure 2.2. *Muscle redundancy. Different combinations of activation of biceps, brachio-radialis, and anterior deltoid are appropriate to oppose force F_1. The force F_2 can be opposed by various combinations of activity in anterior deltoid and biceps.*

activation of shoulder and elbow flexors. Three such muscles are indicated in the figure: anterior deltoid (a shoulder flexor), brachio-radialis (an elbow flexor), and biceps (a shoulder and elbow flexor). The external force could in principle be resisted by activation of only deltoid and brachio-radialis, or by activation of biceps to varying degrees and proportional reduction of the level of activation of the other flexors.

INVARIANTS AND CONSTRAINTS

Bernstein (1967) hypothesized that the central nervous system mastered the problem (and the potential) of redundancy by reducing the number of independent degrees of freedom. Since then, this point of view has been elaborated by, among others, Gelfand, Gurfinkel, Tsetlin, and Shik (1971), Greene (1972), Talbott (1979), Arbib (1984), Lee (1984), and McCollum, Horak, and Nashner (1984).

To make this hypothesis more concrete and to gain some insight into how it can be tested experimentally, it may be useful to see how it might be applied to the preceding examples illustrated in Figures 2.1 and 2.2. As stated, in a pointing task, the path taken by the wrist can in principle be arbitrary. However, one might suppose that it will follow a straight line and that the plane defined by the shoulder, elbow, and wrist always remains vertical. If so, the motion is defined uniquely both at the wrist and in terms of the changes in shoulder and elbow joint angles. Experimentally, then, one can define an invariant of the motion: Movements to targets at different locations and movements performed by different subjects should always obey this rule (straight line motion and vertical plane). Furthermore, the invariant implies that the number of independent degrees of freedom has been reduced and that this reduction is a consequence of neural control (biomechanically, other motions are possible).

A number of other possible invariants can also be envisioned, however. For example, one might suppose that shoulder and elbow angular motion is in tandem, leading to a ratio of angular velocities at the two joints that is constant. In general, such a rule would lead to movements of the wrist that do not follow a straight line (Hollerbach & Atkeson, 1987), and thus it should be possible experimentally to distinguish among different alternatives.

Several other points need to be noted. The identification of an invariant of a particular motor task can lead to insights concerning the parameters utilized by the central nervous system in planning and controlling the movement, that is, the level at which a particular movement is coordinated. The first alternative above would lead to the conclusion that movements are planned in extrinsic coordinates of extrapersonal space (the motion of the end point, wrist), whereas invariant relations among joint angles would imply that movements are organized in this intrinsic space. Experimental details pertinent to this question will be described in detail later on. At this point it suffices to say that, indeed, for arm movements, motion at the shoulder and elbow joints for any particular task is unique, as is the motion of the end point (wrist).

However, one can also cite instances in which invariants of a motion can be defined, and yet the constituent angular motions are highly variable from one instance to another. For example, Cole and Abbs (1986) have described the kinematics of the motion of the index finger and thumb in a rapid pinch. They found that the motion of the metacarpo-phalangeal and interphalangeal joints of the index finger and of the thumb differed from trial to trial. Nevertheless, movement at these three joints was not independent because the point of contact between finger and thumb remained invariant; there are characteristic trade-offs between the angular motions at the three joints. Gracco and Abbs (1985, 1986) have also made observations concerning movements underlying speech. Interlabial distance is an invariant property of such movements in the presence of variant motion of the upper and lower lips and jaw.

The invariants discussed so far have concerned the kinematic characteristics of a movement. The possibility of invariant patterns of muscle activations can also be entertained, such invariants being commonly termed "muscle synergies." For example, Bouisset, Lestienne, and Maton (1977) suggested that all of the elbow flexors were coactivated as a unit—the "flexor equivalent." Similarly, in analyzing the patterns of activity in leg muscles subserving posture following a perturbation, Nashner (1977) and Nashner and McCollum (1985) observed only a few distinct patterns. Although ankle and leg muscles were not activated synchronously, there was an ordered progression in the timing of the activities, and not all combinations of activation were observed.

In principle, muscle synergies and kinematic invariants are not incompatible; that is to say, the presence of muscle synergies could lead to movements that exhibit invariant kinematic characteristics. If so, it would be difficult to deduce experimentally the rules according to which movements are coordinated. However, at least in the case of arm movements, as will be detailed in a subsequent section, it is possible to differentiate between these eventualities.

As the title of this section suggests, it may be useful to distinguish between an invariant and a constraint. An invariant of a movement should be defined by a fixed relation among parameters or by a parameter that remains constant, defined in terms of kinematics or patterns of muscle activation. Furthermore, an invariant, by reducing the number of independent degrees of freedom, should lead to uniqueness of behavior. By contrast, a constraint limits the alternative ways in which a movement is executed without leading to uniqueness. For example, as Hogan (1984) suggested, the smoothness of the motion of the end point (wrist) may be maximized in a pointing movement. This postulate leads to a definite prediction of the path taken by the wrist (straight-line motion) and the velocity profile of that motion, without defining uniquely the angular motion at the shoulder and elbow. Alternatively, one could postulate that angular motion at the joints progresses monotonically; that is, there are no reversals in the direction of motion at the shoulder and elbow (cf. Hollerbach & Atkeson, 1987). Once again, not all motions would be possible, but this constraint by itself does not insure uniqueness. One can also envision other constraints that limit possible patterns of muscle activations.

The possibility exists, then, that coordinated movement derives from the presence of a number of constraints, none of which individually is sufficient to insure uniqueness in motor behavior. This notion has not been pursued extensively, most work to date having been aimed at defining the invariants of a motion. Furthermore, the distinction between invariants and constraints may not always be clear-cut. In the pinch task studied by Cole and Abbs (1986), the point of contact can be considered an invariant. It could equally well be defined as a constraint because by itself, it is insufficient to define uniquely the motion at each of the joints. Nevertheless, there may be some utility in differentiating between invariants and constraints, especially when one attempts to derive models of the neural mechanisms subserving motor behavior. (This point will be taken up in more detail in a subsequent section.)

MOVEMENT DYNAMICS

When motion is restricted to a single joint, force and linear acceleration are proportional to one another, according to Newton's second law, as are torque and angular acceleration. When there is motion at more than one joint, the situation is more complicated. The equation defining the torque at one joint depends on the angular acceleration at other joints as well. In addition, there are terms proportional to the squares of angular velocities that in general are not negligible (Goldstein, 1965; Hollerbach & Flash, 1982; Hoy & Zernicke, 1986). The equations relating torque to movement are nonlinear and not at all simple. For example, for an arm with four degrees of freedom (Figure 2.1), the equation for flexor torque at the elbow contains 25 terms (Soechting, 1983)!

Most importantly, the direction of the changes in torque, dictating the choice of muscles that will produce the movement, need not coincide with the direction of the motion. Consider the movement illustrated in Figure 2.3, involving forward projection of the arm. As shown in the figure, the upper arm is assumed to be initially vertical and the forearm horizontal, and the movement is assumed to require flexion at the shoulder and extension at the elbow. Mathematical details aside, one can predict how torque at the shoulder and elbow should change to propel the arm in the indicated direction from the following considerations. Newton's laws of motion are defined in an inertial frame of reference. With respect to this frame of reference, both upper arm and forearm rotate in the flexor direction (the forearm is inclined upwards at the end of the movement), and therefore flexor torque is required at the shoulder and elbow joints to initiate a movement in this direction. Thus, in the case of multijoint motion, the agonists of a movement, defined as the muscles that need to be activated at the onset of movement, may be muscles that undergo lengthening (brachio-radialis in this example) or shortening (anterior deltoid). In the case of single-joint movements, the agonist will instead always be the muscle that shortens.

Because this analysis has neglected all of the mathematical complexities, it is obviously only approximate. Nevertheless, it points to the computational difficulties any system faces in attempting to control multijointed motion.

SENSORIMOTOR MAPPING

Let us return to the task illustrated in Figure 2.1 and consider the sensory information that may be utilized in performing it. An image of the object falling on the retina provides the primary signal concerning the location of the object, namely, information about the location of the object relative to the gaze of the subject. However, the task ultimately requires knowledge of the location of the object relative to the hand. To determine the latter, the orientation of the

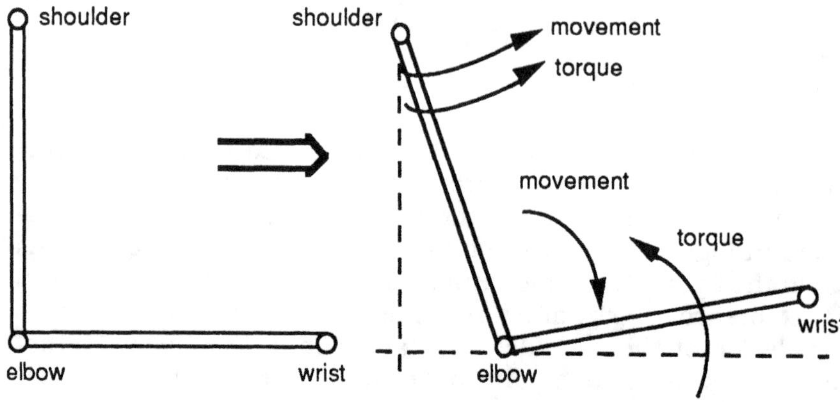

Figure 2.3. Schematic of a movement involving forward projection of the arm. The movement requires flexion at the shoulder but extension at the elbow. To produce it, flexor torque has to be produced at both joints.

eyes relative to the head, the orientation of the head relative to the trunk, and the orientation of the arm relative to the trunk must also be known. For that purpose, information from vestibular receptors and proprioceptors in the neck and arm must be utilized as well.

Thus sensory information from a number of different modalities is utilized in this task. For this information to be utilized effectively, it must be represented in common frames of reference. The need for frames of reference has been stated succinctly by Simpson and Graf:

> . . . the idea of reference frames is central to constructing a definition of sensorimotor integration. First, our description of motor behavior requires the formulations of mechanics, which use the idea of a frame of reference at one or another level of abstraction. Second, reference frames underly the making of measurements. . . . Third, sensorimotor integration . . . can be thought of as a transformation, and the term "transformation" calls for the idea of reference frames. (1985, p. 3)

In the task considered here, a number of frames of reference can be identified readily: a retinocentric one, a head-centered one, and one fixed to the trunk, each used to define the location of the object. Similarly for the limb, one can represent the orientation of one segment relative to the orientation of the adjacent segment (the angle between arm and forearm, for example) or relative to the trunk. Going from one frame of reference to another involves a sensorimotor transformation or mapping. (The term *sensorimotor* is used because some frames of reference may be neither strictly sensory nor strictly motor, as discussed by Grobstein [1988] and Sparks [1988].)

If one is to understand the computational characteristics of such sensorimotor transformations, it is necessary to understand how information is represented in each frame of reference. One possibility is that attached to each frame of reference is a system of coordinate axes. For example, to represent the location of an object relative to the trunk, one might use the Cartesian coordinates x, y and z shown in Figure 2.1. Similarly, to represent the location of the hand relative to the shoulder, one might use a coordinate system of angles.

Several points need to be noted. First, it has proved useful to distinguish between extrinsic coordinates (defining the location of elements extrinsic to the subject—the mug, for example) and intrinsic coordinates (defining the location of elements intrinsic to the subject, such as the arm). Second, it may be useful to introduce the idea of natural coordinate systems: The location of the object could as well be represented in cylindrical or spherical coordinates as in Cartesian coordinates. The term *natural coordinates* is derived from mechanics, where it is used to refer to the coordinate system in which a particular computation is easiest. (For example, straight-line motion is most easily calculated in Cartesian coordinates; the oscillation of a pendulum, in cylindrical or spherical coordinates.) The term adapted to biological systems refers to coordinate systems in which a particular sensorimotor transformation is most readily performed. As a corollary, one would expect to find a neural representation of information in terms of the axes of such natural coordinate systems.

One example of a natural coordinate system is provided by the axes perpendicular to the planes of the semicircular canals and another by the pulling directions of the extraocular muscles. The sensorimotor transformations between these coordinate systems underlying the vestibulo-ocular reflex have been discussed by Robinson (1982, 1985) and by Simpson and Graf (1985). Centrally, if information is encoded in a particular coordinate system, one would expect to find neurons whose firing is tuned to one of the axes of that system. As Simpson (1984) has shown, visually modulated neurons in the dorsal cap of the inferior olive do exhibit such properties.

Finally, Pellionisz and Llinas (1980; Pellionisz, 1984) have proposed a general algorithm, based on tensor network theory, according to which sensorimotor transformations from one coordinate system to another could be accomplished.

Information in a particular frame of reference can also be represented independently of any coordinate system, namely, in terms of a map. A map represents a topographically ordered representation of space, usually on a two-dimensional surface. Its neural correlate would consist of a collection of neurons, each tuned, broadly or narrowly, to one region of space, neighboring neurons being tuned to adjacent regions of space. Some examples of such maps are the representation of auditory and visual space in the optic tectum of the owl (Konishi, 1986; Knudsen, du Lac, & Esterly, 1987) and the representation of echo information in the bat's auditory cortex (Suga, 1984). A sensorimotor transformation from one map to another can be achieved very simply by connecting topographically equivalent points on two maps.

In adapting the idea of maps to the study of limb movements in extrapersonal space, one confronts the immediate problem that there are more than two dimensions in movement space, whereas the concept of maps has been developed, experimentally and conceptually, principally for two-dimensional problems (Grobstein, 1988). For example, in the superior colliculus, the azimuth and elevation of a point in space in some reference frame are represented by a map (Sparks, 1986; 1988); how and where distance is represented is not established.

Second, the fact that information is represented in a map-like fashion in the central nervous system is not necessarily incompatible with the existence of an underlying coordinate system. As mentioned before, neurons tend to exhibit tuning curves that are broad (cf. Knudsen et al., 1987). As a consequence, the location of a point in space, for example, cannot be inferred from the activity of a single neuron; a sampling of the discharge of a population of neurons is required (Georgopoulos, Caminiti, Kalaska, & Massey, 1983; Georgopoulos, Schwartz, & Kettner, 1986; Heiligenberg, 1987). The question then is, How is the information contained in a map "read out"; that is, how is the discharge of the population sampled?

Particular forms of addressing can lead to coordinate representations at subsequent stages of information processing. For example, in the optic tectum of the owl, the tuning of neurons for the azimuth of the location of a point in space varies in the rostro-caudal direction (Knudsen et al., 1987). If neurons in the tectum are addressed by their rostro-caudal and medio-lateral position, a representation of location in terms of azimuth and elevation would evolve automatically. Other forms of addressing would lead to other coordinate representations; and, as mentioned before, point-to-point mappings would lead to representations that are independent of any coordinate system.

COORDINATED LIMB MOVEMENTS

Having outlined some of the problems that the central nervous system faces in producing coordinated limb movements and some of the concepts that may guide their investigation, I will present in this section a hypothesis about the way the task is accomplished (Soechting & Terzuolo, 1986). Subsequent sections will deal with experimental data relevant to the hypothesis as well as with an alternative to it.

The basic hypothesis is that goal-directed limb movements involve a series of sensorimotor transformations, none of which is mathematically exact, executed sequentially. To make the discussion more concrete, consider two tasks: point to a target and draw a figure, such as an ellipse, in a given plane. In both tasks, the information required to perform the movement is represented in an extrinsic frame of reference, provided either by the visual system or cognitively from verbal instructions given to the subject. Successful execution of the tasks ultimately requires the activation of pertinent muscles in an appropriate pattern, that is, a transformation from the extrinsic space to the space of muscle activations.

Such a transformation could in principle be accomplished directly, but the hypothesis is that there are intermediate levels of representation. Making this assumption simplifies the problem conceptually. Furthermore, there is also experimental evidence in its favor. One additional assumption will be made: A coordinate system is associated with each level of representation, and going from one level to another involves coordinate transformations (Benati, Gaglio, Morasso, Tagliasco, & Zaccaria, 1980).

The first such transformation is a kinematic one—from the extrinsic coordinates of extrapersonal space to the intrinsic coordinates describing the orientation of the arm. In the examples considered, this transformation involves the mapping of a point or a curve in space into an appropriate set of joint angles at the shoulder and elbow. As already noted, this transformation is not unique. In the first example, the path taken by the hand is not specified. In neither task is the velocity profile of the movement determinate, nor is the set of joint angles corresponding to a given position of the wrist in space.

For the moment, assume that these parameters are specified. The next transformation is then from kinematics to dynamics, that is, from a trajectory of angular motions at the shoulder and elbow to a variation of the torque at these joints adequate to produce the motion. Note that this particular transformation is unique, but mathematically complex. The last transformation is required to partition the torque among the muscles of the limb. This transformation also is not unique.

Thus two of the postulated three transformations are not unique. Nevertheless, uniqueness in the behavior is observed. The final element of the hypothesis is that uniqueness results because appropriate constraints are introduced and that one utility of these constraints is to simplify the mapping from one coordinate representation to another.

PSYCHOPHYSICS AND COORDINATE SYSTEMS

In the absence of any precise knowledge of the manner in which information about limb position is represented centrally (Costanzo & Gardner, 1981; Mountcastle, Lynch, Georgopoulos, Sakata, & Acuna, 1975; Sakata, Takaoka, Kawarasaki, & Shibutani, 1973), psychophysics can be used to predict what the natural coordinate system might be. The basic premise is that if information is represented in terms of a coordinate system, then the projection of a point in this frame of reference onto one of the coordinate axes should be readily identifiable. A simple example can illustrate this point. Suppose the problem is to identify the location of cities such as Minneapolis, Milan, and Manila. If location is represented in terms of longitude and latitude, then subjects should be able to locate each of the cities in terms of these parameters, longitude independently of latitude and vice versa. If the coordinate system is a different one, say, distance from and compass reading relative to New York (the center of the universe!), then subjects should make smaller errors in locating the other cities in this coordinate system. Finally, if the representation is in terms of a map independent of any coordinate system, subjects would be expected to be able to pinpoint the cities on a globe but to make appreciable errors in either of the two postulated coordinate systems.

A natural coordinate system for representing the orientation of the arm was identified with this approach. Subjects were asked to match one and only one of the four angles that define arm orientation at a time with their right and left arms (Soechting, 1982; Soechting & Ross, 1984). Figure 2.4 illustrates one simple example of the approach. The orientation of the forearm can be defined by the angle it makes with the upper arm (relative orientation) or the angle it makes with the vertical or the trunk (absolute orientation). Subjects were asked to match either the absolute or the relative orientations of their two forearms in situations in which the inclinations of the upper arms differed. Because the error in matching absolute orientation was consistently less than that for matching relative orientation, the conclusion is that the former angle constitutes one of the coordinate axes of the natural coordinate system.

This finding has been corroborated by Worringham and Stelmach (1985) and by Worringham, Stelmach, and Martin (1987). They also found that subjects were very accurate in matching forearm inclinations to the vertical but that when subjects were required to match elbow joint angles, they made errors strongly biased in the direction of matching forearm inclinations. Soechting and Ross (1984) subsequently extended these psychophysical studies to identify the natural coordinates for representing upper arm as well as forearm orientation. The results of this study are shown in Figure

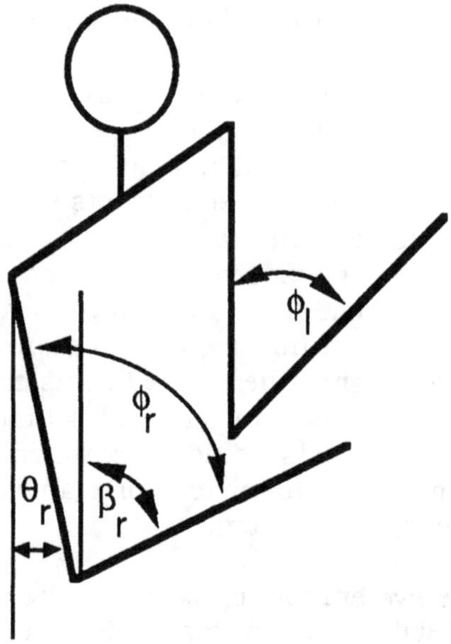

Figure 2.4. Two representations of forearm rotation: elbow angle ϕ and forearm inclination β. They differ unless flexion at the shoulder θ is zero. *Note.* From "Does Position Sense at the Elbow Reflect a Sense of Joint Angle or One of Limb Orientation?" by J. F. Soechting, 1982, *Brain Research, 248,* p. 393. Copyright 1982 by Elsevier Biomedical Press. Adapted by permission.

2.5. Upper arm and forearm orientation are each represented by two parameters: the angular elevation of the limb segment, which is the angle between the segment and the vertical axis measured in the vertical plane, and the yaw angle, which is the angle between the segment and the anterior direction, measured in the horizontal plane.

As will be shown, this particular intrinsic coordinate system has proven useful in understanding how to specify limb angular motion appropriate to producing curvilinear wrist motion in a given plane. A similar approach could provide insight into the preferred extrin-

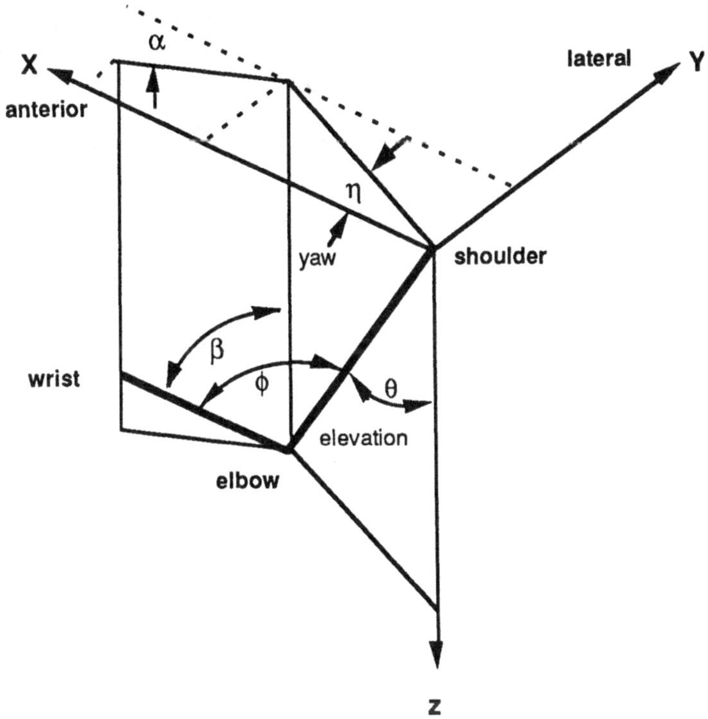

Figure 2.5. The natural coordinate system to represent arm orientation as defined psychophysically. The angles θ and β represent angular elevation of upper arm and forearm; η and α are the corresponding yaw angles. *Note.* From "An Algorithm for the Generation of Curvilinear Wrist Motion in an Arbitrary Plane in Three-Dimensional Space" by J. F. Soechting and C. A. Terzuolo, 1986, *Neuroscience, 19,* p. 1394. Copyright 1986 by IBRO, Pergamon Journals Limited. Adapted by permission.

sic coordinate representation of a point in space and the mapping between the two coordinate systems, but this work has not yet been done.

POINT-TO-POINT ARM MOVEMENTS

Point-to-point arm movements were the first arm movements to be investigated systematically. Initial research efforts were devoted in particular to those arm movements in which movement was confined to the horizontal plane (Abend, Bizzi, & Morasso, 1982; Georgopoulos, Kalaska, & Massey, 1981; Morasso, 1981; Prablanc, Echallier, Komilis, & Jeannerod, 1979) or to the sagittal plane (Soechting & Lacquaniti, 1981). More recently, movements not confined to one of these cardinal planes have also been studied (Georgopoulos et al., 1986; Lacquaniti, Soechting, & Terzuolo, 1986).

Although some details are still unresolved, these and other studies since then all point to one conclusion: Movements are organized at the level of kinematics. Some of the observations that led to this conclusion are (a) there is little trial-to-trial variability in the path taken by the wrist, and the variability decreases with practice (Georgopoulos et al., 1981); (b) the path is independent of the speed of the movement (Soechting & Lacquaniti, 1981) and of any additional inertial loads on the arm (Atkeson & Hollerbach, 1985; Lacquaniti, Soechting, & Terzuolo, 1982); and (c) the path is independent of accuracy constraints placed on the movement (Soechting, 1984).

Thus for movement from one point in space to another, the path taken by the wrist is one invariant of the motion. Angular motion at the shoulder and elbow is also invariant. This invariance is inevitable for movements confined to the horizontal or sagittal plane because only two degrees of freedom of shoulder and elbow motion are involved. However, this conclusion is also true for other point-to-point movements (Lacquaniti et al., 1986), in which all four degrees of freedom of limb motion participate.

The contention that these movements are organized on a kinematic level rests on the fact that during such movements, patterns of muscle activation are not invariant. For example, the pattern of muscle activation depends strongly on the speed of movement (cf. Ghez & Gordon, 1987) and on inertial load. Further support for this contention is that the shoulder and elbow motion is not affected by concomitant wrist motion, such as prono-supination (Lacquaniti & Soechting, 1982) or flexion-extension (Soechting, 1984). For a movement such as the one illustrated in Figure 2.3, biceps is one of the agonists for elbow and shoulder motion. If the movement also requires wrist pronation, one observes less biceps activity because biceps is an antagonist to the wrist motion (Lacquaniti & Soechting, 1982), whereas the kinematics of shoulder and elbow motion remains the same. In summary, kinematic invariants occur in the presence of (predictable) variability in the patterns of muscle activation that give rise to the movement.

One other point can be made. Recall that six parameters are required to specify the location and orientation of an object. The studies of point-to-point arm movement suggest that there is a partitioning of the sensorimotor mapping between intrinsic and extrinsic coordinates, namely, one mapping between object location and the four orientation angles of the arm and a separate one between object orientation and the wrist angles. Such a scheme has also been proposed in the theoretical literature on robotic manipulator control (Benati et al., 1980; Hollerbach, 1985). Also, there is a high degree of variability in the timing of wrist angular motion relative to arm motion (Lacquaniti & Soechting, 1982), as has also been observed for finger and thumb movements. This does not mean that wrist motion is independent of motion at the more proximal joints; it is coordinated to minimize the variability of the position of the finger tips (Lacquaniti, Ferrigno, Pedotti, Soechting, & Terzuolo, 1987; Soechting, 1984).

As for the temporal aspects of the movement, the speed of the wrist also exhibits invariant characteristics, namely, a bell-shaped profile (Abend et al., 1982; Atkeson & Hollerbach, 1985; Morasso, 1981). One advantage of such a temporal invariant may be that it simplifies the transformation between kinematics and dynamics. If the velocity profile obeys scaling laws, then so does torque, provided the gravitational and inertial components are computed separately (Atkeson & Hollerbach, 1985; Hollerbach, 1984). Hogan (1984) and Flash and Hogan (1985) have instead argued that such invariant velocity profiles arise because the system attempts to maximize the smoothness of the movement.

If there are kinematic invariants, the question naturally arises, Are these movements organized in intrinsic or extrinsic coordinates? Initial answers to this question were conflicting, and the question is still not resolved. Morasso (1981) and Abend et al. (1982) noted that for the movements they studied, the path taken by the wrist was close to rectilinear, although they found reversals in angular motion at the shoulder and elbow. They naturally argued in favor of an organization in terms of extrinsic coordinates. Soechting and Lacquaniti (1981) instead found that the amount by which shoulder and elbow angles changed in the terminal phase of the movement was about the same, independent of target location, whereas the motion at the wrist could exhibit a small but consistent curvature. They argued in favor of an organization in terms of intrinsic coordinates.

Both conclusions rested on a limited set of data, and neither holds true for all point-to-point arm movements. For example, in some movements, the path taken by the wrist has a pronounced curvature (Atkeson & Hollerbach, 1985; Lacquaniti et al., 1986). The question whether angular motion at the shoulder and elbow does reverse direction in point-to-point movements is still open; Atkeson and Hollerbach (1985) claimed to have almost never observed such reversals, and the possibility exists that the estimates of joint motion reported by Morasso (1981) were contaminated by translation of the shoulder. Similarly, as suggested by Hollerbach and Atkeson (1987), Soechting and Lacquaniti's (1981) results may hold true only for movements in which elbow extension is close to its limit and may not be a true test of the hypothesis. Hollerbach and Atkeson also suggested that movements are organized in intrinsic coordinates but that a strategy of "staggered joint interpolation" may hold; namely, that the onset of the movement at different joints is staggered, movement beginning first at the joint where the required angular motion is greatest (see also Kaminski & Gentile, 1986). Whether this latest hypothesis will be able to account for all movement trajectories is still an open question.

Another possibility is that neither alternative is correct but that the observed behavior may result from a multiplicity of constraints that, when acting together, lead to a movement kinematically invariant. Some of these constraints may simplify the transformation between kinematics and dynamics; others may act to simplify the transformation between intrinsic and extrinsic coordinates. As will be shown, drawing movements can best be understood from this perspective.

Other constraints may act strictly at the level of dynamics. For example, Lacquaniti et al. (1986) observed that the pattern of torques at the shoulder and elbow for point-to-point movements was always very simple: an initial propulsive phase followed by a reversal in torque to decelerate the limb. For a movement whose wrist trajectory showed a pronounced curvature, they induced subjects to also follow rectilinear trajectories and found that the latter required a reversal in direction of angular motion at the shoulder and a triphasic pattern of torque, that is, two reversals in torque. They thus argued that such curvilinear wrist trajectories may be due to a dynamic constraint acting to simplify the temporal patterns of joint torque.

Concerning the central representation of movements, Georgopoulos and his colleagues (cf. Georgopoulos, 1986) have found cells in motor cortex whose discharge was broadly tuned to the direction of two- or three-dimensional movements. The rate of discharge of these neurons was maximal for movements in one particular direction; for movements away from this preferred direction, the rate of discharge was proportional to the cosine of the angle between the preferred and the actual directions. Given the broad tuning of these cells, they inferred that information about the direction of movement was derived from a population of such neurons. Recently, they (Georgopoulos & Massey, 1988) have approached this problem from an information theoretic point of view and have shown that the information transmitted by the neural population about the intended direction of movement exceeds information about the actual movement. Their results obviously support the contention that there is a central representation of movement kinematics and that the patterns of muscle activations are derived by subsequent sensorimotor transformations.

CURVILINEAR WRIST MOTION

The details of the sensorimotor transformation between extrinsic and intrinsic coordinate systems can be identified more readily from study of movements in which the path of the wrist is specified by the task. For example subjects may be asked to draw circles and ellipses repetitively (Soechting, Lacquaniti, & Terzuolo, 1986; Soechting & Terzuolo, 1986). In such a task, even though the tempo (i.e., the average speed) may be predetermined, the instantaneous speed is a free variable. Nevertheless, there is a characteristic relationship between the spatial and temporal aspects of the motion of the wrist; the radius of curvature has been found to be proportional to the cube of the tangential velocity (or equivalently, angular velocity is proportional to curvature to the 2/3 power), as demonstrated by Viviani and Terzuolo (1982) and by Lacquaniti, Terzuolo and Viviani (1983).

This relationship, which holds true for a large variety of curvilinear wrist movements, represents an invariant of the motion in extrinsic coordinates. Another invariant of the motion can be identified, however, when these movements are described in the intrinsic coordinate system of the orientation angles of the arm. For circles and ellipses drawn in a variety of planes, the modulation in these angles was close to sinusoidal (Soechting et al., 1986). Furthermore, the phase between the modulation in the angular elevations of the upper arm and forearm was close to 180° (Figure 2.6). This was true for ellipses slanted in different directions and drawn in planes ranging from the sagittal plane to the frontal plane and inclined by as much as 45° to the vertical.

Other experimental results and mathematical simulations led to two conclusions: (a) the invariants in extrinsic and intrinsic coordinates have a common origin; and (b) their significance is that they simplify the coordinate transformation between extrinsic and intrinsic coordinates.

Concerning the first point, one can show that in an average sense, a phase difference of 180° between the modulation in the angular elevation of upper arm and forearm minimizes the distortion from sinusoidal motion of the vertical component of wrist velocity. If the motion of the wrist is sinusoidal, the relationship between curvature and tangential velocity is satisfied automatically. Simulations supported this conclusion; random combinations were taken of the amplitudes and mean values of the modulations in upper arm and forearm elevations, and the average distortion resulting for a given value of phase was computed. The significance is that, given a particular combination of amplitudes and means, a phase difference other than 180° could give the least distortion; the minimum at 180° holds true only in an average sense and not necessarily in any particular instance.

As for the coordinate transformation between the two frames of reference, my colleagues and I noted that experimentally (a) the phase difference between upper arm and forearm yaw angles was linearly related to the azimuth of the plane of motion and (b) the phase

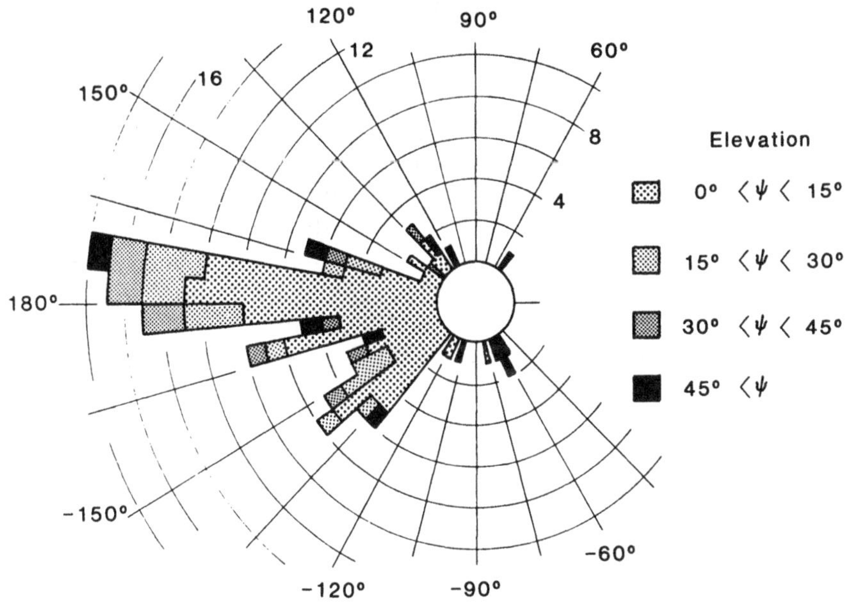

Figure 2.6. Constraint that simplifies the sensorimotor mapping between intrinsic and extrinsic coordinates for curvilinear wrist motion. The sinusoidal modulation in upper arm and forearm angular elevations is close to 180° out of phase for a large range of motions. Shown is the variation of this phase difference with the planar elevation of wrist motion (ψ). *Note.* From "An Algorithm for the Generation of Curvilinear Wrist Motion in an Arbitrary Plane in Three-Dimensional Space" by J. F. Soechting and C. A. Terzuolo, 1986, *Neuroscience, 19,* p. 1398. Copyright 1986 by IBRO, Pergamon Journals Ltd. Adapted by permission.

difference between forearm yaw and angular elevation was linearly related to the slant of the ellipse. Thus the coordinate transformation between these extrinsic and intrinsic parameters is the simplest one imaginable—a linear relation. Simulation studies showed that these two rules were valid, also in an average sense, only if the phase difference between angular elevations was close to

180°; no such simple relations emerged for phases differing substantially (45° or more) from 180°.

Thus in the sense discussed previously, the 180° phase difference between angular elevations constitutes a constraint that leads to a unique solution by reduction of the number of intrinsic degrees of freedom from four to three. Furthermore, its utility is that it simplifies the transformation from extrinsic to intrinsic coordinates; that is, it provides some simple rules whereby curvilinear movement can be produced in an arbitrary plane in space. Note that these rules are only approximate; the mathematically exact transformations are highly nonlinear (Soechting & Terzuolo, 1986). As a consequence, if one in fact uses such rules, one should observe predictable distortions when circles are drawn in some planes, for example, a flattening of the proximal portion of the wrist trajectory in the sagittal plane. Such distortions are in fact observed experimentally (Soechting et al., 1986).

Although the transformation just described is strictly valid only for periodic arm motions that result in elliptical trajectories at the wrist, there is a way to extend its domain by making one assumption: Arbitrary, continuous arm motion is segmented, and each segment traces an arc of an ellipse and is subject to the constraint described previously. Several predictions arise from this assumption (Soechting & Terzuolo, 1986; 1987a; 1987b): (a) The relations described between intrinsic and extrinsic parameters should hold; (b) during each segment, motion of the wrist should be confined to one plane, but the plane of motion can change abruptly from one segment to the next; and (c) subjects should be unable to generate wrist motion whose tangent plane changes smoothly and continuously.

In a test of these predictions, subjects were asked to produce a variety of continuous movements in free space, such as to draw figure eights and stars, arbitrary scribbles, and closed curves with smoothly changing tangent planes, as for example represented by the seams of a tennis ball. In accord with the predictions, in all of these movements motion at the wrist was confined to one plane for periods of time, but the plane of motion could change abruptly. For example, the upper and lower loops of the figure eight were drawn in planes that could differ by 20° or more. For learned movements such as figure eights and stars, the same relations between intrinsic and extrinsic parameters were found as before, as they were for scribbles. Only when subjects were asked to produce novel movements requiring them to generate wrist motion with a smoothly changing tangent plane was there violation of the constraint that the phase between forearm and upper arm elevation be close to 180°. However, even in these tasks wrist motion was piece-wise planar.

In summary, one can also define invariants of continuous arm motion; and these invariants derive from a constraint whose utility is to simplify, in an approximate manner, the coordinate transformation from extrinsic to intrinsic coordinates and vice versa.

TORQUE AND MUSCLE ACTIVATION

In the hypothesized scheme, two more transformations need be considered: the one from kinematics to dynamics and the other from torque to muscle activation. Nothing can be said about the former at this time. The latter has been the subject of several investigations in recent years, with the result that a lawful relationship appears to exist between these two sets of variables; that is, given a level of torques, one can predict the amount by which each of the muscles will be activated.

Figure 2.2 illustrates the general problem. Given the force F_1, which will exert a torque tending to extend the upper arm and the forearm, one can expect activation of elbow and shoulder flexor muscles to resist this torque. Furthermore, one might expect the activity in an elbow flexor such as brachio-radialis to be proportional to elbow torque, that of anterior deltoid to be proportional to shoulder torque, and that of biceps, which spans both joints, to be proportional to both. Now add the force F_2, which acts on the upper arm and produces a torque only at the shoulder. If the initial hypothesis is correct, anterior deltoid and biceps activity should increase because their level of activation was assumed to be proportional to shoulder torque. If so, brachio-radialis activity must *decrease*; otherwise, muscle torque at the elbow would not remain constant. Therefore, brachio-radialis activity should be negatively correlated with shoulder torque even though that muscle does not cross the shoulder joint. In general, the direction in torque space in which a muscle will be maximally activated will not coincide with its pulling direction.

Figure 2.7 illustrates this point for the patterns of activation of shoulder and elbow flexors and extensors under isometric conditions, as reported by Soechting and Lacquaniti (1988). Gielen and van Zuylen (1986) have reported similar results for combinations of torques producing flexion-extension and prono-supination of the forearm, as have Buchanan, Almdale, Lewis, and Rymer (1986) for combinations of flexion-extension and internal-external humoral rotation. Similarly, Baker, Goldberg, and Peterson (1985), investigating the vestibulocollic reflex in decerebrate cats, showed that the direction of head acceleration for which the activation of each of the neck muscles was maximal did not coincide with their pulling direction. All of these results were obtained under isometric conditions.

Relationships between torque and muscle activation during movement remain to be determined. There is the strong likelihood that they may differ from those found under isometric conditions: Keshner et al. (1986) found that the activation vectors during head movements in alert cats could differ substantially from those found under isometric conditions in decerebrate cats.

EQUILIBRIUM POINT HYPOTHESIS

In the preceding sections, I have developed the hypothesis that coordination of arm movements involves a series of sensorimotor transformations and the imposition of a number of constraints and that the constraints lead to uniqueness in the behavior and a simplification of the transformations. An alternate hypothesis, namely, the *equilibrium point hypothesis*, appears to avoid many of the complexities described so far.

This hypothesis, as presented by Bizzi and his colleagues (Bizzi, Accornero, Chapple, & Hogan, 1982, 1984; Hogan, 1985), is a modification of one put forth by Feldman (1986). Its principal elements are as follows. First, in terms of neural control, posture and movement are equivalent, movement being a continuous transition in posture. Second, muscle has spring-like properties. Third, because of these properties, the state of a limb at any time can be characterized by an equilibrium position and a stiffness. Fourth, given sufficient stiffness, movement can be generated by displacement of the equilibrium point. Fifth, during movement the equilibrium point

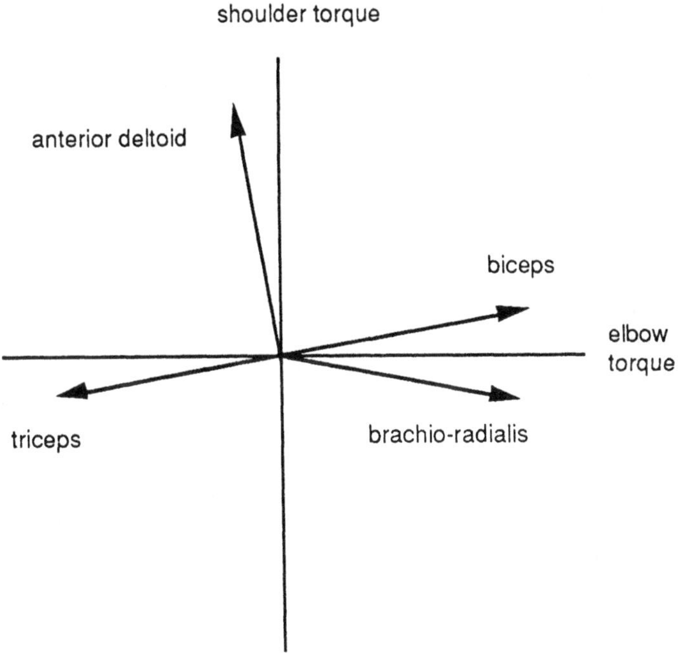

Figure 2.7. Data for some elbow and shoulder flexors and extensors obtained under isometric, steady-state conditions. The direction in torque space for which a muscle is maximally activated does not coincide with the muscle's pulling direction.

(virtual trajectory) is displaced to maximize the smoothness (i. e., minimize jerk) of the movement.

This hypothesis is obviously appealing. By treating statics and dynamics (posture and movement) equivalently, it obviates the need for the transformation between kinematics and dynamics and thus avoids the most difficult part of the problem. Furthermore, Bizzi and colleagues have presented evidence consistent with the hypothesis. For example, at least for small displacements, muscle does exhibit spring-like properties (Mussa-Ivaldi, Hogan, & Bizzi, 1985). Results from deafferentation studies and perturbations imposed during movement (Bizzi et al., 1982, 1984; Hogan, 1984) also support the hypothesis. Finally, Flash (1987) has done simulations studies showing that coordinated multijoint movements can be generated by appropriately displacing the equilibrium point; that is, that the hypothesis is feasible.

There are a number of points that the hypothesis does not address, however. First, a transformation between extrinsic and intrinsic coordinates is still required because the equilibrium point for a given muscle must be set as a function of the desired position of the end point. Second, the hypothesis does not explain how a unique solution emerges, given kinematic redundancy. Third, it is not clear that the solution to the problem as given by this hypothesis is at all simple. For example, as Hasan and Enoka (1985) point out, given muscle geometry, there are instances in which the "equilibrium point" would be displaced during the movement in the direction opposite to the static posture prevailing at the end of the movement. Finally, the equilibrium point is set by an appropriate level of muscle activity. The relationship between muscle activation and the equilibrium point has not been investigated, except under static conditions (Lestienne, Polit, & Bizzi, 1981). As has already been mentioned, the patterns of muscle activity responsible for a given movement can be complicated indeed.

LEARNING

At the close of this chapter, it may be appropriate to consider how the transformations hypothesized to underly coordinated movement come to be established. Loeb (1983) clearly stated the problem and a plausible solution. He noted that two alternative solutions to the problem are generally proposed: "direct computation of the mechanical inverse and the use of look-up tables of preprogramed movements" (p. 203). Because neither solution is very plausible, one resorts to simplifying assumptions, which generally do not work. Loeb then proposed a way out of this dilemma: "When first confronted with . . . an entirely new movement, we look for similarities to previously performed movements and, perhaps, interpolate a reasonable approximation" (p. 203).

The hypothesis put forth in this chapter is consistent with Loeb's proposal. It proposes that coordinated movement results from the imposition of constraints that are acquired through learning. The utility of such constraints is that they simplify particular classes of movement, such as drawing ellipses in different planes, and ensure similarity for a large class of movements. These movements will be only approximately correct. Finally, a particular constraint may be applicable to only one class of movements; different movements may involve different constraints.

ACKNOWLEDGMENTS

Much of the author's work reported in this chapter was done in collaboration with Francesco Lacquaniti and Carlo Terzuolo.

REFERENCES

Abend, W., Bizzi, E., & Morasso, P. (1982). Human arm trajectory formation. *Brain, 105*, 331-348.

Arbib, M. A. (1984). From synergies and embryos to motor schemas. *Advances in Psychology, 17*, 545-562.

Atkeson, C. G., & Hollerbach, J. M. (1985). Kinematic features of unrestrained vertical arm movements. *Journal of Neuroscience, 5*, 2318-2330.

Baker, J., Goldberg, J., & Peterson, B. (1985). Spatial and temporal responses of the vestibulocollic reflex in decerebrate cats. *Journal of Neurophysiology, 54*, 735-756.

Benati, M., Gaglio, S., Morasso, P., Tagliasco, V., & Zaccaria, R. (1980). Anthropomorphic robotics: I. Representing mechanical complexity. *Biological Cybernetics, 38*, 125-140.

Bernstein, N. (1967). *The coordination and regulation of movements*. Oxford: Pergamon.

Bizzi, E., Accornero, N., Chapple, W., & Hogan, N. (1982). Arm trajectory formation in monkeys. *Experimental Brain Research, 46*, 139-143.

Bizzi, E., Accornero, N., Chapple, W., & Hogan, N. (1984). Posture control and trajectory formation during arm movement. *Journal of Neuroscience, 4*, 2738-2744.

Bouisset, S., Lestienne, F., & Maton, B. (1977). The stability of synergy in agonists during the execution of a simple movement. *EEG Clinical Neurophysiology, 42,* 543-551.

Buchanan, T. S., Almdale, D. P. J., Lewis, J. L., & Rymer, W. Z. (1986). Characteristics of synergic relations during isometric contractions of human elbow muscles. *Journal of Neurophysiology, 56,* 1225-1241.

Cole, K. J., & Abbs, J. H. (1986). Coordination of three-joint digit movements for rapid finger-thumb grasp. *Journal of Neurophysiology, 55,* 1407-1423.

Costanzo, R. M., & Gardner, E. P. (1981). Multi-joint neurons in somatosensory cortex of awake monkeys. *Brain Research, 214,* 321-334.

Feldman, A. G. (1986). Once more on the equilibrium-point hypothesis (λ model) for motor control. *Journal of Motor Behavior, 18,* 17-54.

Flash, T. (1987). The control of hand equilibrium trajectories in multi-joint arm movements. *Biological Cybernetics, 57,* 257-274.

Flash, T., & Hogan, N. (1985). The coordination of arm movements: An experimentally confirmed model. *Journal of Neuroscience, 5,* 1688-1703.

Gelfand, I. M., Gurfinkel, V. S., Tsetlin, M. L., & Shik, M. L. (1971). Some problems in the analysis of movements. In I. M. Gelfand, V. S. Gurfinkel, S. V. Fomin, & M. L. Tsetlin (Eds.), *Models of the structural-functional organization of certain biological systems* (pp. 329-345). Cambridge, MA: MIT Press.

Georgopoulos, A. P. (1986). On reaching. *Annual Review of Neuroscience, 9,* 147-170.

Georgopoulos, A. P., Caminiti, R., Kalaska, J. F., & Massey, J. T. (1983). Spatial coding of movement: A hypothesis concerning the coding of movement direction by motor cortical populations. *Experimental Brain Research Supplement, 7,* 327-336.

Georgopoulos, A. P., Kalaska, J. F., & Massey, J. T. (1981). Spatial trajectories and reaction times of aimed movements: Effects of practice, uncertainty and change in target location. *Journal of Neurophysiology, 46*, 725-743.

Georgopoulos, A. P., & Massey, J. T. (1988). Cognitive spatial-motor processes: 2. Information transmitted by the direction of two dimensional arm movements and by neuronal populations in primate motor cortex and area 5. *Experimental Brain Research, 69*, 315-326.

Georgopoulos, A. P., Schwartz, A. B., & Kettner, R. E. (1986). Neuronal coding of movement direction. *Science, 233*, 1416-1419.

Ghez, C., & Gordon, J. (1987). Trajectory control in targeted force impulses: I. Role of opposing muscles. *Experimental Brain Research, 67*, 225-240.

Gielen, C. C. A. M., & van Zuylen, E. J. (1986). Coordination of arm muscles during flexion and supination: Application of tensor analysis approach. *Neuroscience, 17*, 527-539.

Goldstein, H. (1965). *Classical mechanics*. Reading, MA: Addison-Wesley.

Gracco, V. L., & Abbs, J. H. (1985). Dynamic control of the perioral system during speech: Kinematic analyses of autogenic and nonautogenic processes. *Journal of Neurophysiology, 54*, 418-432.

Gracco, V. L., & Abbs, J. H. (1986). Variant and invariant characteristics of speech movements. *Experimental Brain Research, 65*, 156-166.

Greene, P. H. (1972). Problems of organization of motor systems. *Progress in Theoretical Biology, 2*, 304-338.

Grobstein, P. (1988). Between the retinotectal projection and directed movement: Topography of a sensorimotor interface. *Brain Behavior and Evolution, 31*, 34-48.

Hasan, Z., & Enoka, R. M. (1985). Isometric torque-angle relationships and movement-related activity of human elbow flexors: Implications for the equilibrium point hypothesis. *Experimental Brain Research, 59,* 441-450.

Heiligenberg, W. (1987). Central processing of sensory information in electric fish. *Journal of Comparative Physiology A, 161,* 621-631.

Hogan, N. (1984). An organizing principle for a class of voluntary movements. *Journal of Neuroscience, 4,* 2745-2754.

Hogan, N. (1985). The mechanics of multi-joint posture and movement control. *Biological Cybernetics, 52,* 315-332.

Hollerbach, J. M. (1984). Dynamic scaling of manipulator trajectories. *Journal of Dynamical Systems Measurement and Control, 106,* 102-106.

Hollerbach, J. M. (1985). Optimum kinematic design for a seven degree of freedom manipulator. In H. Hanafusa & H. Inoue (Eds.), *Robotics research: The second international symposium* (pp. 215-222). Cambridge, MA: MIT Press.

Hollerbach, J. M., & Atkeson, C. G. (1987). Deducing planning variables from experimental arm trajectories: Pitfalls and possibilities. *Biological Cybernetics, 56,* 279-292.

Hollerbach, J. M., & Flash, T. (1982). Dynamic interactions between limb segments during planar arm movement. *Biological Cybernetics, 44,* 67-77.

Hoy, M. G., & Zernicke, R. F. (1986). The role of intersegmental dynamics during rapid limb oscillations. *Journal of Biomechanics, 19,* 867-877.

Kaminski, T., & Gentile, A. M. (1986). Joint control strategies and hand trajectories in multijoint pointing movements. *Journal of Motor Behavior, 18,* 262-278.

Keshner, E. A., Baker, J., Banovetz, J., Peterson, B. W., Wickland, C., Robinson, F. R., & Tomko, D. L. (1986). Neck muscles demonstrate preferential activation during voluntary and reflex head movements in the cat. *Society for Neuroscience Abstracts, 12,* 684.

Knudsen, E. I., du Lac, S., & Esterly, S. (1987). Computational maps in the brain. *Annual Review of Neuroscience, 10,* 41-65.

Konishi, M. (1986). Centrally synthesized maps of sensory space. *Trends in Neuroscience, 9,* 163-168.

Lacquaniti, F., Ferrigno, G., Pedotti, A., Soechting, J. F., & Terzuolo, C. (1987). Changes in spatial scale in drawing and handwriting: Kinematic contribution by proximal and distal joints. *Journal of Neuroscience, 7,* 819-828.

Lacquaniti, F., & Soechting, J. F. (1982). Coordination of arm and wrist motion during a reaching task. *Journal of Neuroscience, 2,* 399-408.

Lacquaniti, F., Soechting, J. F., & Terzuolo, C. A. (1982). Some factors pertinent to the organization and control of arm movements. *Brain Research, 252,* 394-397.

Lacquaniti, F., Soechting, J. F., & Terzuolo, C. A. (1986). Path constraints on point-to-point arm movements in three-dimensional space. *Neuroscience, 17,* 313-324.

Lacquaniti, F., Terzuolo, C., & Viviani, P. (1983). The law relating the kinematic and figural aspects of drawing movements. *Acta Psychologica, 54,* 115-130.

Lee, W. A. (1984). Neuromotor synergies as a basis for coordinated intentional action. *Journal of Motor Behavior, 16,* 135-170.

Lestienne, F., Polit, A., & Bizzi, E. (1981). Functional organization of the motor processes underlying the transition from movement to posture. *Brain Research, 230,* 121-132.

Loeb, G. E. (1983). Finding common ground between robotics and neurophysiology [Letter to the editor]. *Trends in Neurosciences, 6,* 203-204.

McCollum, G., Horak, F. B., & Nashner, L. M. (1984). Parsimony in neural calculations for postural movements. In J. R. Bloedel (Ed.), *Cerebellar function* (pp. 52-66). Heidelberg: Springer.

Morasso, P. (1981). Spatial control of arm movements. *Experimental Brain Research, 42*, 223-237.

Mountcastle, V. B., Lynch, J. C., Georgopoulos, A., Sakata, H., & Acuna, C. (1975). Posterior parietal association cortex of monkey: Command functions for operations within extrapersonal space. *Journal of Neurophysiology, 38*, 871-908.

Mussa-Ivaldi, F. A., Hogan, N., & Bizzi, E. (1985). Neural, mechanical and geometric factors subserving arm posture in humans. *Journal of Neuroscience, 5*, 2732-2743.

Nashner, L. M. (1977). Fixed patterns of rapid postural responses among muscles during stance. *Experimental Brain Research, 30*, 13-24.

Nashner, L. M., & McCollum, G. (1985). The organization of human postural movements: A formal basis and experimental synthesis. *Behavioral and Brain Sciences, 8*, 135-172.

Pellionisz, A. J. (1984). Coordination: A vector-matrix description of overcomplete CNS coordinates and a tensorial solution using the Moore-Penrose generalized inverse. *Journal of Theoretical Biology, 110*, 353-376.

Pellionisz, A., & Llinas, R. (1980). Tensorial approach to the geometry of brain function: Cerebellar coordination via a metric tensor. *Neuroscience, 5*, 1125-1136.

Prablanc, C., Echallier, J. F., Komilis, E., & Jeannerod, M. (1979). Optimal response of eye and hand motor systems in pointing at a visual target: I. Spatio-temporal characteristics of eye and hand movements and their relationships when varying the amount of visual information. *Biological Cybernetics, 35*, 113-124.

Robinson, D. A. (1982). The use of matrices in analyzing the three-dimensional behavior of the vestibulo-ocular reflex. *Biological Cybernetics, 46*, 53-66.

Robinson, D. A. (1985). The coordinates of neurons in the vestibulo-ocular reflex. *Reviews in Oculomotor Research, 1*, 297-312.

Sakata, H., Takaoka, Y., Kawarasaki, A., & Shibutani, H. (1973). Somatosensory properties of neurons in the superior parietal cortex (area 5) of the rhesus monkey. *Brain Research, 64*, 85-102.

Simpson, J. I. (1984). The accessory optic system. *Annual Review of Neuroscience, 7*, 13-41.

Simpson, J. I., & Graf, W. (1985). The selection of reference frames by nature and its investigation. *Reviews in Oculomotor Research, 1*, 3-20.

Soechting, J. F. (1982). Does position sense at the elbow reflect a sense of elbow joint angle or one of limb orientation? *Brain Research, 248*, 392-395.

Soechting, J. F. (1983). *Kinematics and dynamics of the human arm* (Laboratory of Neurophysiology Report). Minneapolis, MN: University of Minnesota, Laboratory of Neurophysiology.

Soechting, J. F. (1984). Effect of target size on spatial and temporal characteristics of a pointing movement in man. *Experimental Brain Research, 54* ,121-132.

Soechting, J. F., & Lacquaniti, F. (1981). Invariant characteristics of a pointing movement in man. *Journal of Neuroscience, 1*, 710-720.

Soechting, J. F., & Lacquaniti, F. (1988). Quantitative evaluation of electromyographic responses to multidirectional load perturbations of the human arm. *Journal of Neurophysiology, 59*, 1296-1313.

Soechting, J. F., Lacquaniti, F., & Terzuolo, C. A. (1986). Coordination of arm movements in three-dimensional space. Sensorimotor mapping during drawing movement. *Neuroscience, 17*, 295-311.

Soechting, J. F., & Ross, B. (1984). Psychophysical determination of coordinate representation of human arm orientation. *Neuroscience, 13*, 595-604.

Soechting, J. F., & Terzuolo, C. A. (1986). An algorithm for the generation of curvilinear wrist motion in an arbitrary plane in three-dimensional space. *Neuroscience, 19*, 1395-1405.

Soechting, J. F., & Terzuolo, C. A. (1987a). Organization of arm movements. Motion is segmented. *Neuroscience, 23*, 39-52.

Soechting, J. F., & Terzuolo, C. A. (1987b). Organization of arm movements in three-dimensional space. Wrist motion is piecewise planar. *Neuroscience, 23*, 53-61.

Sparks, D. L. (1986). Translation of sensory signals into commands for control of saccadic eye movements: Role of primate superior colliculus. *Physiological Reviews, 66*, 118-171.

Sparks, D. L. (1988). Neural cartography: Sensory and motor maps in the superior colliculus. *Brain Behavior and Evolution, 31*, 49-56.

Suga, N. (1984). The extent to which bisonar information is represented in the bat auditory cortex. In G. M. Edelman, W. E. Gall, & W. M. Cowan (Eds.), *Dynamical aspects of neocortical function* (pp. 315-373). New York: Wiley.

Talbott, R. E. (1979). Ferrier, the synergy concept, and the study of posture and movement. In R. E. Talbott & D. R. Humphrey (Eds.), *Posture and movement* (pp. 1-12). New York: Raven Press.

Viviani, P., & Terzuolo, C. (1982). Trajectory determines movement dynamics. *Neuroscience, 7*, 431-437.

Woltring, H. J. (1974). New possibilities for human motion studies by a real time light spot position measurement. *Biotelemetry, 1*, 132-146.

Worringham, C. J., & Stelmach, G. E. (1985). The contribution of gravitational torques to limb position sense. *Experimental Brain Research, 61*, 38-42.

Worringham, C. J., Stelmach, G. E., & Martin, Z. E. (1987). Limb segment inclination sense in proprioception. *Experimental Brain Research, 66*, 653-658.

Perspectives on the Coordination of Movement
S.A. Wallace (Editor)

SEARCH STRATEGIES AND THE ACQUISITION OF COORDINATION[§]

K. M. NEWELL*, P. N. KUGLER, R. E. A. VAN EMMERIK, and P. V. MCDONALD

Department of Kinesiology
University of Illinois at Urbana-Champaign

ABSTRACT

This chapter synthesizes the extant data on the acquisition of coordination and outlines an approach to examining the exploratory behavior individuals use in mastering the redundant biomechanical degrees of freedom during acquisition of the "form" of movement coordination patterns. A particular focus is the search strategies used to explore and locate stable equilibrium regions in the perceptual-motor work space and the way these strategies relate to the emergent movement form. It is proposed that only a small set of acquisition characteristics may emerge from the search strategies used and that only a small set of search strategies (such as blind, local, and nonlocal) will be identified within and between the equilibrium regions of the work space. Our efforts are oriented toward a theory of skill acquisition that is driven by the learning of the dynamical laws guiding exploratory behavior in the acquisition of skill.

---

*Address correspondence to: K. M. Newell, Department of Kinesiology, University of Illinois at Urbana-Champaign, 906 South Goodwin Avenue, Urbana, IL 61801, U.S.A.

§This work was supported in part by the National Institute of Health Award HD 21212.

One of the most striking features of human action is the commonality of the movement form that arises from the torso and limbs of individuals attempting to satisfy a given set of task constraints. There are, of course, individual differences in the quantitative aspects of response production for a given set of task constraints, but it is the qualitative nature of movement form that provides the basis for defining and labeling physical activities. For example, in activities such as gymnastics and highboard diving, the goal is the production of a set of specified kinematic and geometric movement patterns, although different physical activities clearly afford varying levels of difficulty for identifying and labeling the action. In many other activities, a common movement form appears to emerge from a stable or optimal pattern of movement dynamics that meets the criterion set of outcome task demands, even though these task demands do not directly specify the movement form. The movement form in a variety of tasks can be defined by characterization of the set of kinematic relative motions that emerge from the function for the coordination and control of action (Newell, 1985).

The movement form exhibited by the skilled performer is not, however, generally isomorphic with the movement form exhibited by the beginner or naive performer of a given task. On many occasions, the performance of the beginner appears qualitatively different from that of the skilled performer. It is not merely that the skilled performer is quantitatively better than the beginner on an externally defined task criterion. Rather, the movement pattern or form of the beginner is often nominally distinct from that produced by the skilled performer in a given activity.

In this chapter, we outline a formal approach to characterizing the changes in the qualitative properties of a learner's movement pattern as the learner progresses through the stages of skill acquisition. Previously, we have argued that the very fundamental stage of learning a skill involves acquiring the form of movement or the set of kinematic relative motions that reflects a stable or possibly an optimal pattern of coordination and control for the task constraints at hand (Newell, 1985). This view is consistent with the theoretical proposition that the qualitative dynamics of movement is an emergent property of the constraints imposed on action, rather than the acquisition of prescriptions for action that specify a given movement form or coordination mode (Kugler, Kelso, & Turvey, 1980, 1982; Kugler & Turvey, 1987).

The acquisition of coordination has not been a focus of the extant skill acquisition literature in large part because of the tasks that have been selected for the study of motor skill acquisition. The tasks that have dominated the study of motor skills have required the constraint of a single biomechanical degree of freedom (e.g., key press, linear positioning tasks); or, where more than one biomechanical degree of freedom is involved (e.g., pursuit rotor), the tasks have required the acquisition of only the scaling of the coordination function because the parameters that reflect the essential variables of the coordination function (Gelfand & Tsetlin, 1962) can already be appropriately constrained. In essence, task selection for the study of skill acquisition, no matter what the theoretical viewpoint, has been based on the general principle of studying so-called simple skills, with complex skills loosely characterized as more complicated manifestations of physical activity. The simplicity-complexity dimension for tasks has generally been defined on a unidimensional, quantitative continuum, such as number of muscle groups or body segments, number of alternate stimuli, and number of possible alternate responses.

The one domain in which the acquisition of coordination has been a central concern is that of motor development although, strangely, the development of motor skills has usually been seen as a phenomenon distinct from that of the acquisition of skill. Gesell (1929) proposed that the development of the fundamental movement patterns of posture, locomotion, and prehension follows identifiable directions. However, these behavioral manifestations of directional changes in the qualitative properties of movement kinematics may not be confined to the acquisition of phylogenetic activities. As we show in this chapter, there is mounting evidence that the acquisition of ontogenetic activities follows macrolevel behavioral trends similar to those exhibited by infants in the development of the fundamental movement patterns. Furthermore, we develop the proposition that these trends in the acquisition of coordination generally reflect a dynamical interpretation of the changes in the coordination mode, rather than a maturational perspective, as promulgated by Gesell and other developmentalists, or a prescriptive, cognitive view of action.

In the coordinative structure view (Kugler et al., 1980, 1982; Kugler & Turvey, 1987), the assembly and disassembly of coordination modes is modeled in terms of the generation and annihilation of dynamical attractor or equilibrium regions (such as "preferred" comfort states) in perceptual-motor work spaces. The emerging form of a movement pattern at the behavioral level, which we label here as a coordinative mode, is identified with the stability of the equilibrium regions in the perceptual-motor work space. The exploration and location of these equilibrium regions or attractor states reflects an exploratory behavior we call *search strategies*.

The tenets of the coordinative structure theory have led to a number of empirical tests of (a) the scaling laws associated with the equilibrium manifold for an oscillatory coordination model involving one or two degrees of freedom (Kugler, 1983; Turvey, Rosenblum, Schmidt, & Kugler, 1986; Turvey, Schmidt, Rosenblum, & Kugler, 1988) and (b) the bifurcation laws associated with the breakdown of the equilibrium manifold during transitions between anti-phase and in-phase oscillatory modes (Haken, Kelso, & Bunz, 1985; Kelso, 1981; see also Jeka & Kelso, this volume). However, the identification of the search strategies used to explore the "layout" of stable and unstable properties of the dynamical manifold is an important part of modeling the perceptual-motor work space that heretofore has not been examined systematically. The search strategy orientation provides a principled basis for understanding the macrolevel behavioral changes in the movement form or coordinative mode that occur in the acquisition and development of coordination. Our approach to search strategies and the acquisition of coordination is consistent with a theory of skill acquisition and motor tasks that is driven by the dynamical laws that guide exploratory behavior in the acquisition of skill, rather than by the memory-intensive search procedures that are the hallmark of the traditional laws of perceptual-motor learning.

CHANGES IN MOVEMENT FORM AND THE ACQUISITION OF COORDINATION

Human physical activities are often dichotomized into so-called phylogenetic and ontogenetic activities. Phylogenetic activities are viewed as common to the survival of the species and include such activities as sucking, swallowing, walking, and standing. These fundamental movements have been the primary focus of the field of motor development. In contrast, ontogenetic activities are viewed as a reflection of the shorter term demands of society on individual development and include such activities as piano playing, typewriting, and tennis. These activities have generally been the province of the field of skill acquisition or motor learning.

In this section of the chapter, we begin by synthesizing the literature pertaining to phylogenetic and ontogenetic activities and examine studies that have reported changes in movement patterns as a function of practice and individual experience. In the developmental literature, these changes in movement form are often said to reflect the development of coordination, whereas in the domain of skill acquisition, such changes in movement form are considered a consequence of the acquisition or learning of coordination. Previously, Newell (1986) has proposed that the distinction between phylogenetic and ontogenetic activity is merely a useful, heuristically motivated categorization rather than a fundamental distinction based upon different principles of coordination and control. As further evidence supporting this idea, we show that a number of descriptive parallels exist in the changes of movement form that occur as a function of practice in both phylogenetic and ontogenetic activities.

Phylogenetic Activities

In 1929, Arnold Gesell outlined six principles of maturational growth that characterize the developmental changes observed to be common across children in the first few years of life. These six developmental principles were motor priority and fore-reference, reciprocal interweaving, developmental direction, functional asymmetry, self-regulation, and optimal realization (see also Gesell, 1946, 1952). Of particular relevance to this chapter on the acquisition of coordination is the proposal that the emerging patterns of movement coordination in the developing infant follow the anatomically based directional trends of cephalo-caudal, proximal-distal, and ulnar-radial.

Gesell (1946) had a strong interest in anatomy, embryology, and morphology and believed that the developmental trends in the cephalo-caudal, proximal-distal and ulnar-radial directions, previously identified at a variety of levels of analysis for the structural development of the fetus, were paralleled in the macro-level functional development of the neonate and infant. Furthermore, Gesell recognized that the morphological strategies for characterizing the form of the developing embryo could in principle be applied to the development of movement form in the infant. This link of morphological principles to both the structure and function of biological systems had been articulated previously in considerable detail by D'Arcy Thompson (1917).

Unfortunately, Gesell (1946) never formalized operational approaches to the analysis of movement form, and his anatomical characterization of the directions of movement development in terms of cephalo-caudal and proximal-distal were based on check-list accounts of the earliest chronological age for the onset of a co-ordinated action at a given anatomical unit. Thus, Gesell's developmental trends are anatomically defined and independent of environmental or functional perspectives that also could be formulated to account for the directional changes. The descriptive directional trends of proximal-distal and cephalo-caudal identified by Gesell for the development of postural and locomotory coordination in phylogenetic activities were confirmed in other early characterizations of the emergence of the fundamental movement patterns in infancy (e.g., Bayley, 1935; McGraw, 1935; Shirley, 1931). The approach of using a check list of motor ability to test the development of coordination was, and to a large extent still is, the designated analytical tool for researchers characterizing the milestones in the development of the fundamental movement patterns.

The development of prehension, particularly the grip configurations, has not been studied as extensively as the development of postural and locomotory activities. However, the existing evidence, which is largely based on the work of Halverson (1931), is consistent with the proposition that refined differentiation of finger activity in the grip pattern follows an ulnar-radial directional trend. It should be recognized, however, that a little finger and thumb or a little finger and ring finger two-point grip pattern is not particularly stable or functionally useful for meeting the majority of task constraints. Thus, the developmental ulnar-radial trend (and, in fact, the other developmental directional trends) may be driven as much, if not more, by task constraints as by the biological constraints that reflect maturational development (Newell, 1986; Newell & Scully, 1988a).

The anatomically based directional trends in the development of coordination have also been observed in more sophisticated and detailed analyses of the dynamics of changing movement form exhibited by a particular child as a function of continued practice in the activity. Bernstein (1967), who was also a proponent of using principles from morphology to analyze biological motion, characterized the dynamical changes in the development of locomotion from the onset of independent walking through the 10th year of life. He recorded significant qualitative age-related differences in the leg dynamics during the development of walking and running that are not complete until about the 10th year, although the final sequence in the acquisition of walking generally is evident by the 5th year (see also Grieve & Gear, 1966). For example, Bernstein suggested that it is impossible to observe clear differences between running and walking in the 2nd year of life. Furthermore, the differences in movement form between these locomotory activities begin to emerge in a proximal-distal direction, with the mature form of the leg dynamics always appearing initially in the proximal leg segments.

Bernstein (1967) indicated that anatomical, physiological, and mechanical factors could all potentially contribute to these directional trends in the development of locomotion. However, he argued that it was improbable that the nerve dynamics of the distal musculature are so significantly different from those of the proximal musculature, and thus a mechanical account of the directional trends seemed more appropriate. Bernstein noted that the proximal ends of the legs are surrounded by far more massive muscles than are the distal ends; at the same time, the moments of inertia of the former are much less than the moments of inertia of the latter. For this reason, the muscles of the hip can move the upper sections of the limbs much more easily than they can move the foot because to move the foot, they must oppose the inertia of the entire leg from top to bottom. Thus, the dynamics of the distal leg segments depend on the dynamics of the proximal portions of the legs and ultimately on the functional integration of the dynamics of all the limb link segments. Of course, the "stages" of the qualitative changes in limb dynamics in the development of locomotion that Bernstein outlined are not consonant with the notion of the maturational stage advanced by Gesell and other developmentalists (see Newell & Scully, 1988b, for a discussion of this notion and qualitative changes in the dynamics of motor development).

In summary, there is considerable evidence that the development of the fundamental movement patterns is consistent with a cephalo-caudal, proximal-distal, and ulnar-radial directional description originally advanced by Gesell. Although much of this evidence is limited to the all-or-none check-list characterizations of the development of coordination, there have been some analyses of the response dynamics that also support the developmental directional viewpoint. Gesell (1929) gave a maturational interpretation to these directional trends in the manner of emerging biological prescriptions for action, but the data can also be interpreted from the dynamical perspective of the coordinative structure theory (Kugler et al., 1980; Kugler & Turvey, 1987), a perspective that was developed in part on the basis of Bernstein's (1967) proposals regarding the problem of coordination and control in biological systems. Indeed, the constraints perspective to channeling the response dynamics as opposed to specifying them through prescriptions appears particularly relevant in light of the similar macrolevel directional trends of the qualitative properties of the response dynamics evident in humans of all ages who are learning ontogenetic activities.

Ontogenetic Activities

There has been no systematic study of the acquisition of coordination in ontogenetic activities, but a few isolated studies have reported changes in the qualitative aspects of the response dynamics in a variety of tasks. Most studies of skill acquisition have analyzed only the quantitative changes in a single dimension of response output, such as distance moved or time to move through a given distance, and have thereby provided the field of motor learning and control with a basis for understanding what can be characterized as nonoptimal control (Newell, 1985). In this section, we review studies that have reported qualitative changes as a function of skill acquisition in the response dynamics of ontogenetic activities and then contrast these findings with the developmental data on phylogenetic activities. We begin with the studies that have reported changes in within-limb coordination as a function of practice in a given task.

Arutyunyan, Gurfinkel, and Mirskii (1968, 1969) provided preliminary evidence that with practice in a pistol-aiming task, subjects progress from relatively tightly locked arm links to a cooperative synergism in which compensatory movements occur between wrist and shoulder to minimize variability in pistol movement. Furthermore, it appeared that in skilled marksmen, more biomechanical degrees of freedom are independently introduced into the coordinative mode. Arutyunyan et al.'s experiments also provided evidence that the coordination and control function is very task specific. For example, a change of postural support for pistol shooting, such as standing to sitting, led to a concomitant change in the qualitative aspects of the coordinative structure for the arm action. Unfortunately, these interesting observations on the changes in the qualitative, dynamical aspects of skill acquisition are marred by Arutyunyan et al.'s provision of only very limited data in these pistol-shooting studies to support their theoretical claims.

In a preliminary analysis of adult subjects learning to write with their nondominant limb (Newell & van Emmerik, in press; van Emmerik & Newell, 1987, in press), we found evidence directly relating the number of joint degrees of freedom controlled to the phase of skill acquisition. Furthermore, the nature of the task constraints (e.g., writing orientation, limb dominance, size of writing) strongly influenced the emergence of a given coordination mode. Of particular relevance to the current discussion is the finding that the coordination of limb links in writing appeared to be controlled proximally at the shoulder joint in the nondominant limb, but distally at the wrist joint in the more practiced, dominant limb. An example of these directional coordination trends in the acquisition of an ontogenetic task is displayed in Figures 3.1 and 3.2, which clearly show the distal phase-locking in the pen-wrist and wrist-elbow joint couplings in the unpracticed, nondominant limb while subjects were writing their signature. In contrast, there appears to be independent control of the three limb segments in the practiced, dominant limb.

In this study of the acquisition of handwriting, we also required 3 subjects who were right-hand dominant to practice writing their signature and continuous cursive *e*'s for some 10,000 trials over a period of several weeks. The general effect of practice on the organization of the nondominant left limb in naturally dominant right-handers was small and limited to the proximal shoulder joint motion. This practice effect was consistent with the acquisition principle that control becomes more distal within a given limb segment as a function of practice. However, the findings from this

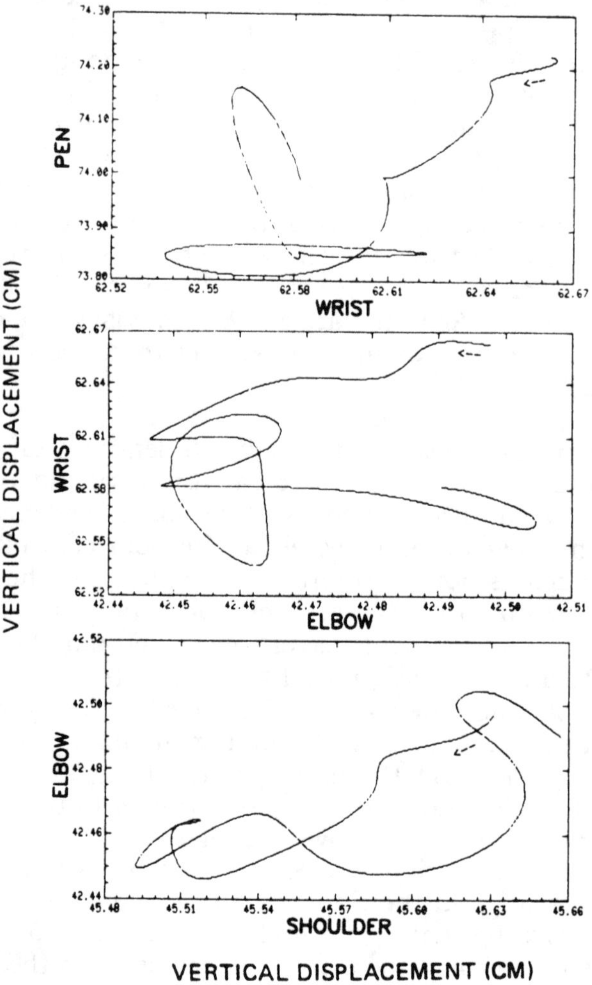

Figure 3.1. Position-position plots of pairs of pen and joint movements for individual trials of dominant signature production. *Note.* From "The Acquisition of Coordination: Preliminary Analysis of Learning to Write" by K. M. Newell and R. E. A. van Emmerik, in press, *Human Movement Science.* Copyright by Elsevier Science Publishers B.V. (North-Holland). Adapted by permission.

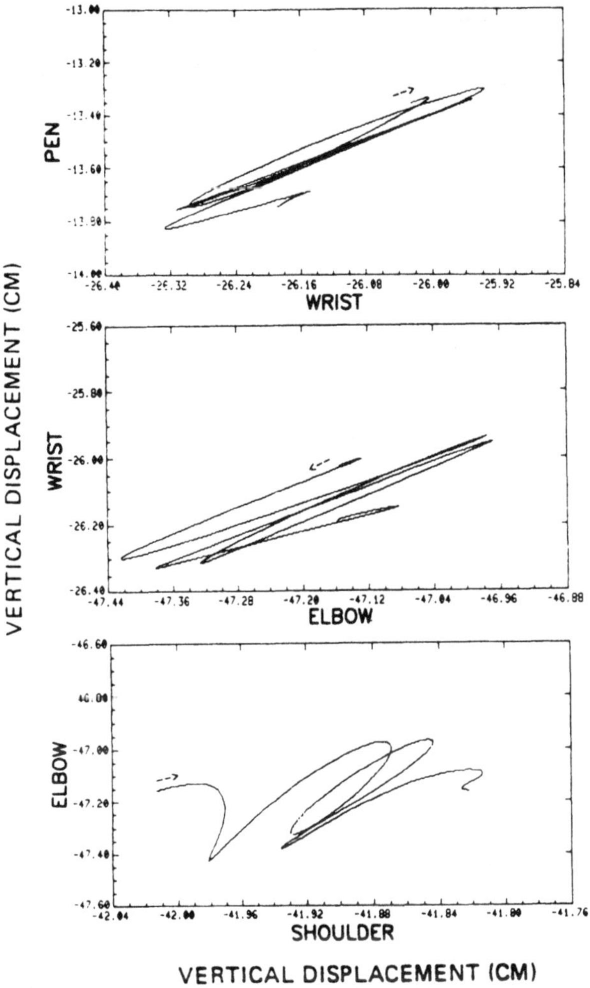

Figure 3.2. Position-position plots of pairs of pen and joint movements for individual trials of nondominant limb signature production. *Note.* From "The Acquisition of Coordination: Preliminary Analysis of Learning to Write" by K. M. Newell and R. E. A. van Emmerik, in press, *Human Movement Science*. Copyright by Elsevier Science Publishers B.V. (North-Holland). Adapted by permission.

practice study suggested that many tasks with multiple degrees of freedom may well require considerably longer periods of practice than have been typically provided in the domain of skill acquisition if subjects are to generate a movement form that reflects skilled behavior. This postulation is consistent with anecdotal observations on the time it takes people to learn to write and generate new movement forms to "solve" motor problems in domains such as music and sport.

A more direct test of the influence of practice on the transitions in directional proximal-distal coordination in a dart-throwing task was recently conducted by McDonald, van Emmerik, and Newell (in press). In this study, adult subjects practiced dart throwing for blocks of 250 trials per session. Practice with the dominant (two sessions) and nondominant (five sessions) limb was spread over a period of 10-14 days. One of the key findings was a significant reduction in the cross-correlations between the angular displacement of the wrist-elbow and wrist-shoulder joint linkages in the dominant arm as a function of practice. Again, this finding is consistent with the proposition that the distal wrist joint was operating more independently as a function of continued practice as a consequence of control shifting from proximal to distal segments. The interlink cross-correlation between joint segments remained high in the nondominant limb, a finding that confirms the earlier findings of Newell and van Emmerik (in press) that the nondominant limb tends to require more practice time than the dominant limb to induce the qualitative changes in limb organization. This contrast between the dominant and nondominant limbs may reflect general differential training arising from prior experience rather than hard-wired, hemispheric asymmetry.

An acquisition of coordination task from our laboratory that examined both within- and between-limb coordination is reported in an experiment by Sparrow and Irizarry-Lopez (1987). This study examined changes in mechanical efficiency and transport efficiency of adult subjects learning to walk on hands and feet (crawl) on a motor-driven treadmill at constant speed (0.76 m/s). The subjects performed twenty 3-min trials spread over a period of a few weeks. Sparrow and Irizarry-Lopez showed considerable changes in the stride duration and step length of the hand and foot movements as a function of practice. Of particular relevance here are the qualitative changes in the hip-knee kinematic organization that occurred over the practice period. There was a progressive decrease in the angle-angle cross-correlations, a result that is consistent with the freeing up of the constraint between the linkages as a function of practice. A sample angle-angle diagram reflecting these coordination changes with practice in some individual subjects from the Sparrow and Irizarry-Lopez study is shown in Figure 3.3.

Van Emmerik and Valk (1984) have also shown that the topological characteristics of torso and limb kinematics change as a function of practice in a slalom-ski simulation task. The inexperienced performer showed relative locking of the upper and lower part of the body whereas the experienced performer moved upper and lower body parts in an out-of-phase coordination mode. These changes are consistent with Bernstein's (1967) intuition that the experienced performer can give direct control to a larger number of biomechanical degrees of freedom than the inexperienced performer can. Furthermore, these qualitative differences in the patterns of coordination were linked to differences in performance level, specifically differences in amplitude and frequency of the ski board motion (see also den Brinker & van Hekken, 1982).

The effects of practice on a between upper limb coordination task were studied by Swinnen, Walter, and Shapiro (in press). They had subjects learn a task requiring simultaneous movements of the left and right hands but imposing different spatial and temporal constraints on each limb. The left limb performed a unidirectional movement; simultaneously, the right limb was required to produce a reversal movement pattern. Swinnen et al. found qualitative changes in the between-limb coordination as a function of practice in that the synchrony constraint was gradually released. Furthermore, the subjects were more successful in decoupling the limbs to meet the independent task demands when the movements of each limb were not initiated simultaneously.

In summary, the studies reviewed on the acquisition of coordination in ontogenetic tasks have revealed a number of parallels to those established for the development of phylogenetic activities. These commonalities relate to the directional trends for shifts in coordination and to the increase in the number of degrees of freedom that appear to be controlled independently with practice. The time frame for the changes in the coordination patterns sometimes

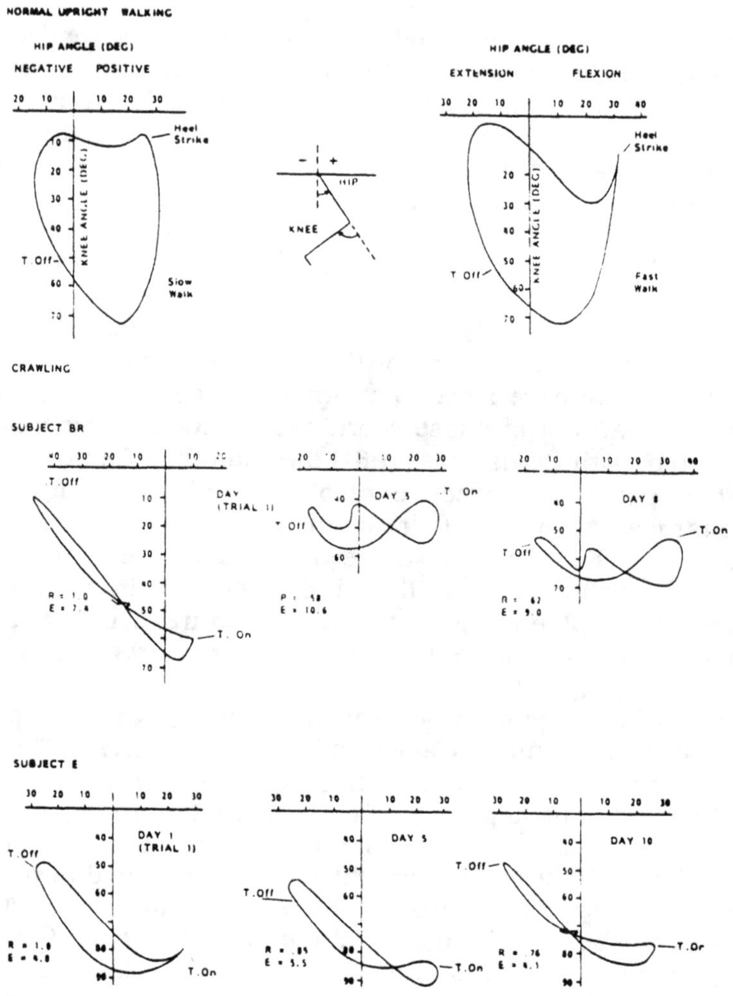

Figure 3.3 Individual subject data for crawling show changes in the pattern of thigh-shank angles as a function of practice. Toe-on corresponds to heel strike in normal walking. *R* is the recognition coefficient for each angle-angle plot comparison with Day 1 (Trial 1). *E* is mechanical efficiency (%) from the linked segment calculations of mechanical power output. *Note.* From "Mechanical Efficiency and Metabolic Cost as Measures of Learning a Novel Gross Motor Skill" by W. A. Sparrow and V. M. Irizarry-Lopez, 1987, *Journal of Motor Behavior, 19*, p. 254. Published by Heldref Publications. Copyright 1987 by the Helen Dwight Reid Educational Foundation. Adapted by permission.

differs between the phylogenetic and ontogenetic classes of activities, but significant differences in the relative time frame for coordination changes as a function of activity class do not necessitate different age-based theories for the acquisition of coordination. These descriptive parallels between coordination changes with practice in both phylogenetic and ontogenetic activities are relevant to a general theory of the acquisition of coordination.

Towards a General Theory of the Acquisition of Coordination

The preceding sections on the acquisition of phylogenetic and ontogenetic activities have revealed some common descriptive principles governing directional changes that occur in macrolevel properties of coordinated movement patterns. These common descriptions of the changes in the coordination mode suggest that common coordination and control principles may drive the acquisition of both phylogenetic and ontogenetic activities. Indeed, as has been mentioned, the constraints perspective on the problem of coordination (Kugler et al., 1980, 1982; Kugler & Turvey, 1987) could in principle accommodate both phylogenetic and ontogenetic classes of activities. By comparison, the traditionally distinct prescriptive theories of motor development and motor skill acquisition are too narrow to hold any potential as a general theory for the acquisition of coordination (see Newell, 1986).

It is also of interest to note that the data reviewed previously on the acquisition of phylogenetic and ontogenetic activities are generally consistent with Bernstein's (1967) proposal that through practice the performer learns to coordinate independently an increasing number of degrees of freedom. This principle is reflected in the often quoted characterization of the unskilled performer as initially freezing out many of the biomechanical degrees of freedom by locking joint angles and, hence, having limb segments (for example) operate as a single unit. Except for the two studies reviewed earlier (Newell & van Emmerik, in press; van Emmerik & Newell, 1987), there have been no systematic tests of Bernstein's proposition, although there are isolated data sets in support of the proposition and relevant anecdotal observations of subjects learning whole body actions, such as gymnastics. The mastery of an increasing number of degrees of freedom during skill acquisition is another principle linking phylogenetic and ontogenetic activities, especially when viewed in light of Gesell's (1929, 1946) directional principle of proximal-distal development, which is always accompanied by an increase in the number of biomechanical degrees of freedom directly controlled.

The observation of common cephalo-caudal, proximal-distal, and ulnar-radial directional trends in the acquisition of coordination in ontogenetic activities provides prima-facie evidence against a developmental maturational interpretation of the development of the fundamental movement patterns (e.g., Gesell, 1929, 1946) because a maturational argument would not or could not be invoked in explanation of the changes previously reported for adults learning ontogenetic activities. One reason that the maturational argument does not apply is that in adults learning motor skills, the time frame over which some coordination changes occur is relatively short and outside the relatively long time frame consistent with the changes projected by maturational developmental theory. Thus, the scientific principle of parsimony alone should encourage one to look beyond the traditions of maturational theory, even for an account of the development of coordination in phylogenetic movement patterns.

It is our position that the prevalence of anatomical directional trends in the acquisition of coordination in both phylogenetic and ontogenetic activities is due to the general imposition of a very narrow set of task constraints to action (Newell, 1986). The logical implication is that if the various sources of constraint to action were to be manipulated more strongly than they have been heretofore, then a different set of trends in coordination changes would emerge as a consequence of practice in both phylogenetic and ontogenetic activities. One implication of this proposal is that the anatomically based directional trends evident for the acquisition of coordination in both phylogenetic and ontogenetic activities should not be viewed as a reflection of hard-wired neural constraints as projected, for example, by maturational theory, but rather as a reflection of soft coupling due to the constraints imposed on action (Kugler, 1986). Thus, strong manipulations of certain physical variables could lead to significant departures from anatomically based directional trends in the development of both phylogenetic and ontogenetic activities.

It should also be recognized that the directional trends observed in the changes of coordination patterns of phylogenetic and ontogenetic activities are merely a *description* of changing events. Thus, even if one calls these directional changes in coordination modes principles of action, as Gesell did, they are still only descriptive principles. As a consequence, if maturational theory is eliminated as a viable explanation of the acquisition of coordination on the grounds stated previously, then we are left with no theoretical construct to account for the changes in coordination mode evident in the dynamics of either phylogenetic or ontogenetic activities. The general prescriptive accounts to action, such as the motor program, logically cannot accommodate the findings on the acquisition of coordination (Kugler et al., 1980; Newell & van Emmerik, 1987), and proposals derived from them have generally been restricted to skill changes within a given coordination mode.

Although additional descriptions of the behavioral changes in movement form that accompany the acquisition of skill are still required, it is our position that a concomitant research focus should develop a characterization of the dynamical processes that support the observed changes in the macrolevel behavioral movement forms. How subjects explore the perceptual-motor work space to bring the redundant biomechanical degrees of freedom on-line during acquisition is the focus of the next section of the chapter. It is proposed that characterization of the search strategies used by subjects to explore and locate the gradient and equilibrium regions of the dynamical properties of the perceptual-motor work space will provide a principled basis for a theory of the acquisition of coordination that is general to all activity classes. According to coordinative structure theory, the physical processes underlying changes in the layout of the equilibrium regions of the response dynamics are strategically continuous with the physical design principles of nonconservative, self-organizing systems (cf. Haken, 1977; Iberall, 1972; Prigogine, 1980; Yates, 1987).

SEARCH STRATEGIES AND PERCEPTUAL-MOTOR WORK SPACES

We use the term *search strategies* to characterize the way in which an organism explores the perceptual-motor work space of the organism-environment interaction to "solve" the motor problem. Our approach is to link directly the macrolevel changes in movement patterns as a function of the acquisition of skill in phylogenetic and ontogenetic activities with the exploratory strategies subjects use to search through the "layout" of gradient and equilibrium regions defining the perceptual-motor work space. This layout is perceptually explored in terms of low-energy kinematic field descriptors (Gibson, 1979; Kugler & Turvey, 1987). The structure of the kinematic field may be described with standard techniques from nonlinear dynamics and stability theory (Abraham & Shaw, 1982, 1984, 1985; Thom, 1972).

Kinematic Field Morphologies, Information, and the Perceptual Motor Work Space

The perspective advanced here is that exploration of perceptual-motor work spaces is guided by constraints that are specific to the structured energy distribution of evolving field processes (Kugler & Turvey, 1987; Turvey & Kugler, 1984). It is hypothesized that the information organisms use to search a perceptual-motor work space is a macroscopic property defining the "form" of the layout of gradient and singular regions. These forms are revealed in the kinematics of the evolving energy distributions. The topological properties of this kinematic field provide information that constrains the high-energy force-output of the neuromuscular system. In this regard, it is proposed that the kinematic fields carried by the central nervous system "specify" the kinetics (see Kugler, 1983; Runeson, 1983). Figure 3.4 provides a schematic of this perception-action loop.

It was Gibson's (1966, 1979) fundamental insight that the layout of gradient and critical (singular) properties of the optic flow field, arising from the organism-environment interaction, provide information for perceiving and acting. This field-based theory of perception was not limited to optical information but was viewed as fundamental to other sensory systems (haptic, acoustical, etc.) as well. Since Gibson's formulation of a field-oriented approach, considerable research on the invariant properties of optical flow has been forthcoming (e.g., Koenderink, 1985; Lee, 1980). We are attracted by the prospect that the muscle-joint work spaces are likewise organized by perceptual-motor fields that have similar invariant properties.

The invariant flow properties of the perceptual-motor field are not to be viewed as static structures but rather as dynamic patterns. These patterns continually undergo change as a consequence of movements about the work space. These patterns are organized by

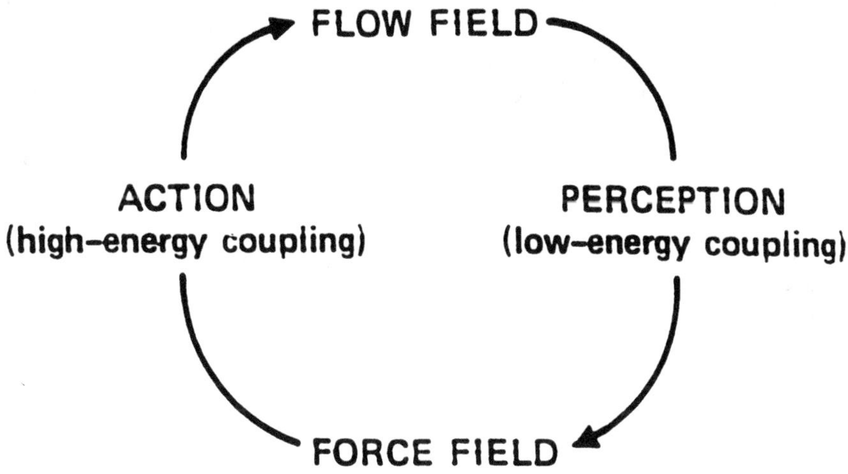

Figure 3.4. A schematic of the perception-action cycle. *Note:* From *Information, Natural Law, and Self-Assembly of Rhythmic Movement: Theoretical and Experimental Investigations* by P. N. Kugler and M. T. Turvey, 1987, p. 88. Copyright 1987 by Lawrence Erlbaum Asscciates, Inc. Adapted by permission.

the critical (singular) regions in the field. It is these critical regions that we view as providing a major source of information for orientation in a biomechanical work space. The motion about the work space both creates and annihilates critical regions in the field; that is to say, the field is "open" to the creation and annihilation of information states (Kugler et al., 1982; Kugler & Turvey, 1987). The perception-action loop is an open, self-organizing information system that can be approached by the mapping of the creation and annihilation of these informational properties and the lawful relations underlying their creation and annihilation.

We seek to understand the perception-action loop by examining the gradient and singular properties of the fields defining the perceptual-motor work space and by determining how these properties evolve as a function of motion through the space. In turn, our understanding of the way subjects explore this work space is very dependent on an analysis of the space. In fact, the description of exploratory behavior through the perceptual-motor work space must be complementary to the description of the work space. We now provide some detail about the types of attractors or critical regions that act as the basic building blocks for the field properties in perceptual-motor work spaces.

Building Blocks for Field Properties in Perceptual-Motor Work Spaces

Attractors (equilibrium regions) are states on which a set of different starting conditions in the state space settles. A major characteristic of attractors is their stability to external perturbations. There are four fundamental types of attractors: the fixed point, the limit cycle, the quasi-periodic, and the chaotic or strange attractor. Figure 3.5 depicts geometrical building blocks of these four classes of attractors that are viewed as the fundamental and general organizers of field processes in both physical and biological systems. A brief description of each of these attractor types follows with some examples from the motor control literature that directly relate to these different attractors. (For a more detailed discussion of attractors, see Abraham & Shaw, 1982, 1984; Shaw, 1981).

The first type is the fixed-point attractor (Figure 3.5, upper left). With a fixed-point attractor, the motion in phase space eventually comes to a stop, and regardless of initial position, the system is attracted to one point and stays there. An example is a displaced, damped pendulum whose amplitude decreases with time as a function of frictional forces. Regardless of the displacement at the initiation of motion, the amplitude always converges onto a point attractor defined when velocity is damped out. A point attractor has a dimension of zero because the state of the system can be completely specified in terms of a single point in state (or phase) space. An example of the point attractor notion in the movement domain is the equilibrium point concept in the mass-spring model as developed by Feldman (cf. Feldman, 1986). In this model, changes in body posture are the result of changes in the equilibrium points, which are in turn the result of the interaction of centrally oriented activation thresholds and external loads.

A second type of attractor is the stable limit cycle, namely, a steady closed oscillation (in phase space) that attracts all adjacent motions. To assemble a single stable limit cycle, it is necessary

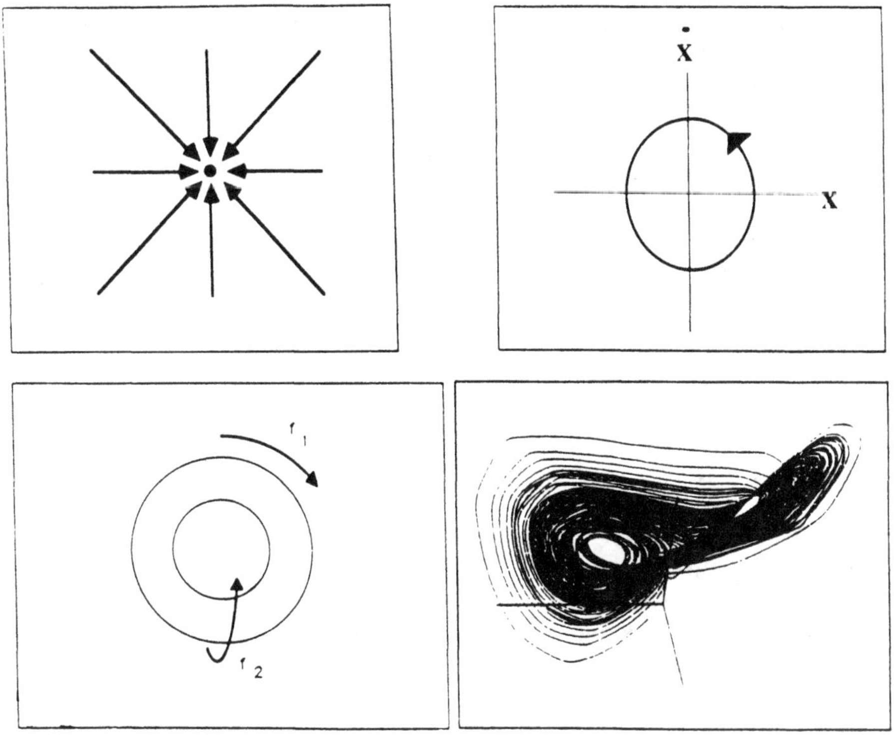

Figure 3.5. Schematic representations of the basic attractors. Upper left: point attractor; upper right: limit-cycle attractor; lower left: torus attractor where f_1 and f_2 represent the frequencies of two component oscillators; lower right: the Lorenz attractor as an example of a strange or chaotic attractor with folding in direction one, stretching in another.

that the origin (0,0) of phase space be unstable such that the motion is driven away from the steady state associated with the damped limit point attractor previously identified. This is accomplished by providing an escapement process that injects metabolically converted potential (chemical) energy into time-dependent kinetic energy. This process drives the trajectory through phase space in an outwardly spiraling manner. The outward spiraling motion is prevented from growing indefinitely large by a second energy-converting process that converts the outwardly spiraling kinetic energy into a micro heat mode. The second process is a dissipative process commonly associated with coloumb (zero order) or viscous (first order) friction. For a limit cycle to be possible, this second process must convert the energy at a different rate (first process, first order; second process, second order) and in an opposite direction (first process, micro to macro; second process, macro to micro).

If an oscillator is riding a limit cycle, then the number of coordinates required to locate the oscillator in phase space is one: The specification of phase determines both velocity and displacement of the oscillator. In this regard, the dimension of the limit cycle attractor is one.

In the coordinative structure perspective (Kugler et al., 1980), muscle ensembles are taken to reflect dissipative structures in which energy flows allow the system to operate in a limit cycle fashion. Experimental tests have been provided of the scaling laws that relate to the equilibrium manifold of these attractors (Kugler, 1983; Kugler & Turvey, 1987) and their stability to external perturbations (Kelso, Holt, Rubin, & Kugler, 1981; Kay, 1986).

A third type of attractor can emerge when two or more oscillators are weakly coupled. The attractor can be represented by a torus of two or more dimensions (see Figure 3.5, lower left). If the ratio between the frequencies is an irrational number, the trajectory of states will completely fill out a torus-like region of this extended phase plane. This class of solutions is termed *quasi-periodic*. If the ratio between the component frequencies is rational, a closed (limit cycle) orbit will occupy a specific region of the torus, and the solution is called *periodic*. The dominance of the periodic or quasi-periodic solution emerges as a function of the coupling strength between the oscillators. If the coupling strength is low, quasi-periodic solutions dominate. If the coupling strength is increased, periodic solutions or phase locking starts to dominate. The latter situation is exemplified by two clocks, which, when rigidly connected, tend to synchronize their motion.

The fourth type of attractor is the chaotic or strange attractor. In contrast to the attractors previously discussed, chaotic attractors have noninteger or fractal dimensions (Mandelbrot, 1983). Chaotic attractors exhibit the apparent paradox of global confinement in phase space and continuous local divergence of stages. The confinement is associated with the attraction of phase trajectories to an attractor occupying a finite region of phase space. The evolving trajectory of states exhibits a slow stretching and folding as the states locally diverge and are globally folded back to maintain a confinement to a bounded region of phase space. In three-dimensional space, these folding and stretching operations are shown through the concept of hyperbolicity: Attraction takes place in one direction, divergence in the other.

The importance of chaotic attractors is particularly relevant to the emergence of complex dynamics from a relatively simple organization (for a review of chaotic attractors, see Berge, Pomeau, & Vidal, 1984; Thompson & Stewart, 1986). Figure 3.5 (lower right) shows the Lorenz attractor, a chaotic attractor that was used to represent the dynamic evolution of weather patterns. This attractor evolves in three-dimensional space, where the axes in the illustration represent the interaction between heat, viscous, and conductive processes. Chaotic attractors have also been identified in chemical reactions, fluid mechanics, and neuronal interactions in biological systems (Mpitsos, Creech, Cohan, & Mendelson, 1988). One of us (PNK) is currently investigating the existence of quasi-periodic and chaotic attractors in oscillatory limb movements with two or more degrees of freedom.

The description of the dynamics of the perception-action loop in terms of nonlinear dynamics and stability theory provides a methodology for developing a physical biology perspective on movement control (cf. Kugler & Turvey, 1987). This orientation also provides, however, some very specific advantages in the consideration of the problem of the acquisition of coordination. First, attractors offer a simplification of the control problem by reducing the dimensionality or degrees of freedom in the system through the convergence of neighboring trajectories (Kay, 1988). Thus, the dimensionality of the perceptual-motor work space is usually lower than the dimensionality of the biomechanical degrees of freedom at the macro level. Second, complex dynamics can emerge from a relatively simple organization (e.g., May, 1976). Thus, the apparent complexity of gross motor activity evident in phylogenetic and ontogenetic activities may be due more to the experimenter than to the perceptual-motor system. Third, the description of the field processes in terms of attractors provides a formal basis for considering the search through the layout of the equilibrium and gradient regions of the perceptual-motor work space. Thus, the problems of the acquisition of skill and motor development may be formally linked to general field descriptions of the perceptual-motor work space in biological systems.

Search Strategies

The characterization of exploratory strategies in the acquisition of coordination in biological systems was introduced by Gelfand and Tsetlin (1962), who outlined a number of possible search strategies for solving, for example, a simple postural problem requiring the interaction of an organism with its environment. Solving the postural problem is here expressed in terms of finding the equilibrium manifold associated with the postural problem. This equilibrium manifold can be described by a potential well, the preferred state or solution to the postural problem being given by the minimum potential or the valley of the well. The search strategies proposed by Gelfand and Tsetlin (1962) include (a) *a method of blind search* in which the space points of the working parameters of the system may be looked at in random or in well-defined order; (b) *methods of local search* such as the gradient, the relaxational, the method of steepest descent, and the different varieties of all of these methods; (c) *methods of nonlocal search*, in which the movement of the working point in the space of the parameters is not continuous. These search strategies, of which some examples will be presented subsequently, are in essence procedures for optimizing the attainment of the solution to the coordination and control function. These procedures are consistent with optimum-seeking methods designed to guide the search for the optimum of any function about which full knowledge is not available (Wilde, 1964).

Each of these methods of searching for a solution to the coordination and control function has advantages and disadvantages (Gelfand & Tsetlin, 1962). For example, in a blind search, only a single value of the function is evaluated at any moment, and no memory characteristics are available. Thus, information from a previous search is not utilized in the current search, and the intrinsic organization of the evaluation function is utilized only minimally. In contrast, the various methods of local search are continuous and use knowledge from a previous search in relation to a small local neighborhood of the potential working space available. However, local searches are based only on properties that are related to a particular local space, a feature that creates the possibility of confining the search to a given equilibrium region. The nonlocal search methods are characterized by noncontinuous or intermittent searches, and this feature allows considerably more of the work space to be explored for a given unit of time. The discontinuity of the nonlocal search leaves open the possibility of passing over a given attractor or equilibrium region, and this search strategy is therefore often linked to work in conjunction with a local search.

Gelfand and Tsetlin (1962) defined as well-organized those functions that can be decomposed into both essential and nonessential variables. Nonessential variables are those variables whose manipulation is associated with gradient changes in the evaluation values of the function for coordination and control and which lead to scaling changes in the coordinative mode (movement pattern) at the behavioral level. Essential variables, in contrast, are those variables whose manipulation produces no change in the evaluation value of the function for coordination and control as they define the behavioral structure of the coordinative mode. In contrast to nonessential variables, essential variables are few in number and are likely to be instantiations of the elementary parameters in the system, much like the order parameter concept (Haken, 1977). The number of essential variables defines the dimensionality of the equilibrium region, or as Gelfand and Tsetlin (1962) refer to it, the dimensionality of the ravine. A number of methods exist for assessing the dimensionality of the equilibrium region or attractor. The correlation dimension indicates the number of dimensions that govern the equilibrium region (Grassberger & Procaccia, 1983). The lyapunov exponent provides a means for determining the topological growth of an attractor. These and other methods have recently been reviewed by Kay (1988) and applied to rhythmic movement patterns by Kay (1986) and Mpitsos et al. (1988).

There are two general experimental methods available for investigating the nature of search strategies in the acquisition of coordination. In one method, the natural perceptual-motor work spaces that arise from the response dynamics during the execution of a task are studied. The location and layout of the equilibrium regions in these work spaces can be manipulated, for example, through variations in inertial and frictional loadings as a function of task constraints and modeled with the techniques of nonlinear dynamics. The problem with this approach to studying the acquisition of coordination is that the equilibrium regions of the perceptual-motor work space are manipulated indirectly and can be very difficult to characterize even a posteriori, particularly when a large number of biomechanical degrees of freedom are being coordinated. The problem facing experimenters in this modeling approach is to create a task that requires the constraint of a sufficient number of degrees of freedom to make the task rich and interesting from a dynamical viewpoint, but not to have so many degrees of freedom that the dynamics cannot be modeled formally.

In the other general experimental method, the perceptual-motor work space is structured analytically from computer-generated motor algorithms and displays (Fowler & Turvey, 1978; Krinskii & Shik, 1964). This approach allows the explicit a priori control of the structure of the perceptual-motor work space and, as a consequence, the direct location and layout of the equilibrium regions of the response dynamics and flow properties of the organism-environment interaction. Of course, this approach cannot specify a priori the nonlinearities that the organism brings to the task, such as the variety of biological thresholds. In both the empirically and analytically driven methods, the independent measures relate to the layout of the equilibrium regions in the perceptual-motor work space, and the dependent measures relate to the macro- and microlevel changes in coordination and control associated with the search strategy.

In the remainder of this chapter, we emphasize the analytical approach to search strategies and the acquisition of coordination because this approach provides a more direct framework for examining models of search strategies and the acquisition of coordination. Our long-term goal, however, is to link the analytical, computer-driven motor algorithm approach to the analysis of the perceptual-motor work space of natural actions, and to the coordination and control exhibited in other tasks whose dynamics are not specified a priori by the experimenter. In summary, we believe that the modeling of search strategies provides the basis for a general theory of the acquisition of coordination; but at this juncture in the theoretical and methodological development of the problem, the analysis of movements generated to satisfy a motor problem defined a priori may be the most fruitful experimental method.

The Krinskii and Shik (1964) Experiment

The prototypical experiment on the study of search strategies in structured dynamical perceptual-motor work spaces is that of Krinskii and Shik (1964). In their task, subjects were required to coordinate and control the joint angles (wrist, elbow, and shoulder pairs) so as to minimize the error (E) in solving a motor algorithm:

$$E = [x - y - (a - b)] + Z(x - a) + Z(y - b) \quad (1)$$

where, for example, x and y represent the respective elbow angles at any moment, Z is a control variable that can be modified to change the scaling of the relationship between the joint angles, and a and b are constants. These control variables and constants directly vary the flow field properties of the perceptual-motor work space and can be manipulated both within or between trials, although Krinskii and Shik varied these properties only between trials. The motor algorithm of Equation 1 produces a point attractor equilibrium region for the configuration space of limb angle relationships analogous to that depicted in Figure 3.5 (upper left). This point attractor in the Krinskii and Shik protocol has, in addition, a number of local minima (six) on the major gradient. The slope of the local equilibrium manifold to be explored by the subject and the depth of the local minima can be heightened or flattened when the experimental constant Z of Equation 1 is changed. Figure 3.6 depicts the attractor region of the Krinskii and Shik task with both a high and low Z.

A sample output of the coordination relationship between two independently controlled degrees of freedom is shown in Figure 3.7. In this example, the two degrees of freedom are formed by two independently moving computer mice. The upper left and right panels of the figure show searches from one subject for a low and high Z, respectively. For the low Z (shallow gradient) in Figure 3.7, upper left, the search got stuck in a local minimum without ever getting close to the absolute minimum (center point). For the high Z (upper right), the highly systematic orthogonal search ended straight on target. The lower left and right panels of Figure 3.7 show the two example outputs of the search behavior for another subject. In the

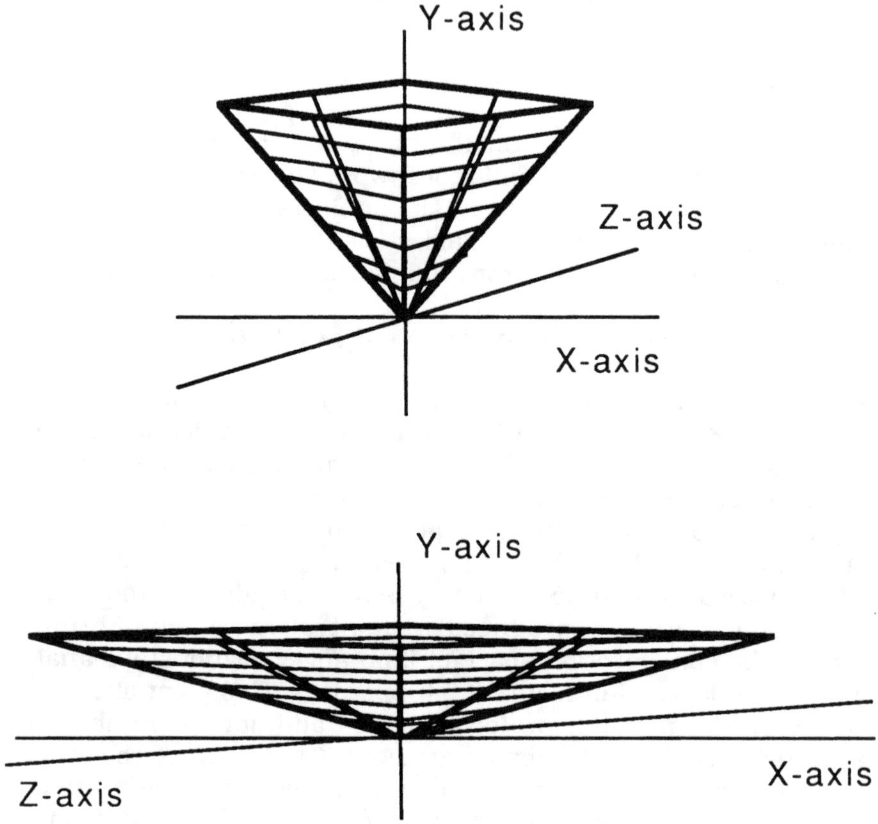

Figure 3.6. Schematic of the point attractor region formed from the algorithm of the Krinskii and Shik (1964) task with both a high (upper panel) and low (lower panel) *Z*.

lower left panel, the diagonal motion of the line representing the relationship between the two degrees of freedom indicates that as movement occurred away from the initial starting position, there was simultaneous motion of approximately the same rate at both degrees of freedom (mice). This coordinated motion gave way eventually to independent motion (characterized by straight, orthogonal lines) as the subject further minimized the error to the motor algorithm (Figure 3.7, lower right). This discrete change in search strategy reflects a shift from a nonlocal to a local search strategy

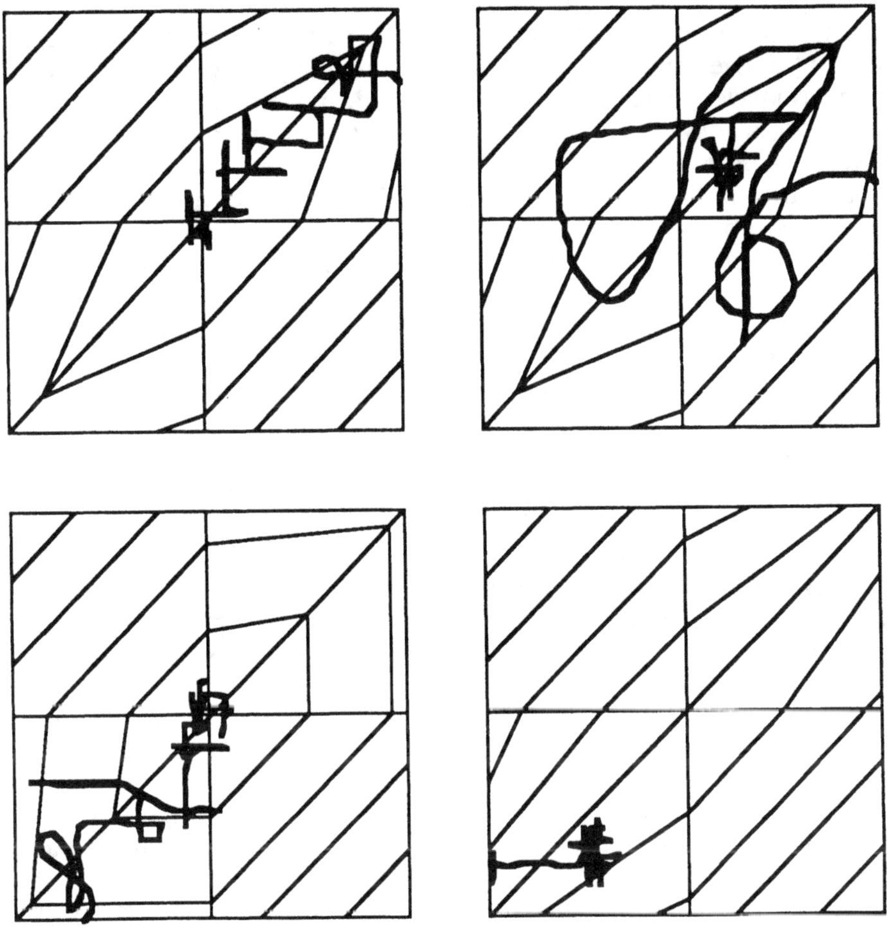

Figure 3.7. Sample outputs of search strategies of 2 subjects engaged in finding the solution to the Krinskii and Shik (1964) algorithm.

and is consonant with what Gelfand and Tsetlin (1962) labeled the ravine method. Subjects can and do generate a number of distinct search strategies in this and related tasks. A major problem for this orientation is the meaningful characterization of the output of the evaluation function (*E*) and the detection of structure in that output, which can easily be passed off as random behavior or noise.

The subject did not see the configuration space output of the coordination task represented in Figure 3.7, but rather the output E of the mapping of the limb organization to the motor problem as reflected by the relative position of a moving scale pointer displayed on a screen. This moving pointer provided information feedback to the subject to help solve the motor problem created by Equation 1. Fowler and Turvey (1978) showed that there was not an unambiguous solution between the value of E, as reflected in the scale pointer feedback given subjects in a Krinskii and Shik (1964) type of task, and the relative joint angle position. In other words, a variety of joint angle combinations could produce the same value of E in Equation 1. Fowler and Turvey proposed that it is the higher order properties of the optical stimulation afforded by the scale pointer that provide the unambiguous joint-related information and prescribe what the subject should do next in attempting to minimize error in Equation 1. Specifically, they suggested that it is the minimizing of the value of $\Delta E/E$ that implies the minimization of error with the current search strategy and the need for a shift (ravine step) to a new search strategy to minimize error still further. Furthermore, they claim that the higher order values of E (namely ΔE and Δ (ΔE) not only tell the subject when to shift strategies but also prescribe how the subject should alter his or her strategy, a feature that traditional knowledge of results or other forms of information feedback lack (see also Newell, Morris, & Scully, 1985).

The Krinskii and Shik (1964) motor task may seem to impose a very artificial set of task constraints, but the task demands provide all the essential ingredients for understanding how subjects search the perceptual-motor work space to optimize the coordination and control function. Furthermore, the perceptual-motor environment is formally defined a priori by the motor algorithm so that the experimenter has a firm grasp on the environment that the subject is exploring. These analytical tasks are very rich in terms of the variables that can be manipulated to help characterize the search strategy a subject uses in seeking the solution to coordination and control tasks. The Krinskii and Shik experimental protocol was confined to mimicking the perceptual-motor work space as a point attractor. This methodology can be extended, however, to the other types of attractors either individually or in some combinational form.

A significant problem for the search strategy orientation is the development of the appropriate methods for characterizing the output, such as that displayed in Figure 3.7. A variety of discrete output measures, such as time to minimize E and number of discrete steps in the search, could be formulated, as they have been in tracking tasks, for example. All these output measures of the response, however, are based on the assumptions of a linear system—which, of course, is the very antithesis of the approach proposed here to modeling the perceptual-motor environment and the exploratory behavior of subjects searching the gradient and equilibrium regions of the work space layout (Kugler & Turvey, 1987). Thus, considerable work will need to be done to develop and effectively utilize appropriate dependent variables to characterize the search strategy.

CONCLUDING REMARKS

The general claim advanced here is that the acquisition of coordination in both phylogenetic and ontogenetic activities may be modeled in terms of the way subjects search the gradient and equilibrium regions of the perceptual-motor work space. The critical points that help to organize and change the observed coordination patterns at the macro level of behavior exist in both the kinematic and kinetic field properties of the organism-environment interaction. Indeed, it is the mapping of the kinematic and kinetic field properties that constrains and organizes the perception-action cycle (Kugler & Turvey, 1987). Thus, our approach attempts to map the coordination modes and changes in coordination modes evident in practice to the equilibrium and gradient properties of the perceptual-motor work space.

The ability of biological systems to optimize the solution that constrains the coordination and control function is self-evident. Our understanding of the exploratory behavior used to optimize the search of the variables requiring constraint in coordination is, however, very limited. Modeling the search strategies provides explicit physical models for understanding the changes in the coordination patterns with practice, and the Krinskii and Shik (1964) analytically driven experimental strategy provides a very useful protocol for testing this modeling approach.

The orientation proposed here to the acquisition of coordination is a new way of examining many of the traditional issues of skill learning, such as learning, retention, transfer, and information feedback (see Newell, 1985 for a further account of these arguments). The research program attempts to understand the learning of motor skills through models that minimize the representational requirements or logical constraints of the system and that maximally exploit the free dynamics emerging from the organism-environment synergism. Shaw & Alley (1985) have labeled this ecological approach to the issues of learning as the "learning of the laws" in contrast to the traditional learning theory approach, which has attempted to define the "laws of learning."

REFERENCES

Abraham, R. H., & Shaw, C. D. (1982). *Dynamics: The geometry of behavior: Part 1. Periodic behavior.* Santa Cruz, CA: Aerial Press.

Abraham, R. H., & Shaw, C. D. (1984). *Dynamics: The geometry of behavior: Part 2. Stable and chaotic behavior.* Santa Cruz, CA: Aerial Press.

Abraham, R. H., & Shaw, C. D. (1985). *Dynamics: The geometry of behavior: Part 3. Bifurcation behavior.* Santa Cruz, CA: Aerial Press.

Arutyunyan, G. H., Gurfinkel, V. S., & Mirskii, M. L. (1968). Investigation of aiming at a target. *Biophysics, 13*, 536-538.

Arutyunyan, G. H., Gurfinkel, V. S., & Mirskii, M. L. (1969). Organization of movements on execution by man of an exact postural task. *Biophysics, 14*, 1162-1167.

Bayley, N. (1935). Development of motor abilities during the first three years. *Monograph of the Society for Research in Child Development, 1.*

Berge, P., Pomeau, Y., & Vidal, C. (1984). *Order within chaos: Towards a deterministic approach to turbulence.* New York: Wiley.

Bernstein, N. (1967). *The co-ordination and regulation of movements.* New York: Pergamon.

den Brinker, B. P. L. M., & van Hekken, M. F. (1982). The analysis of slalom-ski type movements using a ski simulator apparatus. *Human Movement Science, 1*, 91-108.

Feldman, A. G. (1986). Once more on the equilibrium-point hypothesis (Lambda Model) for motor control. *Journal of Motor Behavior, 18*, 17-54.

Fowler, C. A., & Turvey, M. T. (1978). Skill acquisition: An event approach with special reference to searching for the optimum of a function of several variables. In G. E. Stelmach (Ed.), *Information processing in motor control and learning* (pp. 1-40). New York: Academic Press.

Gelfand, I. M., & Tsetlin, M. L. (1962). Some methods of control for complex systems. *Russian Mathematical Surveys, 17*, 95-116.

Gesell, A. (1929). Maturation and infant behavior pattern. *Psychological Review, 36*, 307-319.

Gesell, A. (1946). The ontogenesis of infant behavior. In L. Carmichael (Ed.), *Manual of child psychology* (pp. 295-331). New York: Wiley.

Gesell, A. (1952). *Infant development*. New York: Harper and Row.

Gibson, J. J. (1966). *The senses considered as perceptual systems.* London: Allen and Unwin.

Gibson, J. J. (1979). *The ecological approach to visual perception.* Boston: Houghton Mifflin.

Grassberger, P., & Procaccia, I. (1983). Characterization of strange attractors. *The American Physical Society, 50*, 346-349.

Grieve, D. W., & Gear, R. J. (1966). The relationships between length of stride, step frequency, time of swing and speed of walking for children and adults. *Ergonomics, 5*, 379-399.

Haken, H. (1977). *Synergetics*. Heidelberg: Springer.

Haken, H., Kelso, J. A. S., & Bunz, H. (1985). A theoretical model of phase transitions in human hand movements. *Biological Cybernetics, 51*, 347-356.

Halverson, H. M. (1931). An experimental study of prehension in infants by means of systematic cinema records. *Genetic Psychology Monographs, 10*, 107-283.

Iberall, A. L. (1972). *Towards a general science of viable systems.* New York: McGraw-Hill.

Kay, B. A. (1986). *Dynamic modeling of rhythmic limb movements: Converging on a description of the component oscillators.* Unpublished doctoral dissertation, University of Connecticut.

Kay, B. A. (1988). The dimensionality of movement trajectories and the degrees of freedom problem: A tutorial. *Human Movement Science, 7*, 343-364.

Kelso, J. A. S. (1981). On the oscillatory basis of movement. *Bulletin of the Psychonomic Society, 18*, 63.

Kelso, J. A. S., Holt, K. G., Rubin, P., & Kugler, P. N. (1981). Patterns of human interlimb coordination emerge from the properties of non-linear, limit cycle oscillatory processes: Theory and data. *Journal of Motor Behavior, 13*, 226-261.

Koenderink, J. J. (1985). Space, form and optical deformations. In D. J. Ingle (Ed.), *Brain mechanisms and spatial vision* (pp. 31-58). Boston: Martinus Nijhoff.

Krinskii, V. I., & Shik, M. L. (1964). A simple motor task. *Biophysics, 9*, 661-666.

Kugler, P. N. (1983). *A morphological view of information for the self-assembly of rhythmic movement.* Unpublished doctoral dissertation, University of Connecticut.

Kugler, P. N. (1986). A morphological perspective on the origin and evolution of movement patterns. In M. G. Wade & H. T. A. Whiting (Eds.), *Motor development in children: Aspects of coordination and control* (pp. 459-525). Boston: Martinus Nijhoff.

Kugler, P. N., Kelso, J. A. S., & Turvey, M. T. (1980). On the concept of coordinative structures as dissipative structures: I. Theoretical lines of convergence. In G. E. Stelmach & J. Requin (Eds.), *Tutorials in motor behavior* (pp. 3-47). Amsterdam: North-Holland.

Kugler, P. N., Kelso, J. A. S., & Turvey, M. T. (1982). On the control and coordination of naturally developing systems. In J. A. S. Kelso & J. E. Clark (Eds.), *The development of movement control and coordination* (pp. 5-78). New York: Wiley.

Kugler, P. N., & Turvey, M. T. (1987). *Information, natural law, and self-assembly of rhythmic movement: Theoretical and experimental investigations.* Hillsdale, NJ: Erlbaum.

Lee, D. N. (1980). The optic flow field: The foundation of vision. *Philosophical Transcripts of the Royal Society of London. B, 290*, 169-179.

Mandelbrot, B. (1983). *The fractal geometry of nature.* San Francisco: Freeman.

May, R. M. (1976). Simple mathematical models with complicated dynamics. *Nature, 261*, 459-467.

McDonald, P. V., van Emmerik, R. E. A., & Newell, K. M. (in press). The effects of practice on limb kinematics in a throwing task. *Journal of Motor Behavior.*

McGraw, M. B. (1935). *Growth: A study of Johnny and Jimmy.* New York: Appleton-Century.

Mpitsos, G. J., Creech, H. C., Cohan, C. S., & Mendelson, M. (1988). Variability and chaos: Neurointegrative principles in the generation of rhythmic motor patterns. In J. A. S. Kelso, A. J. Mandell, & M. F. Shlesinger (Eds.), *Dynamic patterns in complex systems* (pp. 162-190). World Scientific Press.

Newell, K. M. (1985). Coordination, control and skill. In D. Goodman, R. B. Wilberg, & I. M. Franks (Eds.), *Differing perspectives in motor learning, memory, and control* (pp. 295-317). Amsterdam: North-Holland.

Newell, K. M. (1986). Constraints on the development of coordination. In M. G. Wade & H. T. A. Whiting (Eds.), *Motor development in children: Aspects of coordination and control* (pp. 341-360). Boston: Martinus Nijhoff.

Newell, K. M., Morris, L. R., & Scully, D. M. (1985). Augmented information and the acquisition of skill in physical activity. In R. J. Terjung (Ed.), *Exercise and sport sciences reviews* (Vol. 13, pp. 235-261). New York: Macmillan.

Newell, K. M., & Scully, D. M. (1988a). *The development of prehension: Constraints on grip patterns*. Manuscript submitted for publication.

Newell, K. M., & Scully, D. M. (1988b). Steps in the development of coordination: Perception of relative motion? In J. E. Clark & J. H. Humphrey (Eds.), *Advances in motor development research* (Vol. 1, pp. 153-170). New York: AMS Press.

Newell, K. M., & van Emmerik, R. E. A. (1987). Can schema account for the generation of novel action patterns? *Cahiers de Psychologie Cognitive, 7*, 177-180.

Newell, K. M., & van Emmerik, R. E. A. (in press). The acquisition of coordination: Preliminary analysis of learning to write. *Human Movement Science.*

Prigogine, I. (1980). *From being to becoming: Time and complexity in the physical sciences.* San Francisco: Freedman.

Runeson, S. (1983). On visual perception of dynamic events. *Acta Universitatis Upsaliensis: Studia Psychologia*, Serial No. a (originally published 1977 as a dissertation).

Shaw, R. (1981). Strange attractors, chaotic behavior, and information flow. *Zeitschrift fur Naturforschung, 36a*, 80-112.

Shaw, R. E., & Alley, J. R. (1985). How to draw learning curves: Their use and justification. In T. D. Johnston & A. T. Pietrewicz (Eds.), *Issues in the ecological study of learning* (pp. 275-304). Hillsdale, NJ: Erlbaum.

Shirley, M. M. (1931). *The first two years: A study of twenty-five babies: Vol. 1. Postural and locomotor development.* Minneapolis, MN: University of Minnesota Press.

Sparrow, W. A., & Irizarry-Lopez, V. M. (1987). Mechanical efficiency and metabolic cost as measures of learning a novel gross motor skill. *Journal of Motor Behavior, 19*, 240-264.

Swinnen, S., Walter, C. B., & Shapiro, D. C. (in press). The coordination of limb movements with different kinematic patterns. *Brain and Cognition.*

Thom, R. (1972). *Structural stability and morphogenesis.* (D. H. Fowler, Trans.). Reading, MA: Benjamin.

Thompson, D. (1917). *On growth and form.* Cambridge: Cambridge University Press.

Thompson, J. M. T., & Stewart, H. B. (1986). *Nonlinear dynamics and chaos.* New York: Wiley.

Turvey, M. T., & Kugler, P. N. (1984). An ecological approach to perception and action. In H. T. A. Whiting (Ed.), *Human motor actions: Bernstein reassessed* (pp. 373-412). Amsterdam: North-Holland.

Turvey, M. T., Rosenblum, L. D., Schmidt, R. C., & Kugler, P. N. (1986). Fluctuations and phase symmetry in coordinated rhythmic movements. *Journal of Experimental Psychology: Human Perception and Performance, 12*, 564-583.

Turvey, M. T., Schmidt, R. C., Rosenblum, L. D., & Kugler, P. N. (1988). On the time allometry of co-ordinated rhythmic movements. *Journal of Theoretical Biology, 130*, 285-325.

van Emmerik, R. E. A., & Newell, K. M. (1987). *The influence of task and organismic constraints on intralimb kinematics in a drawing task.* Manuscript submitted for publication.

van Emmerik, R. E. A., & Newell, K. M. (in press). The relationship between pen-point and joint kinematics in handwriting and drawing. In C. Y. Suen, R. Plamondon, & M. L. Simner (Eds.),

Computer and human recognition of handwriting. Singapore: World Scientific.

van Emmerik R. E. A., & Valk, P. J. L. (1984). *Movement pattern formation for a skilled and unskilled performer in a slalom-ski type task.* Unpublished manuscript.

Wilde, D. J. (1964). *Optimum seeking methods.* Englewood Cliffs, NJ: Prentice-Hall.

Yates, F. E. (Ed.). (1987). *Self-organizing systems: The emergence of order.* New York: Plenum Press.

Perspectives on the Coordination of Movement
S.A. Wallace (Editor)

ABSOLUTE COORDINATION: AN ECOLOGICAL PERSPECTIVE§

R. C. SCHMIDT* and M. T. TURVEY

Center for the Ecological Study of Perception and Action
University of Connecticut

and

Haskins Laboratories

ABSTRACT

The ecological perspective on the coordination of movement is discussed with regard to the most basic, pervasive form of coordination, namely, absolute coordination. The working hypothesis of the ecological perspective is that coordinations are largely due to general laws and principles. Dynamical explanations of phenomena such as von Holst's magnet effect and maintenance tendency, as well as locomotory time allometries of both large and small organisms, are reviewed. The role of information in the functioning of dynamically based action systems is discussed, where information is understood in the Gibsonian (1979) specificational sense.

---

*Address correspondence to: R. C. Schmidt, Center for the Ecological Study of Perception and Action, Department of Psychology, U-20, 406 Cross Campus Road, University of Connecticut, Storrs, CT 06269, U.S.A.

§The writing of this chapter was supported in part by a NIH grant (BRS-RR-05596) awarded to Haskins Laboratories, a grant from NSF (BNS-8811510), a University of Connecticut dissertation fellowship awarded to the first author, and a James McKeen Cattell Fellowship awarded to the second author.

INTRODUCTION

In this chapter we illustrate the ecological view of coordination of movement by reviewing research and theory on absolute coordination (von Holst, 1937/1973, 1939/1973). Two or more limbs or two or more body segments are in absolute coordination when they oscillate at the same period and maintain, thereby, a steady phase relation between them. This state has been contrasted by von Holst with relative coordination, in which the period of oscillation of two coupled oscillators is not equal and, hence, the phase relation between the two oscillators is changing constantly. The research and theory in question has focused on the absolute coordination of what might be termed "pendular clocking movements." As instanced in walking and running, the components in absolute coordination are limbs whose individual motions are pendulum-like (they are raised and lowered with respect to gravity) and clock-like (the raising and lowering is done at regular intervals and to approximately the same degree each cycle). This type of absolute coordination is by far the most prevalent. Not only does it characterize terrestrial locomotion but also it is common to many other everyday activities. The principles behind this type of absolute coordination remain, however, largely unknown.

Neurophysiological theories of absolute coordination assume that the nervous system is causally responsible for the coordinating. Theories of arthropod and vertebrate locomotion refer to multiple central rhythmic-pattern generators, at least one per limb (Grillner, 1975; Shik & Orlovskii, 1976; Stein, 1977) and possibly one per joint (e.g., Edgerton, Grillner, Sjogstrom, & Zangger, 1976). In neurophysiology, the problem of absolute coordination is to describe how the central rhythmic-pattern generator of one limb is coordinated with the central rhythmic-pattern generator of another limb. One proposed neural mechanism is a coordinating neuron that sends a neural copy of the motor discharge of one limb to the other (Stein, 1971, 1976, 1977). Another proposed neural mechanism is the mutual inhibition of one rhythmic unit by another rhythmic unit (or every other rhythmic unit), without mutual excitation (Stafford & Barnwell, 1985).

In the ecological approach to absolute coordination, the nervous system is not regarded as solely responsible, in causal terms, for the interaction. Rather, the nervous system is seen as the *medium* supporting the causal agencies of coordination. The burden of explanation is shifted from modular anatomical units that are ascribed a special causal function a priori to very general physical principles that coordinate both animate and inanimate things. Biological things have not only biological properties but also physical ones, and it would be expeditious for biological things to exploit the organizing that takes place, a postcriori, as a consequence of having physical properties. The purpose of an ecological analysis is to determine the degree to which coordination phenomena can be understood in terms of lawful regularities and principles at the ecological scale, the scale at which the physical entities of animal and environment are defined (Gibson, 1979; Turvey & Carello, 1986).

In this way of proceeding, the high dimensional biological situation (a number of limbs, each with their detailed neural, vascular, and muscular microstructures) can be given a low dimensional, physical redescription. Much work has been done from this perspective in both theoretical (Beeke & Beek, in press; Feldman, 1986; Kelso, 1986; Kugler & Turvey, 1987; Schöner & Kelso, 1988) and empirical (Feldman, 1966, 1980; Kay, Kelso, Saltzman, & Schöner, 1987; Kugler & Turvey, 1987; Turvey, Rosenblum, Schmidt, & Kugler, 1986) domains. This work is aimed at Bernstein's problem of degrees of freedom (Bernstein, 1967; Turvey, 1977; Turvey, Shaw, & Mace, 1978), namely, how components comprising very many degrees of freedom are regulated to yield behaviors of very few degrees of freedom. The many degrees of freedom of an organism are not all controlled by the organism. Many, perhaps most, are controlled by physical constraints that form the context for a movement. In this chapter, we try to illustrate how the presence of physical properties and their functional relations explain many facts about the coordination of limbs in locomotion.

From the ecological perspective, absolute coordination is to be understood in terms of the macroscopic observables of limb complexes (comprising bones, joints, muscular, vascular, and neural components) organized to function as oscillatory units. As such, the individual units of analysis are continuous with the rhythmic "automatisms" identified in von Holst's (1937/1973, 1939/1973) investigations of the interactions among the rhythmically moving fins of the fish *Labrus*. Von Holst discovered three macroscopic phenomena that oscillating fins often exhibit in absolute and relative coordination, namely, *superposition*, *the magnet effect*, and *the maintenance tendency*. Von Holst noticed that one fin's oscillation sometimes showed up as a second periodicity in another fin's oscillation; the one oscillation was superimposed upon the other. Additionally, he noticed a magnet effect. Each fin oscillator tries to draw the other fin (or fins) to its characteristic period, the tempo it exhibits when oscillating alone. Hence, two coupled oscillators with different endogenous tempos would settle ultimately on a cooperative tempo that was in between the tempos each preferred individually. Further, each oscillator tries to maintain its identity when participating in such a coupling; that is, even though it is operating at the cooperative period, there are residual effects of its preferred period, as indexed by the fluctuations around the mean cooperative period. A physical basis for the latter two features of absolute coordination is suggested later in this chapter.

It must be pointed out that ecological investigations of the physical principles underlying coordination are concerned with biological coordination in general. That is, these investigations are interested not only in the coordination between limbs (as in locomotion) but also in the coordination between an animal and its environment (as in fulfilling an intent). Examples of the latter research are provided by Warren (1984), Lee (1980), and others and are summarized in Turvey and Carello (1986). This chapter is limited to a review of the theory and research within the ecological framework that address the absolute coordination of limbs. We consider how this coordination arises as an emergent property through the operation of physical principles that organize parts into wholes. We then show that this physical modeling has implications for understanding the scalings of locomotory cycles to body proportions found in species as diverse as quadrupeds and insects. Finally, we demonstrate that the linkage of the oscillators in absolute coordination has an informational basis and suggest what form this information might take. We begin with the experimental paradigm used and a physical model for the component pendular clocking movements.

THE "PENDULAR CLOCKING MODE" METHODOLOGY

The pendular clocking movements of a single limb of a human subject have been examined using hand-held pendulums that are swung at the wrist joint with the forearm parallel to the ground (Kugler & Turvey, 1987; Turvey et al., 1986; Turvey, Schmidt, Rosenblum, &

Kugler, 1988). The subject was instructed to swing the pendulum at a comfortable, preferred tempo—the tempo that felt suitable for continuous work. As a rule, people are very good at finding and sustaining the tempo at which they prefer to perform a given task (e.g., Frischeisen-Köhler, 1933; Smoll, 1975a, 1975b). There is evidence to suggest that a preferred tempo corresponds to the working pace that represents a near minimum value of energy per cycle (e.g., Corlett & Mahadeva, 1970). For the task of swinging hand-held pendulums, a person will settle on different characteristic periods for pendulums of different masses and lengths. The smaller the pendulum's moment of inertia, the smaller the period.

How is the preferred period of oscillation of a single wrist-pendulum system to be understood? Physically speaking, the answer ought to follow from the fact that such a system comprises a body oscillating as a function of two potentials, namely, gravity and the restorative forces formed by the limb musculature and its associated metabolism. There is evidence that these restorative forces in rhythmic movements are based partly upon the inherent elasticity of muscle (Cavagna, 1977). Further, the elastic stiffness assembled in a given movement can be varied by manipulation of the co-contraction of the flexors and extensors at the joint (Feldman, 1980).

The behavior of a wrist-pendulum system can be modeled most simply as a pendulum in the gravitational field with a linear spring attached, the spring corresponding to the restorative contribution of the musculature and the attendant metabolism (Figure 4.1). The equation governing the period of this system is

$$\tau = 2\pi[ML^2/(gML + kb^2)]^{0.5} \tag{1}$$

where M is the mass of the pendulum, L is the length of the pendulum, g is the acceleration due to gravity, k is the elastic stiffness of the attached spring, and b is the distance from the spring to the center of rotation of the pendulum (see Turvey et al., 1988). One can calculate a subject's contribution to the period from this model by measuring the period of oscillation and assuming that the mass and length involved are those of the simple pendulum equivalent of a compound pendulum, consisting of the mass of the

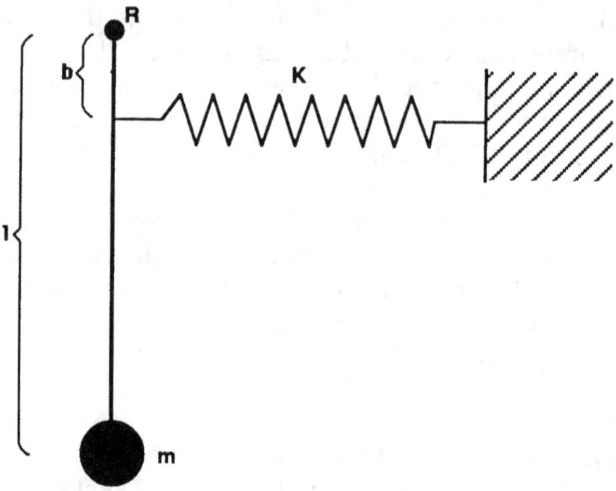

Figure 4.1. The hybrid mass-spring simple pendulum model, where *m* is the mass of the pendulum, *l* is the length of the pendulum, *K* is the elastic stiffness of the attached spring, *b* is the distance from the spring to the center of rotation of the pendulum, and *R* is the center of rotation.

wrist, the mass of the rod, and the mass of the attached weight, each at different lengths from the point of rotation.

The value of the elastic contribution is not constant across different pendulums but increases nonlinearly with the pendulum's inertia (Figure 4.2). An increase in the elastic stiffness of the musculature as a function of the load has been observed similarly in the calf muscle of humans landing from a fall without bending their knees (Cavagna, 1970). Assuming that the muscular co-contraction of the flexors and extensors gives rise to the macroscopic physical property of an elastic potential, the characteristic periods (τ) of single wrist-pendulum system movements can then be understood as a consequence of the physical properties comprising the system. A mass is rotating about an axis under the influence of two forces, namely, gravity and an assembled elastic potential whose magnitude depends upon the load involved. This understanding of the basis of a preferred tempo contrasts with the view that preferred tempos in rhythmic tasks are controlled exclusively by physiologi-

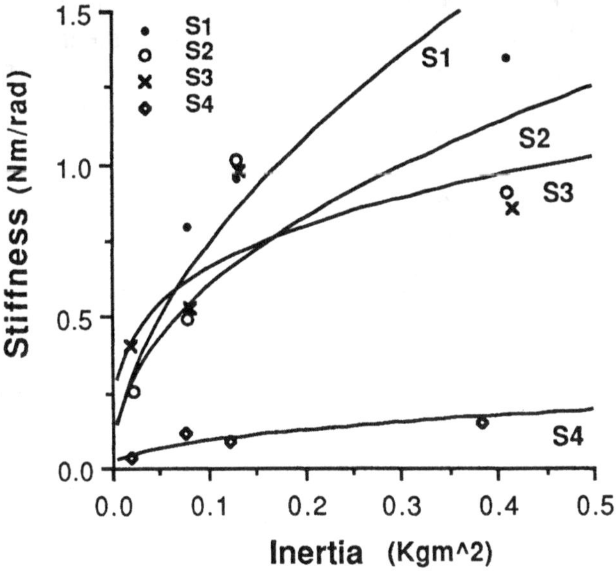

Figure 4.2. Assembled elastic potential (K) versus inertia of a wrist-pendulum system for 4 subjects of Kugler & Turvey's (1987) main experiment (S1: $K = 2.65I^{0.56}$, $R^2 = .92$; S2: $K = 1.728I^{0.46}$, $R^2 = .80$; S3: $K = 1.20I^{0.28}$, $R^2 = .72$; S4: $K = 0.266I^{0.48}$, $R^2 = .83$).

cal or cognitive processes (Smoll, 1982). The only variable that needs to be controlled to produce different characteristic periods in a wrist-pendulum system is the elastic potential provided by the musculature (see Feldman, 1980, for a theory of how the control is effected).

THE MAGNET EFFECT IN THE ABSOLUTE COORDINATION OF WRIST-PENDULUM SYSTEMS

The wrist-pendulum methodology has been used to examine the absolute coordination of pendular clocking movements. The task is to swing two wrist-pendulums (one in each hand) in absolute coordination at the same period and either in the same direction (0° phase difference) or in opposite directions (180° phase difference) (Kugler

& Turvey, 1987; Turvey et al., 1986). These two versions of absolute coordination resemble the organization of symmetric and asymmetric gaits, respectively, in quadrupeds. The subject is asked to find the period that is most comfortable for swinging the pendulums simultaneously. Significantly, the two wrist-pendulum systems to be coordinated can be either identical or different in mass and length parameters. When they have identical parameter magnitudes, the characteristic periods of the wrist-pendulum systems when swung alone are equal; when they have different parameter magnitudes, the characteristic periods of the wrist-pendulum systems when swung alone are different. The ability to manipulate the characteristic periods of the component oscillators to be coordinated permits a replication of the circumstances of von Holst's (1937/1973, 1939/1973) experiments with fish, in which fin oscillators of different inherent periods were coordinated. Given von Holst's work, a central question is whether or not the cooperative period follows from each oscillator drawing the other oscillator toward its own characteristic period. That is, Is there a magnet effect?

Figure 4.3 replots data from Kugler and Turvey (1987, main experiment). This new plot gives (a) the difference between the cooperative period ($\tau_{cooperative}$) and the characteristic period of the left system (τ_{left}) as a function of the difference between the two characteristic periods (τ_{right}-τ_{left}) and (b) the difference between the cooperative period ($\tau_{cooperative}$) and the characteristic period of the right system (τ_{right}) as a function of the difference between the two characteristic periods (τ_{right}-τ_{left}). To a first approximation, as the absolute difference in the characteristic periods of the component oscillators increases, the deviation of the cooperative period from a component period increases. What dictates the cooperative period and its relation to the characteristic periods of the component oscillators? One answer might be that it is a simple function of the periods of the two pendulums in isolation, as suggested by von Holst (1939/1973) and as suggested by a central pattern-generator analysis. A different answer follows from entertaining the possibility, as Kugler and Turvey (1987) did, that the two wrist-pendulum systems form a common or cooperative period in a manner analogous to ordinary physical pendulums that are coupled rigidly. If the coupling between wrist-pendulum systems were rigid—more exactly, if the nervous system simulated a rigid coupling—then the cooperative

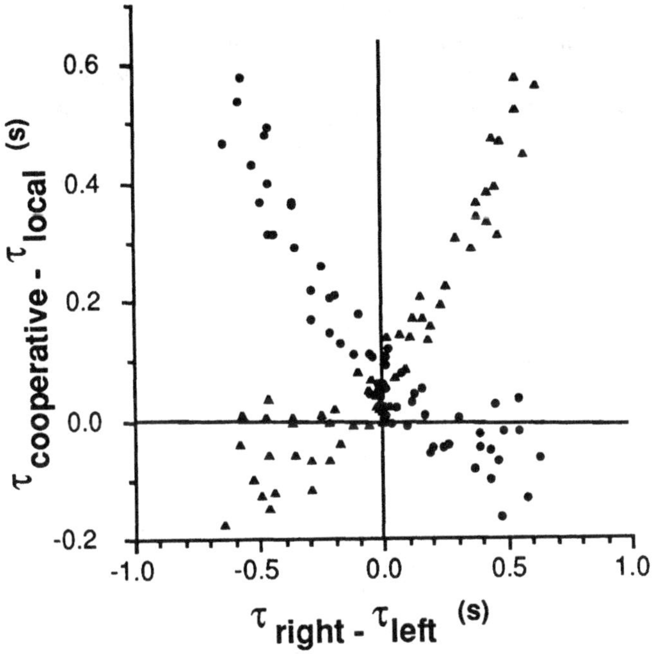

Figure 4.3. The magnet effect in the coordination of wrist-pendulum systems. The triangles represent the deviation of the cooperative period from the preferred period of the left pendulum, and the circles represent the deviation of the cooperative period from the preferred period of the right pendulum. These deviations increase as the differences in the two preferred periods ($\tau_{right}-\tau_{left}$) increase.

period at which they settle would follow from the theory of the compound pendulum as advanced by Huygens in the 17th century (Bell, 1950). For two simple pendulums coupled rigidly to form a compound pendulum, the characteristic period of the resultant system depends upon the length

$$L_v = (M_1L_1^2 + M_2L_2^2)/(M_1L_1 + M_2L_2) \quad (2)$$

where M_1, L_1 refer to the mass and length, respectively, of one of the simple pendulums and M_2, L_2 refer to the mass and length, respectively, of the other simple pendulum. The length L_v is the distance to the compound pendulum's center of oscillation (at which the masses are essentially summated and concentrated) from its rotation axis. In the case of so-called rigid coupling of wrist-pendulum systems, the resultant "virtual single system" parameter magnitudes of L_V (distance to center of oscillation) and M_V (total mass of component parts) are physical properties that arise as a consequence of the physical principles behind Huygens's (Bell, 1950) derivation, as those principles are manifest in biological tissue (see Kugler & Turvey, 1987, for the details).

There is a simple test of whether or not the cooperative period is a function of some local linear combination of the periods of the individual oscillators (e.g., some weighted mean of the two) or a function of a physical cooperativity of the individual oscillators as suggested by the Huygens (Bell, 1950) analysis. The test is whether or not the variation of the cooperative period accounted for by the squared virtual length parameter (L_V) is significantly greater than that accounted for by the characteristic periods of the two pendulums (τ_{left} and τ_{right}). As shown in Table 4.1 with the data from Kugler and Turvey's (1987) main experiment, a multiple regression of the cooperative period on L_V, τ_{left} and τ_{right} revealed that L_V accounted for most of the variation of the cooperative period, with the characteristic periods of the individual systems accounting for relatively less overall. In short, the virtual length (which is, roughly speaking, an emergent physical quantity) dictated the cooperative period. This result, it may be argued, points to a lawful basis for the common period, or magnet effect, exhibited by two pendular systems in comfortable, absolute coordination, a basis that depends, presumably, upon purely physical properties of biological systems, under certain boundary conditions.

IMPLICATIONS OF THE VIRTUAL SYSTEM ANALYSIS FOR TIME ALLOMETRIES

The analysis of Kugler and Turvey's (1987) data demonstrates that two limb-like oscillators in absolute coordination can be given a macroscopic redescription as a single virtual oscillator. The cooperative period at which the two limb-like systems settled was linked to the length L_V of this virtual oscillator. As noted, the virtual oscillator has a virtual single concentration of mass defined as the sum of the individual masses and located at the end of this length. Further, it can be assumed that the virtual oscillator has a virtual single restorative potential to represent the contribution of the left and right limb neuromusculatures and metabolisms needed to sustain

Table 4.1

Relative Contribution of Virtual and Local Oscillator Systems to the Cooperative Period

Subject	Multiplc regression standardized coefficients (β)		
	Virtual length (L_v)	Characteristic period of left system (τ_l)	Characteristic period of right system (τ_r)
1	.792**	.219*	.062
2	.557**	.149	.453**
3	.495**	.427**	.267
4	.930*	.042	.004

*$p < .05$
**$p < .01$

the oscillation. Hence, with this virtual system analysis, the two wrist-pendulums in absolute coordination can be redescribed as the single spring-pendulum system depicted in Figure 4.1, and the equation $\tau = 2\pi\ [ML^2/(gML + kb^2)]^{0.5}$ can be predicted to govern them.

Terrcstrial locomotion typifies absolute coordination, as noted. An important question to raise is whether or not the physical model depicted in Figure 4.1 can explain the periodic timing of the absolute coordination exhibited in locomotion. Surprisingly, the model rationalizes the different time allometries (that is, how an animal's locomotory cycle time scales to the magnitudes of its body) found over the ranges of different lengths occupied by insects, hummingbirds, small birds, large birds, medium-sized quadrupeds, and large quadrupeds.

If a quadruped's set of limbs is regarded as a compound pendulum, convertible into a virtual pendulum length L_v, then for quadrupeds in excess of 10 kg, the period found in walking, trotting, and cantering approximates closely the cycle times of a freely oscillating pendulum at length $L_V/2$, $L_V/7$ and $L_V/10$, respectively (Kugler & Turvey, 1987). What this means is that all quadrupeds' cycle times for a particular gait are related regardless of their differences in size and shape. For example, the warthog at 85 kg and the giraffe at 1,000 kg walk at step cycle times (0.784 s and 2.083 s, respectively) that are very much the same proportion of each animal's virtual limb length. How does this empirical result fit with the equation for the pendular clocking mode? The denominator of the equation can be separated into the torsional stiffness due to gravity G (= gML) and the torsional stiffness K (= kb^2) due to the elastic and metabolic contribution from the individual. It proves to be the case (Turvey et al., 1988) that the scalings of the periods of the gaits to L_V follow from the pendular clocking equation if *the elastic restoring torque* K *assembled by each animal is a constant multiple of the gravitational restoring torque* G *affecting each animal*, with a different multiple for each gait. If $K/G = 1$ then the equation reduces to $\tau = 2\pi(L/2)^{0.5}$, which characterizes walking; if $K/G = 6$ and $K/G = 9$, then the equation reduces to $\tau = 2\pi(L/7)^{0.5}$ and $\tau = 2\pi(L/10)^{0.5}$, which characterize trotting and cantering, respectively (Turvey et al., 1988). Further, the time allometries exhibited for quadruped locomotion (across all gaits) are $\tau \propto M^{0.125}$ or $M^{0.166}$ and $\tau \propto L^{0.5}$ (see Kugler & Turvey, 1987; Pennycuick, 1975; Turvey et al., 1988). The pendular clocking equation can be translated into biological terms on the assumption that there is geometric similarity in the size and shape of the organism (that is, $L = M^{0.3}$; Peters, 1983). If this is done, and one calculates periodic times for quadruped-sized masses and lengths at constant K/G ratios from the equation, then a simple regression analysis reveals that the scaling of time to mass and length is $M^{0.166}$ and $L^{0.5}$, respectively. This scaling agrees with observation.

The pendulum clocking mode equation $\tau = 2\pi[ML^2/(gML + kb^2)]^{0.5}$ also seems to explain the diverse time allometries found for other creatures across a great range of sizes. This feature is revealed if one takes into account both geometric similarity and the fact that the elastic potential K scales, on the average, to the mass of the animal to the first power (Turvey et al., 1988). The elastic system is constructed in different ways across the range of animals bounded by the smallest insect and the largest quadruped. In insects, it is built from at least three elastic materials: (a) the solid skeletal cuticle of the thoracic box, (b) a typical elastomere in the form of protein resilin, and (c) an elastic component in the myofibrils (Buchtal & Weis-Fogh, 1956). Insects have no mechanism for adjusting the magnitude of the elastic potential. Vertebrates, in contrast, can manipulate the magnitude of the stiffness by varying the co-contraction of the flexors and the extensors at the joint.

This difference in elastic structures has behavioral consequences. The mechanism used by insects restricts them to moving their wings at an almost constant period, irrespective of flight maneuvers and speed of travel (Sotavalta, 1952, 1954). The same appears to be true of birds (Greenewalt, 1960, 1975). In contrast, the elastic mechanism used by quadrupeds lets them walk, trot, and canter at a number of different frequencies although each gait does exhibit a preferred period that appears to be metabolically least costly (Hoyt & Taylor, 1981). Various analyses suggest, however, that despite differences in the construction of elastic machinery, there is a constant relation across animals between the elastic potential assembled at the characteristic frequency and body mass; that is, K increases as the first power of body mass (Turvey et al., 1988).

If the pendular clocking equation is translated into biological terms and periodic times are calculated (with length scaling as mass to the one-third power and K scaling as mass to the first power) across the range of lengths from the smallest insect to the largest quadruped, then the regression of log period on log length and log period on log mass exhibit a number of interesting results. The regressions are significantly quadratic and significantly linear (see Figures 4.4 and 4.5). For length < 0.1 m, period scales as length to the 0.913 power and mass to the 0.305 power; for length > 0.1 m, period scales as length to the 0.543 power and mass to the 0.183 power. These are the magnitudes of the time allometrics for insects (< 0.1 m) and quadrupeds (> 0.1 m), respectively. The region of the transition between the two slopes, that is, the region of length scales in the vicinity of 0.1 m, is the region occupied by large birds, small birds (passerines), and hummingbirds. Whereas the wing periods of large birds scale to length close to the 0.5 power, those of the small birds scale close to the 1.0 power (Turvey et al., 1988). Moreover, for hummingbirds, the relation in double logarithmic coordinates between period and length is largely quadratic, yielding what has been traditionally an odd scaling (Greenewalt, 1975). Collectively, this pattern of results for large birds, small birds, and hummingbirds is

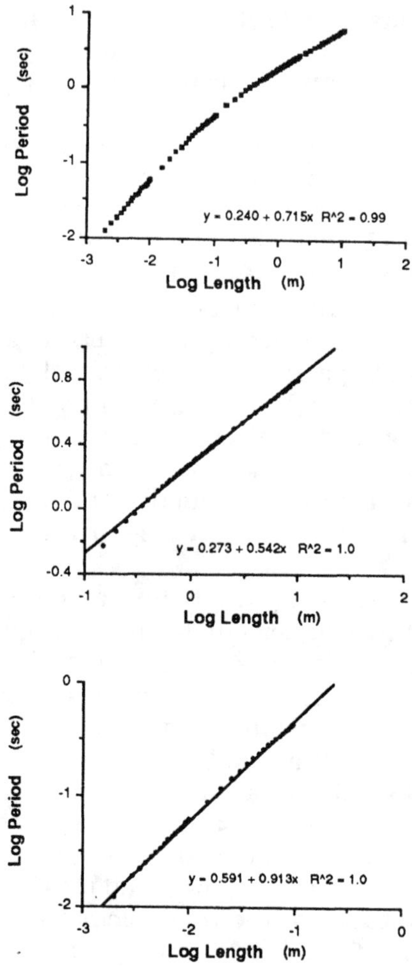

Figure 4.4. Log period plotted against log length for the pendular clocking mode equation over the length scales of insects, birds and quadrupeds. The equation is constrained biologically: $m \propto l^3$, and $K \propto m$. Upper panel is total range of lengths. Middle and lower panels are restricted to length > 0.1 m and length < 0.1 m, respectively. *Note.* From "On the Time Allometry of Coordinated Rhythmic Movements" by M. T. Turvey, R. C. Schmidt, L. D. Rosenblum, and P. N. Kugler, 1988, *Journal of Theoretical Biology, 130,* p. 332. Copyright 1988 by Academic Press. Reprinted by permission.

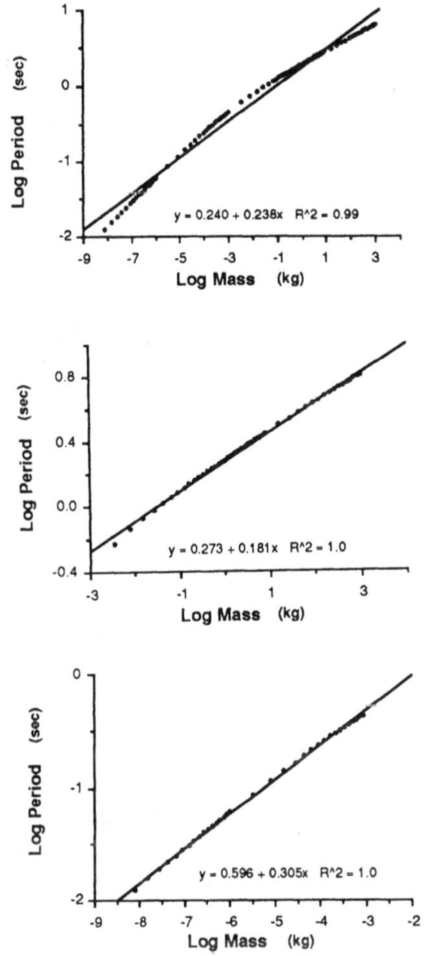

Figure 4.5. Log period plotted against log mass for the pendular clocking mode equation over the length scales of insects, birds and quadrupeds. The equation is constrained biologically: $m \propto l^3$, and $K \propto m$. Upper panel is total range of lengths. Middle and lower panels are restricted to length > 0.1 m and length < 0.1 m, respectively. Top panel regression line shows how the universal biological time scale ($\tau \propto M^{0.25}$) is obtained from the pendular clocking equation. *Note.* From "On the Time Allometry of Coordinated Rhythmic Movements" by M. T. Turvey, R. C. Schmidt, L. D. Rosenblum, and P. N. Kugler, 1988, *Journal of Theoretical Biology, 130,* p. 323. Copyright 1988 by Academic Press. Reprinted by permission.

rationalized by the biological version of the pendular clocking equation. These creatures are to be found near the inflection of the functional relation (in Figures 4.4 and 4.5). In sum, the pattern of time allometries does not point to shifts in biological design for different creatures but to the physical consequences of the pendular clocking mode at different lengths.

THE MAINTENANCE TENDENCY IN THE ABSOLUTE COORDINATION OF WRIST-PENDULUM SYSTEMS

So far, we have shown how the cooperative frequency exhibited in absolute coordination can be the consequence of very general physical principles rather than specific anatomical devices. In so doing, we have given a physical basis for the time allometries found in animal locomotion and von Holst's (1937/1973, 1939/1973) magnet effect. We now turn to von Holst's (1939/1973) maintenance tendency. We wish to show that the maintenance tendency (a) is found in the absolute coordination of wrist-pendulum systems and (b) can be explained by dynamic modeling of action systems.

Von Holst (1939/1973) saw evidence for the maintenance tendency in the relative coordination of rhythmic fin movements. For example, whenever one fin of a coordinated pair was stopped from moving, the other would move from the cooperative period to its characteristic period. Further, he found that when two fins had different characteristic periods, the faster one would reach its peak excursions before the slower one did. Stein (1973, 1974) corroborated this finding in the interappendage coordination of crayfish swimmerets. He found that the degree of phase lag between the two swimmerets was correlated with the difference between the characteristic periods of the two swimmerets. Other evidence for the maintenance tendency is an increase in the variability of periodic timing of a rhythmic unit as it is moved away from its characteristic period by coupling with another rhythmic unit. This result has been found in human finger tapping (Scripture, 1899) and in human arm movements (Smoll, 1982).

Do wrist pendulum systems show the tendency to maintain their intrinsic steady states when in absolute coordination; that is, do they show a maintenance tendency? A series of six experiments comprising a total of 54 different wrist-pendulum systems was designed to answer this question. The subjects were instructed to swing the pendulums in an alternating fashion (180° out of phase) and at the most comfortable tempo (see Rosenblum & Turvey, 1988, for a complete description of the methodology.) In each of these experiments, three subjects had to coordinate the same right-hand pendulum (the "target" system) with a number of different-sized, left-hand pendulums. The effect of the different-sized, left-hand pendulums was to pull the target system away from its characteristic period by varying amounts. Because of the magnet effect, a pendulum larger than the target pendulum will cause the target system to move more slowly than its characteristic period whereas a pendulum smaller than the target pendulum will cause the target system to speed up. This design allowed the deviation from preferred period to be indexed by τ_i/τ_0, where τ_i is the period of oscillation exhibited by the target wrist-pendulum in a coupling and τ_0 is the period of the target wrist-pendulum when swung alone. To measure fluctuations, von Holst's (1939/1973) relativized root mean square variance measure was used:

$$t_{fluc} = \sqrt{[((\Sigma(t_{pv}\ \ meant_{pv})^2/meant_{pv})/n_{pv} - 1) + ((\Sigma(t_{vp} - meant_{vp})^2/meant_{vp})/n_{vp} - 1)]} \qquad (3)$$

where *pv* means peak-to-valley half cycles, *vp* means valley-to-peak half cycles, and *n* is the number of cycles. Figure 4.6 shows how the relative periodic fluctuations change as a function of τ_i/τ_0. As the target pendulum is pulled away from its preferred period ($\tau_i/\tau_0 = 1$), the instability of the period increases. This is true whether the target pendulum is made to slow down ($\tau_i/\tau_0 > 1$) or to speed up ($\tau_i/\tau_0 < 1$). There is a leveling off, however, in the range $\tau_i/\tau_0 \geq 1.5$. Up to this point, the increase of periodic fluctuations with deviation from preferred tempo is symmetrical on both sides of $\tau_i/\tau_0 = 1$. Further, the same basic pattern is true for relativized amplitude fluctuations (computed from the amplitude version of the equation for the variance measure) except that they increase faster when the target pendulum is made to move faster than preferred (Figure 4.7).

The maintenance tendency in coordinated wrist-pendulum movements is indexed not only by fluctuations around the mean states of period and amplitude but also by the phase lag between the wrist-pendulum systems. The phase difference between systems increases with the magnitude of the difference between their charac-

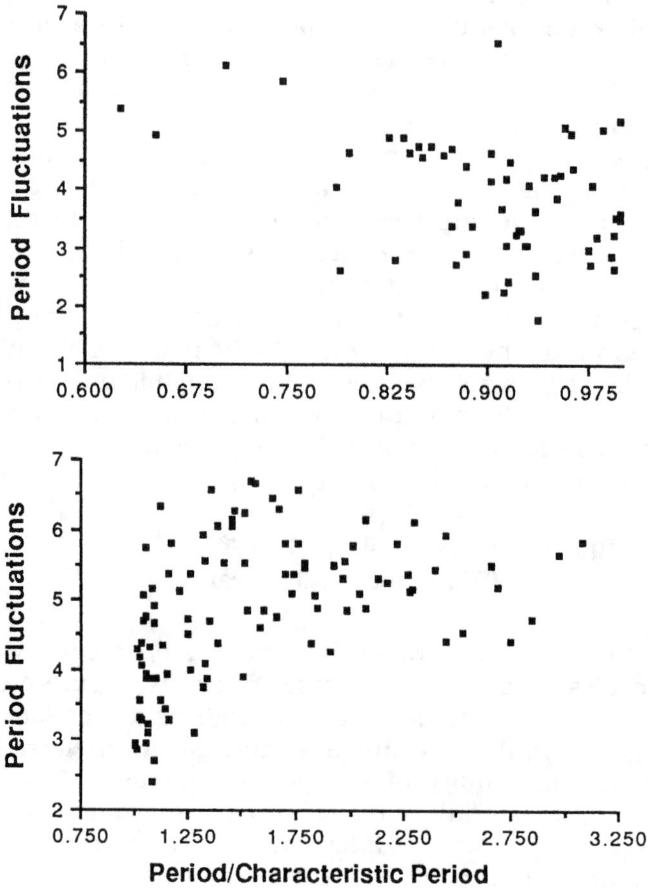

Figure 4.6. The maintenance tendency in wrist-pendulum coordination as revealed by periodic fluctuations (expressed as percentage of mean period). The upper panel shows the fluctuations as a function of τ/τ_0 (period/characteristic period) ≤ 1, and the lower panel shows them as a function of $\tau/\tau_0 \geq 1$. Data are pooled over the 3 subjects and six experimental sessions of Rosenblum and Turvey (1988). *Note.* From "Maintenance Tendency in Coordinated Rhythmic Movements: Relative Fluctuation and Phase" by L. D. Rosenblum and M. T. Turvey, 1988, *Neuroscience, 27*, p. 293. Copyright 1988 by IBRO. Reprinted by permission.

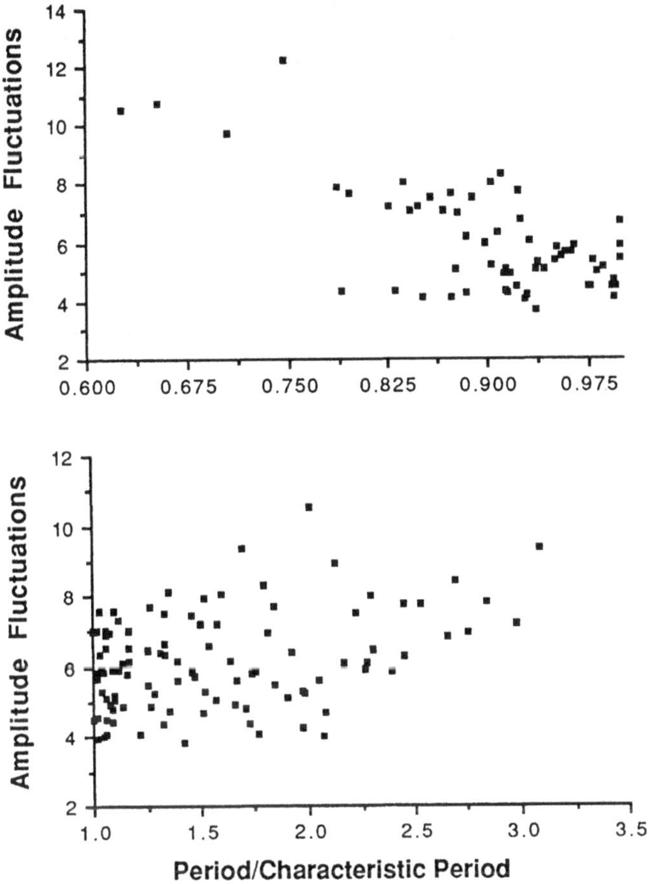

Figure 4.7. The maintenance tendency in wrist-pendulum coordination as revealed by amplitude fluctuations (expressed as percentage of mean amplitude). The upper panel shows the fluctuations as a function of τ/τ_0 (period/characteristic period) ≤ 1, and the lower panel shows them as a function of $\tau/\tau_0 \geq 1$. Data are pooled over the 3 subjects and six experimental sessions of Rosenblum and Turvey (1988). *Note.* From "Maintenance Tendency in Coordinated Rhythmic Movements: Relative Fluctuation and Phase" by L. D. Rosenblum and M. T. Turvey, 1988, *Neuroscience*, 27, p. 295. Copyright 1988 by IBRO. Reprinted by permission.

teristic periods. Because the preferred period (τ_0) was not measured for each pendulum in a wrist-pendulum pairing (it was measured only for the right-hand system), the metric used for the difference between the left and right pendulums' characteristic frequencies was L_T - L_l (that is, the length of the right system minus the length of the left system). The mean phase difference between the two oscillating systems was calculated with a point estimate at the peaks of each cycle (Yaminishi, Kawato, & Suzuki, 1979):

$$\phi = 360\ [\Sigma\ [(\tau_{1i} - \tau_{2i})/(\tau_{2i} - \tau_{2i+1})]/n\ ] \tag{4}$$

where τ_{1i} is the time of peak i of the target pendulum, τ_{2i} the time of peak i of the reference pendulum, and n is the number of cycles in a trial. Hence, phase differences greater than 180° indicate that the right pendulum is leading the left pendulum, and phase differences less than 180° indicate that the left pendulum is leading the right pendulum. Figure 4.8 demonstrates that when L_T is less than L_l (and L_T is the faster pendulum in isolation), the target pendulum leads the left pendulum; when L_T is greater than L_l, the target pendulum lags the left pendulum. This result is just what one would predict for a system demonstrating the maintenance tendency; namely, the intrinsically faster pendulum always leads the intrinsically slower pendulum in phase.

What kind of physical system would manifest these characteristics? It can be argued that the physical system formed in the absolute coordination of two wrist pendulums is a stratified system with three levels of stratification (Kugler & Turvey, 1987; Turvey et al., 1986). The highest level is the intentional level at which the subject has chosen to perform a pendular clocking behavior. Intentional states act as extraordinary boundary conditions on the dynamical principles underlying human movements (Kugler & Turvey, 1987). The middle level, the cooperative level, is a coordinative structure or functional unit (Kugler, Kelso, & Turvey, 1980) formed as a consequence of the interaction of the individual's intentional state and the dynamical principles governing rhythmic limb movements. This formative process consists, in part, of the freezing out of biomechanical degrees of freedom (for example, setting the postural context for the movement and setting the antagonist co-contraction) to allow the harnessing of dynamics so that absolute coordination can be obtained with minimal moment-to-moment intervention.

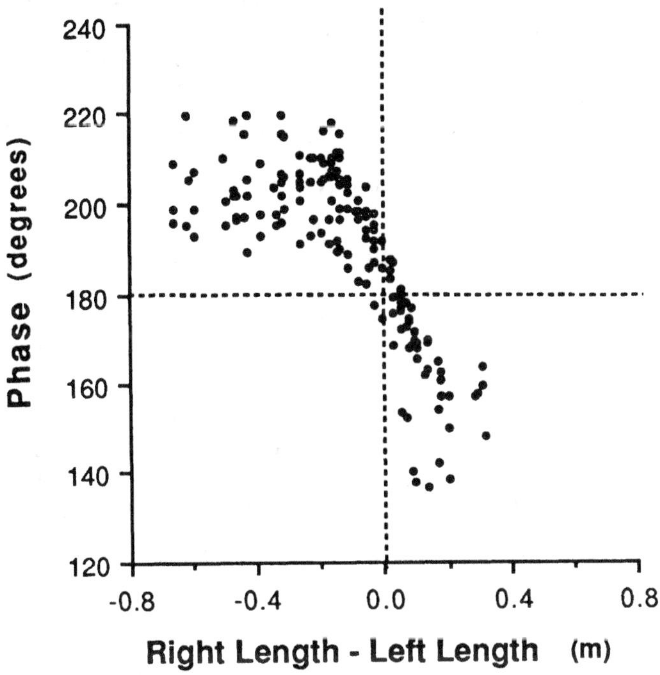

Figure 4.8. The maintenance tendency in the coordination of wrist-pendulum systems as revealed by phase differences (in degrees). Phase differences are presented as a function of the difference between the lengths of the right and left systems. Data are pooled over the 3 subjects and six experimental sessions of Rosenblum and Turvey (1988). *Note.* From "Maintenance Tendency in Coordinated Rhythmic Movements: Relative Fluctuation and Phase" by L. D. Rosenblum and M. T. Turvey, 1988, *Neuroscience, 27,* p. 297. Copyright 1988 by IBRO. Reprinted by permission.

The lowest level is that of the component oscillators. When two wrist pendulums are in absolute coordination, neither of the component oscillators is at its preferred period. Each is pulled towards the cooperative period defined by the natural laws underlying the virtual system. The interaction between the coordinative level and the component level is very much like that of self-organizing systems found elsewhere in nature (Kugler & Turvey, 1987). An example is the oft-cited Bénard convection. Here, a vat of oil is heated from the bottom and cooled from the top. The energy injected into the system by heating causes the component oil molecules to collide randomly with one another. At a critical temperature of heating, the molecules that have been pulled far from their equilibrium states form hexagonal columns which, upon further heating, begin to roll vertically. This type of self-organizing system has been called a dissipative structure (Prigogine, 1980) because the new macroscopic, coordinative level is formed in response to the need to dissipate the energy injected into the system. The conditions forming the cooperative virtual system are not only the metabolic energy put into the system to create the movements but also the informational-functional linking of degrees of freedom as designated by the subject's intentional state. As a consequence of forming a cooperative level of organization, the component microstructures (the individual molecules in the oil convection experiment and the component pendulums in the wrist-pendulum experiment) are driven far from the state they would settle at if left to their own exigencies and not subject to external influences. The individual wrist-pendulum systems in absolute coordination index the maintenance tendency through fluctuations because they have been marshalled temporarily to cooperate in an organization that has a dynamical basis. In order for stability to be in evidence at the cooperative level, the components must be made unstable temporarily. One can then expect the kinematics (period and amplitude) of the components to index their unstable states. The more the cooperative state demands that the components compromise their preferences, the larger will be the fluctuations seen in their behavior.

In sum, the fluctuations that index the maintenance tendency can be explained by the supposition that the system formed to produce the coordinated movement has a dynamical basis and a stratified structure of cooperating levels in common with many self-organizing systems in nature. In order to explain the phase difference result, Rosenblum and Turvey (1988) presented two hypotheses. The various phase differences between the wrist-pendulum systems are: (a) consequences of neural innervations of the musculature at those phase difference, or (b) consequences of a single phase difference of neural innervation (in this case 180°) and resultant pendulum phase differences produced by a self-organizing property of the musculature comprising each component system (Partridge, 1966,

1967). The second hypothesis is a strong candidate from the ecological point of view. Loaded muscle, for a given frequency of neural innervation, varies the force of its output with the load it carries very much as a spring would. This property leaves amplitude of contraction constant across different inertial loads but produces a variable phase lag between the input neural signal and the output movement for pendulums of different inertial magnitudes. On the assumption that a neural signal is sent to the coordinated limbs exactly at a 180° phase difference, the observed deviation in phase difference from 180° between the two pendulums can be attributed to this compensatory property of the component musculature. The degree of this compensation depends upon each wrist-pendulum system's moment of inertia (I_l or I_r), which correlates with its length (L_l or L_r). If the wrist-pendulum systems have the same inertial loadings ($I_r = I_l$), then the phase lag between innervation and movement is equal for both, and their phase relation when coupled will be the intended 180° (out of phase). The wrist-pendulum system with the lower moment of inertia will always lead, and the wrist-pendulum system with greater moment of inertia will always lag, as is seen in Figure 4.8.

THE INFORMATIONAL BASIS FOR ABSOLUTE COORDINATION

The physical interactions between limbs that underly absolute coordination must be nonstandard from a physical point of view. At issue here is the fact that the two oscillating wrist-pendulum systems are not interacting in a classical mechanical way by a transportation of mechanical energy from one unit to the other. The reason is simply that they are not forcefully linked as would be the case if the two pendulums were connected by a metal rod. Patently, the physical linkage, or medium for the interaction, is the nervous system. The problem that this fact poses for a physical account is that nervous system interactions are *informational*. That is, even though an electromagnetic potential is the basis for the function of the nervous system, it is not the electromagnetic energy transported over the nervous system that is the basis for interactions among limbs. Rather, it is the *patterning* of action potentials (e.g., rate) over neural lines, not the amount of energy each line contains, that supports these interactions. When a patterning of some medium is the basis for the interaction of units (e.g., an animal and its environment via structured ambient light), rather than energy transported along this medium, then the interaction may be termed informational (Kugler, Turvey, Carello, & Shaw, 1985; Kugler & Turvey, 1987). If physical principles indeed shape absolute coordination, and if they are functioning across the nervous system, then some physical principles may apply equally in either energy or informational interactions. This intriguing possibility is corroborated by experimental results.

If it is true that the principles of organization underlying the magnet effect and the maintenance tendency are physical and operational across an informational medium, then the magnet effect and the maintenance tendency should be manifest in absolute coordination across an informational medium different from the central nervous system. In research now underway, the wrist-pendulum system methodology is being used to study the *visual* absolute coordination of movements between two people. In this research, two people sit facing the same direction but turned slightly towards each other to enable each person to see the other's outside wrist. Each person swings a wrist-pendulum with the outside hand in two conditions: (a) coordinated synchronously with the other person at 180° out of phase; and (b) with the eyes closed. Four wrist-pendulum systems of different sizes (the same pendulum magnitudes and coupling as in Kugler & Turvey, 1987) are combined in 16 ways. The characteristic periods of the individual pendulums are measured in the eyes closed condition. With this design we can investigate whether or not the phenomena of absolute coordination that were found in the within-person or neuro-anatomical coupling of wrist-pendulum systems will be found in the between-person or visual coupling of wrist-pendulum systems.

Preliminary analyses suggest that both the maintenance tendency and the magnet effect occur. The analyses are complicated, however, by the heterogeneity of the elastic potential brought to the task by different subjects (see Figure 4.2). (In the within-person case, the elastic stiffness seems to be uniform across the two wrists.) The data for the magnet effect are presented in Figure 4.9 for two pairs of subjects and replicate the within-person coupling data presented in Figure 4.3. The period of a pendulum decreases when that pendulum is coupled with a faster pendulum but increases when the pendulum is coupled with a slower pendulum. The virtual system analysis, which assumes that a lawful cooperative state is formed to dictate the coupled period, seems to apply equally well to this visual instance of the magnet effect. Under the assumption that the coupled

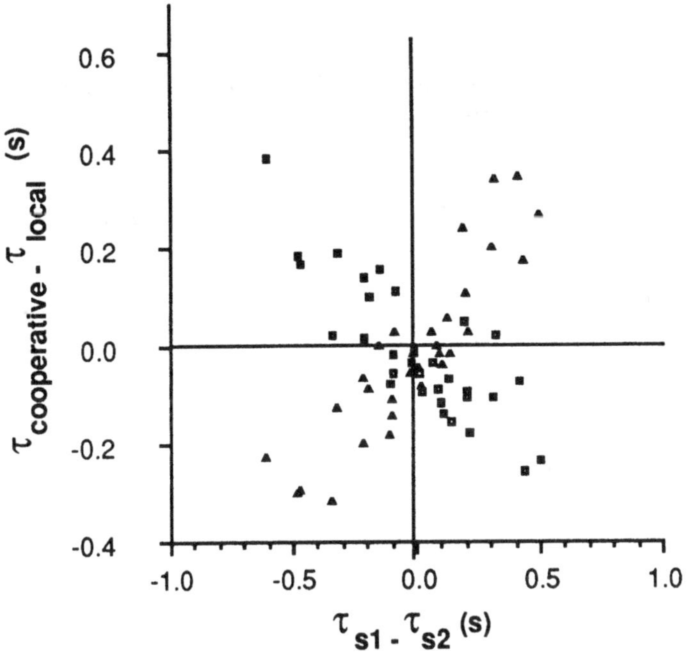

Figure 4.9. The magnet effect in the visual coordination of wrist-pendulum systems between two people. The squares represent the deviation of the cooperative period from the preferred period of the 1st subject's pendulum and the triangles represent deviation of the cooperative period from the preferred period of the 2nd subject's pendulum. These deviations increase as the differences in the two preferred periods ($\tau_{first} - \tau_{second}$) increase.

wrist-pendulums are performing as if rigidly coupled, the length to the center of oscillation L_V (the length of the virtual simple pendulum) can be calculated. In a multiple regression of the cooperative period on L_V, τ_{s1} and τ_{s2} (where s1 refers to the first subject of a pair and s2 refers to the second), the virtual length L_V was found to account for more variation in the cooperative period than did either of the individual pendulums' characteristic periods (Table 4.2).

In the visual case of absolute coordination, the cooperative period cannot be a consequence of signals from a central pattern generator interacting via interneuronal connections because there is no nervous system connecting the two individuals. It can, however, be the result of a very general physical organizing strategy—one that

Table 4.2

Relative Contribution of Virtual and Local Oscillator Systems to the Visual Cooperative Period

	Multiple regression standardized coefficients (β)		
Subject pair	Virtual length (L_v)	Characteristic period of left system (τ_l)	Characteristic period of right system (τ_r)
1	.542*	.306	.379*
2	.644*	.016	.185

* $p < .05$

confines the comfortable period of absolute coordination to that determined by the coupled systems' center of oscillation—that is manifest equally in energy and information contexts. How is it possible for information to link the dynamical states of two components of a physical organization? To answer this question, we must say something about J. J. Gibson's notion of information and its function in the behavior of wrist-pendulum systems.

In Gibson's (1979) ecological theory of optical information, the light ambient to a point of observation, light that has been structured by multiple reflections from surfaces, is referred to as the optic array. Gibson argued that macroscopic qualitative properties of the optic array and, in particular, the transforming optic array, are unique and specific to surface layout, changes in surface layout, and displacements of the point of observation. As such, the transforming optic array is a low-energy field that contains information about the environment and about movements of the perceiver, by virtue of its kinematic form (Kugler et al., 1980). Visual perception studies by Runeson and Frykholm (1981) and Bingham (1987) have demonstrated that dynamical properties of an event (for example, momentum and mass) are perceivable from merely the kinematic morphologies in the optic array. They referred to this phenomenon as the *kinematic specification of dynamics*.

Kugler and Turvey (1987) suggested that the subject discovers the preferred period of wrist-pendulum behavior through a process similar to kinematic specification of dynamics. They maintain that the deformation of tissue during a movement induces a patterning of mechanoreceptors that is unique and specific to the dynamical properties of the movement. The dynamical property that needs to be perceived is the minimum of the potential energy function associated with a particular single or coupled wrist pendulum system. Finding the most comfortable period of oscillation is tantamount to finding the period at which the subject contributes as little energy to the cycle as possible. By exploring different tempos at the initiation of the behavior, the subject is exploring the potential space of that wrist-pendulum system. Where he or she is in the space with respect to the potential minimum is specified by the gradient of the space (that is, the magnitude and direction of the rate of change of the potential). The gradient of the potential needs to be perceived in order for the subject to lengthen or shorten his or her cycle to obtain the preferred period. Kugler and Turvey (1987) suggested that there is information for this dynamical property in the patterning of the deformation of tissue produced by the movement. Biomechanically, the shortening or lengthening of a cycle can be understood as the varying of the assembled neuromuscular elastic potential, previously discussed. In short, the physiology is being tuned by the perceived dynamical properties of the activity.

Armed with this ecological notion of information, we are in a better position to try to understand how "virtual" quantities emerge from an interaction defined across an informational medium such as the optic array or the nervous system. The dynamics of each wrist pendulum's movement creates a neural patterning and an optical patterning that are unique and specific to the dynamics. The subject's task (or each subject's task, in the visual case) is to find the period that is dynamically least costly for coordinating both pendulums isochronously and 180° out of phase. The solution that emerges is interpretable as the harnessing of the two different dynamics by means of *the period dictated by the center of oscillation of the two components*. At this period, the potential energy function of the cooperative state comprising the two wrist-pendulum systems is at its minimum. In sum, for single wrist-pendulum systems, and for two wrist-pendulum systems connected neuroanatomically or visually, the basis for perceiving the dynamical state of an oscillatory system in order to find the least energetically costly cycle period is identical.

CONCLUDING REMARKS

The ecological perspective on coordination maintains that one should look for a low-dimensional redescription of the high-dimensional (viz., neural, vascular, etc.) system in question and see whether the patterning of the macroscopic components can be explained by physical principles at the macroscopic scale. Special anatomical mechanisms that are posited to explain a particular phenomenon should be proposed only after more general forms of explanation have been proved unusable.

The magnet effect is such a patterning of the macroscopic components that can be explained by physical principles. It is a general characteristic of coordinated rhythmic movements: It is observed for limbs, fins, and wings; it is observed for rhythmic units that relate neuroanatomically, as in the within-person coordination of wrist-pendulum systems, and for rhythmic units that relate visually, as in the between-people coordination of wrist-pendulum systems. The commonalities among these very different circumstances must be the result of laws operating over the physical properties that all the circumstances share. The implication is that organisms use the natural organizing dispositions of their physical attributes to facilitate the coordination of their movements. Thus, the predictions from the pendular clocking mode equation, concerning the time allometries involved in the locomotion of animals of all sizes, are evidence for the operation of physical constraints in the coordination of wrist-pendulum systems and in the coordination of limbs in locomotion generally. Further, the fluctuations that are the hallmark property of the maintenance tendency can be understood as an index of the tension between two levels of the dynamical organization formed to coordinate two or more rhythmic units.

A major assumption behind the arguments in this chapter is that absolute coordination is a lawful process involving information. This assumption is in keeping with the understanding that organisms are physical entities that relate to their surroundings through informational contact, an understanding that poses significant challenges for the development of a physical biology (Kugler et al., 1985; Kugler & Turvey, 1987; Yates, 1987). An organism's ability to perceive the dynamical properties of its limbs on the basis of information about those dynamics underlies the operation of physical strategies of organization in locomotion. The surprising outcome of this fact is that physical laws are manifest in situations that are governed more by information than by forces. A thoroughgoing ecological perspective necessitates exploring in detail the role of information in the dynamical constraints on coordinated movements, where information is understood in Gibson's (1979) specificational sense.

REFERENCES

Beek, P. J., & Beek, W. J. (in press). Tools for constructing dynamical models of rhythmic movement. *Human Movement Science.*

Bell, A. E. (1950). *Christian Huygens and the development of science in the seventeenth century.* London: Arnold.

Bernstein, N. A. (1967). *The control and regulation of movements.* London: Pergamon Press.

Bingham, G. P. (1987). Kinematic form and scaling: Further investigations on the visual perception of lifted weights. *Journal of Experimental Psychology: Human Perception and Performance, 13,* 155-177.

Buchtal, F., & Weis-Fogh, T. (1956). Contribution of the sarcolemma to the force exerted by resting muscle of insects. *Acta Physiologica Scandanavia, 35,* 345-364.

Cavagna, G. (1970). Elastic bounce of the body. *Journal of Applied Physiology, 29,* 297-282.

Cavagna, G. (1977). Elastic energy in muscle. *Exercise and Sports Science Reviews, 5,* 89-129.

Corlett, E. N., & Mahadeva, K. (1970). A relationship between a freely chosen working pace and energy consumption curves. *Ergonomics, 13*, 517-524.

Edgerton, V. R., Grillner, S., Sjogstrom, A., & Zangger, P. (1976). Central generation of locomotion in vertebrates. In R. M. Herman, S. Grillner, P. S. G. Stein, & D. G. Stuart (Eds.), *Neural control of locomotion* (pp. 439-464). New York: Plenum Press.

Feldman, A. G. (1966). Functional tuning of the nervous system with control of movement or maintenance of a steady posture: III. Mechanographic analysis of execution by man of the simplest motor tasks. *Biophysics, 11*, 766-775.

Feldman, A. G. (1980). Superposition of motor programs: I. Rhythmic forearm movements in man. *Neuroscience, 5*, 81-90.

Feldman, A. G. (1986). Once more on the equilibrium-point hypothesis (lambda model) for motor control. *Journal of Motor Behavior, 18*, 17-54.

Frischeisen-Köhler, I. (1933). The personal tempo and its inheritance. *Character and Personality, 1*, 301-313.

Gibson, J. J. (1979). *The ecological approach to visual perception.* Boston: Houghton-Mifflin.

Grillner, S. (1975). Locomotion in vertebrates: Central mechanisms and reflex interactions. *Physiological Review, 55*, 247-304.

Greenewalt, C. H. (1960). The wings of insects and birds as mechanical oscillators. *Proceedings of the American Philosophical Society, 104*, 605-611.

Greenewalt, C. H. (1975). The flight of birds. *Transactions of the American Philosophical Society, 65*, 1-67.

Hoyt, D. F., & Taylor, C. R. (1981). Gait and the energetics of locomotion in horses. *Nature, 292*, 239-240.

Kay, B. A., Kelso, J. A. S., Saltzman, E. S., & Schöner, G. (1987). The space-time behavior of single and bimanual rhythmical movements. *Journal of Experimental Psychology: Human Perception and Performance, 13* , 564-583.

Kelso, J. A. S. (1986). Pattern formation in speech and limb movements involving many degrees of freedom. In H. Heuer & C. Fromm (Eds.), *Experimental Brain Research: Supplementum: Vol. 15. Generation and modulation of action patterns* (pp. 107-128). Berlin: Springer.

Kugler, P. N., Kelso, J. A. S., & Turvey, M. T. (1980). On the concept of coordinative structures as dissipative structures: I. Theoretical lines of convergence. In G. E. Stelmach & J. Requin (Eds.), *Tutorials in motor behavior* (pp. 3-47). Amsterdam: North-Holland.

Kugler, P. N., & Turvey, M. T. (1987). *Information, natural law and the self-assembly of rhythmic movement.* Hillsdale, NJ: Erlbaum.

Kugler, P. N., Turvey, M. T., Carello, C., & Shaw, R. (1985). The physics of controlled collisions: A reverie about locomotion. In W. H. Warren & R. E. Shaw (Eds.), *Persistence and change* (pp. 195-229). Hillsdale, NJ: Erlbaum.

Lee, D. N. (1980). Visuomotor coordination in space-time. In G. E. Stelmach & J. Requin (Eds.), *Tutorials in motor behavior* (pp. 281-295). Amsterdam: North-Holland.

Partridge, L. D. (1966). Signal-handling characteristics of load-moving skeletal muscle. *American Journal of Physiology, 210*, 1178-1191.

Partridge, L. D. (1967). Intrinsic feedback factors producing inertial compensation in muscle. *Biophysics Journal, 7*, 853-863.

Pennycuick, C. J. (1975). On the running of the gnu (Connochaetes taurinus) and other animals. *Journal of Experimental Biology, 63*, 775-799.

Peters, H. P. (1983). *The ecological implications of body size.* Cambridge, MA: Cambridge University Press.

Prigogine, I. (1980). *From being to becoming: Time and complexity in the physical sciences*. San Francisco: Freeman.

Rosenblum, L. D., & Turvey, M. T. (1988). Maintenance tendency in coordinated rhythmic movements: Relative fluctuation and phase. *Neuroscience, 27*, 298-300.

Runeson, S., & Frykholm, G. (1981). Visual perception of lifted weight. *Journal of Experimental Psychology: Human Perception and Performance, 7*, 733-740.

Schöner, G., & Kelso, J. A. S. (1988). Dynamic pattern generation in behavioral and neural systems. *Science, 239*, 1513-1520.

Scripture, E. W. (1899). Observations on rhythmic action. *Studies from the Yale Psychology Laboratory, 7*, 102-108.

Shik, M. L., & Orlovskii, G. N. (1976). Neurophysiology of locomotor automatism. *Physiological Review, 56*, 465-501.

Smoll, F. L. (1975a). Between-days consistency in personal tempo. *Perceptual and Motor Skills, 41*, 731-734.

Smoll, F. L. (1975b). Preferred tempo in the performance of repetitive movements. *Perceptual and Motor Skills, 40*, 439-442.

Smoll, F. L. (1982). Accuracy of rhythmic motor behavior in response to preferred and nonpreferred tempos. *Journal of Human Movement Studies, 8*, 123-138.

Sotavalta, O. (1952). The essential factor regulating the wing stroke frequency of insects in mutilation and loading experiments and in experiments in subatomospheric pressure. *Annals of the Zoological Society "Vanamo," 15*, 1-66.

Sotavalta, O. (1954). The effect of wing inertia on the wingstroke frequency of moths, dragonflies, and cockroach. *Annales Entomologici Fennici, 20*, 93-101.

Stafford, F. S., & Barnwell, G. M. (1985). Mathematical models of central pattern generators in locomotion: I. Current problems. *Journal of Motor Behavior, 17*, 3-26.

Stein, P. S. G. (1971). Intersegmental coordination of swimmeret motor neuron activity in crayfish. *Journal of Neurophysiology, 34*, 310-318.

Stein, P. S. G. (1973). The relationship of interlimb phase to oscillator activity gradients in crayfish. In R. B. Stein, K. G. Pearson, R. S. Smith, & J. B. Redford (Eds.), *Control of posture and locomotion* (pp. 621-623). New York: Plenum Press.

Stein, P. S. G. (1974). The neural control of interappendage phase during locomotiom. *American Zoologist, 14*, 1003-1016.

Stein, P. S. G. (1976). Mechanisms of interlimb coordination. In R. M. Herman, S. Grillner, P. S. G. Stein, & D. G. Stuart (Eds.), *Neural control of locomotion* (pp. 465-487). New York: Plenum Press.

Stein, P. S. G. (1977). A comparative approach to the neural control of locomotion. In G. Hoyle (Ed.), *Identified neurons and behavior of arthropods* (pp. 227-239). New York: Plenum Press.

Turvey, M. T. (1977). Preliminaries to a theory of action with reference to vision. In R. E. Shaw & J. Bransford (Eds.), *Perceiving, acting and knowing* (pp. 211-265). Hillsdale, NJ: Erlbaum.

Turvey, M. T., & Carello, C. (1986). The ecological approach to perceiving-acting: A pictorial essay. *Acta Psychologica, 63*, 133-155.

Turvey, M. T., Rosenblum, L. D., Schmidt, R. C., & Kugler, P. N. (1986). Fluctuations and phase symmetry in coordinated rhythmic movements. *Journal of Experimental Psychology: Human Perception and Performance, 12*, 564-583.

Turvey, M. T., Schmidt, R. C., Rosenblum, L. D., & Kugler, P. N. (1988). On the time allometry of coordinated rhythmic movements. *Journal of Theoretical Biology, 130*, 285-325.

Turvey, M. T., Shaw, R. E., & Mace, W. M. (1978). Issues in a theory of action: Degrees of freedom, coordinative structures and coalitions. In J. Requin (Ed.), *Attention and performance, VII* (pp. 557-595). Hillsdale, NJ: Erlbaum.

von Holst, E. (1973). On the nature of order in the central nervous system. In R. Martin (Ed. and Trans.),*The collected papers of Erich von Holst: Vol. 1. The behavioral physiology of animals and man* (pp. 3-32). Coral Gables, FL: University of Miami Press. (Original work published 1937)

von Holst, E. (1973). Relative coordination as a phenomenon and as a method of analysis of central nervous function. In R. Martin (Ed. & Trans.), *The collected papers of Erich von Holst: Vol. 1. The behavioral physiology of animals and man* (pp. 33-135). Coral Gables, FL: University of Miami Press. (Original work published 1939)

Warren, W. H. (1984). Perceiving affordances: Visual guidance of stair climbing. *Journal of Experimental Psychology: Human Perception and Performance, 10*, 683-703.

Yaminishi, J., Kawato, M., & Suzuki, R. (1979). Studies on human finger tapping neural networks by phase transition curves. *Biological Cybernetics, 33*, 199-208.

Yates, F. E. (1987). Quantumstuff and biostuff: A view of patterns of convergence in contemporary science. In F. E. Yates (Ed.), *Self-organizing systems: The emergence of order*. New York: Plenum Press.

Perspectives on the Coordination of Movement
S.A. Wallace (Editor)

MOTOR COORDINATION FOR FUNCTIONAL HUMAN BEHAVIORS: PERSPECTIVES FROM A SPEECH MOTOR DATA BASE

James H. ABBS*

Departments of Neurology and Neurophysiology
University of Wisconsin Medical School

Speech Motor Control Laboratories, Waisman Center
University of Wisconsin

and

Nadine P. CONNOR

Department of Neurophysiology
University of Wisconsin Medical School

Speech Motor Control Laboratories, Waisman Center
University of Wisconsin

ABSTRACT

The term *coordination*, justifiably, is almost as common in discussions of motor systems as the word *control*. In this vein, speech motor coordination provides a potential model system for examining this phenomenon. For speech and for most natural motor tasks, generating properly timed and measured multiple muscle contractions is a primary role of the motorsensory system. However, the understanding of the neurobiological processes underlying motor coordination has

---

*Address correspondence to: James H. Abbs, Ph.D., Speech Motor Control Laboratories, Waisman Center, 1500 Highland Avenue, University of Wisconsin, Madison, WI 53706, U.S.A.

been hampered by several difficulties. First, single joint or oversimplified behaviors have been emphasized; it is difficult to examine coordination if the number of elements being coordinated is unnaturally restricted. Second, many investigators have utilized superficial measures of behavior (keystrokes, speech acoustic signals, cursive writing patterns), thus largely ignoring the multiple underlying motor actions that are the crux of coordination. Finally, several lines of study, apparently aimed at coordination, have consisted largely of blanket searches for invariant patterns in certain aspects of system output—primarily without scientific hypotheses regarding biological processes. Fortunately, recent analyses of speech motor actions not only address some of these difficulties but also offer some more concrete data on underlying neurobiological mechanisms.

INTRODUCTION

The objective of this chapter is to examine motor coordination based upon knowledge of perhaps our most human motor function, speech. However, this discussion is not intended to be a description of speech as an isolated behavior; rather, we assume that the coordination of speech motor actions is similar in principle to most other actions in which we engage. Indeed, consideration of speech motor function will illustrate that many important sequential motor behaviors such as food gathering, tool using, locomotion, typewriting, handwriting, and musical instrument performance share many features with speech; research findings on some of these varied motor functions will thus be drawn upon, when relevant, to provide some critical directions. Although knowledge of coordination associated with speech is not fully described, more information is available on multiple movements and muscle actions than for most other motor functions of this complexity.

At this juncture, it is necessary to raise the long-standing issues of motor programs and motor programming. To a large extent, one cannot discuss motor coordination, and particularly the underlying biological correlates of motor coordination, without considering what a motor program is and how it might be formed. Specifically, one must address how the temporal-spatial organization of the motor output for a given task is generated. The alternatives range from a stereotypic pattern generator (a hard-wired programmer), to a more flexible planning process that is implemented completely prior to the initiation of that motor output sequence (event-by-event preprogramming), to very flexible interactions between ascending sensory input and general motor objectives, which operate on a moment-to-moment basis to shape the interactions among the constituent movements. Until recently, it has been popular to examine certain features of the motor output (e.g., relative timing, intermovement intervals), or conversely variations therein, to discern the nature of the underlying programming process. Unfortunately, for reasons to be addressed subsequently, regularities in certain limited aspects of the motor output are not generally prima facie evidence for any of the above-noted alternatives for motor coordination or programming, despite claims to the contrary. As will be apparent, however, recent data of a different kind do address this issue at least for coordination of speech motor output and some other sequential motor functions of a like nature.

AN HISTORICAL COMMENT ON STUDIES OF MOTOR COORDINATION

Early kinematic analyses of movement involved techniques such as cinematography, cyclography, and variations thereof. These analyses typically yielded graphic representations of movement patterns that were examined visually in their natural and albeit complex form as a means to develop general concepts and overall hypotheses (cf. Bernstein, 1967). Because of the form of these data and the complexity of the motor behaviors observed in that early work, the statements and theories regarding motor coordination were largely qualitative. Ironically, the perspectives offered by these early workers were generally very insightful. Almost all scientists of motor control have been impressed if not inspired by the testimony of Nikolai Bernstein in his conceptualization of motor coordination and the function of the nervous system in that process. Likewise, Gordon Holmes's (1939) observations on impairments of coordination in World War I patients with gunshot wounds of the cerebellum, conducted with rudimentary techniques over 60 years ago, remain the best of their kind; almost no modern discussion of cerebellar function is without reference to these works. Perhaps these early insights were particularly valuable because observations were made largely on relatively unconstrained and functional motor tasks. At the least, because quantitative details were limited, investigators were forced to focus more generally on overall patterns, perhaps yielding insights not discernible from the microanalyses of more restricted behavior that have characterized much of the modern work.

Consideration of efforts in the last 15-25 years, since the active work of Holmes and Bernstein, is also very revealing. There is a time-worn metaphor concerning a drunk, searching for his keys under the lamppost on a darkened street, who explains that although he lost the keys in a nearby dark alley, he prefers to look where it is easier to see. This story is relevant to our understanding of coordination and to more recent investigations of this phenomenon, including coordination of speech. In general, much of the recent work on the issue of coordination has taken one of two avenues, both of which appear to suffer the problems of the proverbial imbiber. One approach has focused on the patterns of timing in the final motor output, for example, the key strokes in typing or acoustic segments in speech. These superficial outputs were relatively easy to transduce and measure, especially in their temporal patterning; hence the applicability of the metaphor. Some concepts have emerged from these analyses regarding nervous system representation of sequential motor behaviors, particularly in the cognitive psychology literature (re: memory storage and retrieval processes, search algorithms, etc.). However, observations of the output side of behaviors such as speech and typewriting (cf. Kent, 1976; Kent & Moll, 1972; Sternberg, Monsell, Knoll, & Wright, 1979; Terzuolo & Viviani, 1979), have also led to the arguments that motor control of sequencing is (a) the major aspect of coordination in these behaviors, and (b) largely dependent upon predetermined representations, stereotypically generated with certain temporal invariances. As will become apparent, there are serious problems with these perspectives.

A second recent approach relevant to the present chapter is exemplified in studies using various modern techniques for the transduction of human movement; these studies have benefited from being able to examine the detailed movements underlying the more superficial outputs, with considerable potential for quantification. Because of the complexity of natural human movement, many of these instrumentally advantaged studies have focused on rather simple aspects of behavior. Technical factors also caused some of this work to be aimed at simpler motor acts, including some difficulty in transducing movements around multiple joints and the intractability of the multivariant data obtained.

Not surprisingly, many hypotheses concerning motor control and coordination were derived from studies of single joint movements on the one hand or measures of terminal output on the other. The present authors are of the opinion that these two approaches both yield limited views of coordination, but for different reasons.

Problems in interpretation will be dealt within a separate section. However, regarding the analysis of simpler, single joint movements to discern patterns of control and coordination, one fundamental flaw is that the mammalian nervous system may not have evolved to optimally generate such outputs. Experimental constraint of such actions is likely to force the organism to perform tasks for which the nervous system was not designed. In the case of primates, including naive human subjects, it is possible only with considerable care and usually force of personality to ensure that isolated single joint movements are generated; very often precautions must be taken to minimize "unwanted" contributions from adjacent arm segments. In this vein, several researchers have argued that movements constrained to a single degree of freedom may be more difficult to learn and perform than natural movements (Abbs, Gracco, & Cole, 1984; Fentress, 1984; Hogan, 1985; Hollerbach, 1981). One could argue that single joint movements are a basic unit of more complex behaviors; but this argument, however appealing, is without empirical support and most probably is overreductionistic. Considering this issue from the standpoint of underlying biological processes, it is also apparent that even the most abstract definition of coordination must include the operation of functionally significant relations among multiple motor output variables: muscle contractions, movements, forces. A meaningful view of coordination is not likely to be forthcoming if the number of such variables is artificially limited. Even in studies of brain structures asserted to have a primary role in motor coordination, such as the cerebellum, many of the motor tasks studied have involved single joint movements (cf. Brooks & Thach, 1981).

Most investigations of even multiarticulate or multijoint movement have been focused on single trials of reaching, pointing, or grasping. However, the problems of analyzing simpler movements are not solved fully by examination of isolated multijoint actions, although this is a step in the right direction. Most actions in nature are not produced in isolation but rather as part of functional, progressive sequences. Food gathering, manual communication, speech, locomotion, tool using, and so on typically involve the accomplishment of significant objectives in sequence; only rarely are these important actions useful in singular or isolated form. Thus a general definition of coordination must also include relations among sequential output variables. Without proper sequencing and in some cases fairly precise timing of those sequences, the functional objectives of the many motor behaviors cannot be achieved.

As should be apparent, considerations of motor coordination in natural motor behaviors must include (a) the codependent adjustment and multiple muscle actions and movements as these are combined concurrently for each element of a sequential behavior (e.g., reaching, object grasping, engagement, and transport or alternatively key presses for a typed word) as well as (b) timing of the initiation and completion of that element with those elements that follow or precede it. In fact, even a single reaching movement involves the coadjustment of the magnitudes of the angular rotations and translations at the torso, shoulder, elbow, and wrist along with the shaping of the hand (and possibly torso) and the timing of these movements with the subsequent actions of object grasp, engagement, and transport (see Athenès & Wing, this volume, for a review). As is apparent in even this qualitative example, these two forms of coordination may be fundamentally inseparable in natural behaviors. That is, for a given action (e.g., reaching) some components (i.e., shoulder joint rotation) are superimposed across several elements in a sequence, whereas others (the wrist or hand) are adjusted discretely for each. Certain components of such natural multi-element actions manifest anticipatory movements, such as preshaping of the hand during the reach. These considerations thus demonstrate the limitation of simply considering motor coordination as a sequencing or timing issue on the one hand or as the orchestration of concurrent actions for a single element on the other. Not coincidentally, the patterns of actions associated with speech motor functions are a prime example of coordination so defined.

THE NATURE OF SPEECH MOTOR COORDINATION

We do not view speech motor actions as strikingly different in complexity from other motor behaviors. As such, it is illuminating to explicate the nature of these particular motor phenomena as a basis for appreciating the task of the nervous system in coordination generally. Table 5.1 lists the approximately 70 muscles that appear to be involved in generating even the simplest speech motor output. Because these muscles are innervated by five cranial nerves (V, VII, IX, X, and XII), as well as nerves from the spinal anterior roots, it is not expected that local neural pattern generators can easily explain their integration, at least for speech behavior. The 70 or so muscles for speech actively control (a) respiratory movements of the diaphragm, rib cage, and abdomen; (b) the laryngeal movements of the trachea, thyroid, cricoid, and arytenoid cartilages; and (c) the upper airway actions of the pharynx, tongue, soft palate, lips, and jaw. In general, although it has been popular to separate speech movements into these respiratory, laryngeal, and "articulatory" (upper airway) components, these divisions are not justified by their neural control or biomechanics.

The function of all of these movements for speech is multiple, and even individual movements often contribute in several different ways. Figure 5.1 illustrates the different contributions of speech movements, including

> *aerodynamic functions*: classic pulmonary-like manipulations of air volumes to control air pressure;
>
> *sound generation*: creation of constrictions and transient occlusions across which air pressures equalize to generate the tonal vibrations, buzzes, and pops associated with speech; and
>
> *acoustic resonance manipulation*: time-varying patterns of changes in the shape and size of the vocal tube extending from the larynx to the lips whereby the generated sound sources are selectively filtered to yield distinctive spectral characteristics.

Although Figure 5.1 oversimplifies the processes involved in speech control, the importance of coordination among the multiple

Table 5.1.

Muscles of Speech

RESPIRATORY	VELAR	LARYNGEAL
Diaphragm	Levator palatini	Sternohyoid
Pectoralis major	Tensor palatini	Omohyoid
Pectoralis minor	Musculus uvulae	Thyrohyoid
Subclavius	Palatoglossus	Throarytenoid
Serratus anterior	Palatopharyngeus	Thyroyocalis
External intercostals		Posterior crico-aryntenoid
Levator costalis	FACIAL	
Serratus posterior superior		Lateral crico-aryntenoid
Sternocleidomastoid	Buccinator	
Scalenus anterior	Risorius	Cricothyroid
Scalenus medius	Quadratus labii superior	
Scalenus posterior	Quadratus labii inferior	LINGUAL
Latissimus dorsi	Zygomaticus	
Sacropinalis	Mentalis	Superior longi-tudinal
Triangularis sterni	Triangularis	
Subcostals	Caninus	Inferior longitudinal
Serratus posterior inferior	Incisivis labii superior	Transverse
Iliocostalis dorsi	Incisivis labii inferior	Vertical
Iliocostalis lumborum	Platysma	Genioglossus
Quadratus lumborum		Styloglossus
External oblique	PHARYNGEAL	Palatoglossus
Internal oblique		Hyoglossus
Transversus abdodminus	Inferior constrictor	
	Middle constrictor	MASTICATORY
	Superior constrictor	
	Stylopharyngeus	Digastricus
	Salpinopharyngeus	Mylohyoid
	Pharyngopalatinus	Geniohyoid
		External pterygoid
		Masseter
		Temporalis
		Internal pterygoid

muscles and movement is apparent, even superficially. First, each of the three kinds of actions must be timed properly in relation to the others. A constriction or occlusion of the oral opening (for an *s* or a *p*, respectively) would be of no value if the appropriate air pressures had not been generated so as to create a basis for air flow. This timing is more critical than simply sequencing the respiratory volume reductions to occur prior to the upper airway actions. In the

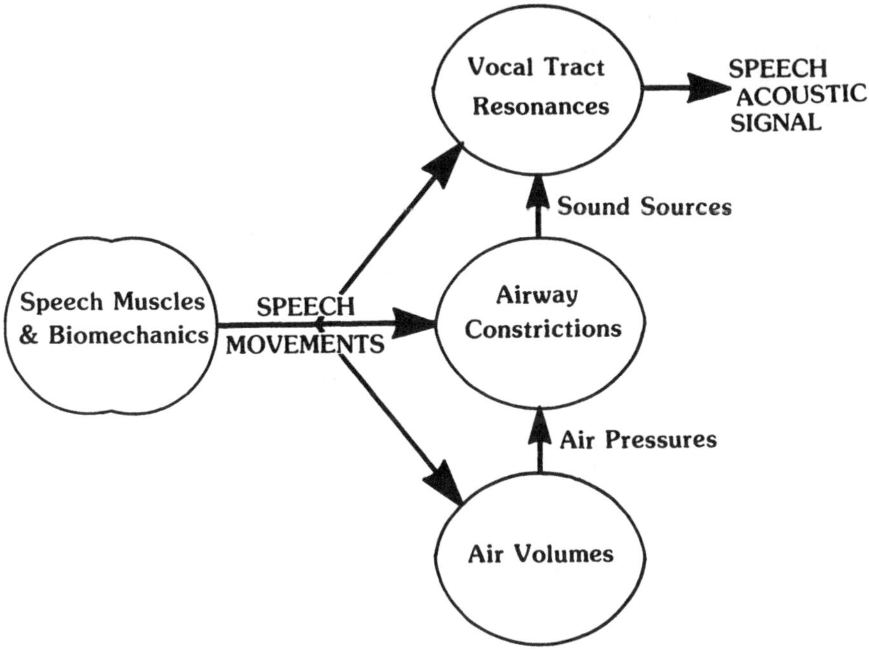

Figure 5.1. Schematic illustration of the contribution of movements to the generation of the speech acoustic signal.

generation of the consonant *p*, aside from overall provision of subglottal air pressure, the sequence of events is as follows: (a) approximately simultaneous vocal fold abduction and oral closure leading to a buildup of intraoral air pressure, (b) oral opening releases that pressure, and (c) subsequent vocal fold adduction. If the vocal folds abduct as little as 20 ms early prior to labial closure or 20 ms later following release of that closure, the actual speech sound "heard" by a listener would likely be different from the desired *p*. Specifically, early vocal fold abduction, with the oral cavity not quite occluded, yields a *f* pattern acoustically, and early vocal fold adduction after or at the time of release of the oral opening can change a *p* into a *b*. In addition to timing and sequencing, the magnitudes of the actions likewise must be coadjusted appropriately. For the *p*, the oral cavity is sealed at the lips and at the opening between the oral and nasal cavities. Electromyographic studies have shown that the lip and velopharyngeal muscle activity increases in direct proportion to the magnitudes of intraoral air pressure for the *p* (Lubker, Fritzell, & Lindqvist, 1970, Lubker & Parris, 1970). Obviously, the air pressures must be matched by the occlusal forces of the lip muscles; if the pressures are too high, the labial or oral-nasal seal will be ruptured, with detrimental consequences for the necessary acoustic pattern.

Although this example demonstrates the necessity of coordination among the classic three parts of the speech motor system, it may leave the impression that the neural control problem is confined to just a three-element integration. But for each of the actions described, a number of subactions also must be orchestrated. Although the oral occlusion at the lips is a single action in terms of its acoustic and aerodynamic consequences, it involves several potentially independent movements, minimally including the muscles of the upper lip, lower lip, and jaw (cf. Table 5.1) as well as active closure of the oral-nasal cavity passage via combined actions of the velar (soft palate) and pharyngeal muscles. The laryngeal actions include elevation of the larynx (presumably to modify the oral-pharyngeal cavity size) and coordinated contraction of the muscles that move the arytenoid cartilages. Similarly, the generation of the positive air pressure by the respiratory apparatus involves actions of the muscles that move the abdomen, rib cage, and diaphragm. These movements also must be accomplished in a complementary fashion; paradoxical movements among these components of the chest cavity, which interestingly have been observed in the respiratory patterns of patients with cerebellar disorders, would be expected to yield unacceptable fluctuations in the acoustic intensity of speech.

Although these general descriptions by themselves do not provide immediate insights into underlying mechanisms, it is glaringly obvious that focusing solely upon sequencing is an inappropriate oversimplification. Any perspective on coordination, even for serial behaviors, must deal with the orchestration of multiple concurrent or overlapping events, as well as with such sequencing.

IN SEARCH OF MECHANISMS OF SPEECH MOTOR COORDINATION

Recognizing what must be coordinated in complex motor functions such as speech (and other like sequential behaviors) sharpens the perspective of what must be addressed in regard to underlying processes. However, the key issue is the potential neurobiological mechanisms utilized by the nervous system in controlling sequencing, timing, and magnitude of activity in all of the muscles involved. Theories regarding these mechanisms have varied widely, particularly in terms of the role of central processes versus peripheral sensory mechanisms (cf. Gentner, 1987; cf. Schmidt, 1988).

Classical views of speech motor sequencing, as a subprocess of coordination, have viewed the rate of speech as too rapid (4-7 syllables/s) to permit any moment-to-moment adjustments to the timing or sequencing of the movements involved once a sequence is initiated (cf. Lashley, 1951). The sequencing and magnitude of muscle contractions was thought to involve some sort of preprogrammed central commands with an output that is produced essentially open-loop. To determine the invariant characteristics of these hypothesized central motor programs, and the "parameters" of more generalized motor programs, a number of investigators have undertaken large-scale searches for such invariances in the motor output. It has been argued that these invariances, which ideally reflect constants despite experimental manipulation (e.g., variations in the rate of movement), reflect components of the underlying central programs. An alternate view, more recently advanced, is that these coordinative processes are accomplished by a hierarchical organization whereby moment-to-moment modifications are produced by on-line sensorimotor mechanisms. According to this alternate view, sensory input is argued to be critical in achieving coordinated movements for speech and other skilled sequential actions. In the discussions that follow, these two positions are evaluated.

Relative Timing of Motor Sequences: Invariances in Limb and Speech Motor Control as Evidence for Central Motor Programs

When the overall duration of a movement sequence is varied, certain analyses have suggested that the durations of the constituent elements are changed proportionately (Armstrong, 1970; Carter & Shapiro, 1984; Glencross, 1973; Pew, 1974; Summers, 1975; Terzuolo & Viviani, 1979; Tuller & Kelso, 1984; Tuller, Kelso & Harris, 1981, 1982, 1983; Weismer & Fennell, 1985; cf. Atkeson & Hollerbach, 1985; Schmidt, 1985). Namely, some of these experiments suggested that manipulations of overall movement durations do not alter the timing relations among the submovements. It therefore has been argued that this so-called *relative timing* is indicative of a predetermined motor program for coordination of those constituent elements. For example, in one frequently cited work, individual interkeystroke intervals appeared to vary in proportion to overall duration (Terzuolo & Viviani, 1979). Similar studies of relative timing for speech have attempted to determine whether such patterns could be discerned in electromyographic, kinematic, or acoustic segment durations of speech utterances produced at different rates (Tuller et al., 1981, 1982, 1983; Weismer & Fennell, 1985). The apparent proportional timing reported was interpreted to suggest that the relative time dedicated to a given element was an invariant aspect of the motor output and that such elements are preprogrammed for the coordination of a wide range of human movements, including speech. Similarly, such relative timing (or alternatively referred to as time rescalability) became an essential component of the generalized motor program theory (Schmidt, 1975, 1985; Zelaznik, Schmidt, & Gielen, 1986). Further examination indicates that these interpretations may have been premature, and at the least far too simple to explain the complexities of sequencing timing of human movement (Barry, 1983; Gentner, 1987; Schmidt, 1985; Zelaznik et al., 1986).

Because these relative timing results have significant implications, they have been carefully scrutinized. The most recent interpretations of the earlier data are that fundamental procedural and measurement problems and/or misinterpretations occurred in most or all of the studies reporting such relative timing (also see Gentner, 1982; 1987). *Procedural* problems in these studies included (a) widespread post hoc elimination of data from certain subjects, tasks, or parts of sequences, (b) the choice of motor systems or tasks in which peripheral physical factors may have constrained timing patterns, and (c) paradigms in which relative sequence timing measures were not meaningful because absolute timing variations were not properly manipulated. In parallel, *measurement or statistical* limitations included (a) analyses confined to data averaged across subjects or experimental trials or both, (b) the mis- or overinterpretation of correlations between part/whole sequence durations, and (c) "liberal" setting of inappropriately stringent significance levels to minimize the identification of differences in relative segment durations. Inasmuch as all of the above-cited studies suffer from one or more of these problems, they fail to support claims of timing regularity. However, because these concepts are pervasive and potentially significant for the mechanisms underlying coordination of motor sequences, a more detailed critique is warranted.

One major methodological concern in studies of relative timing was selection and elimination of data. For example, in one study of relative timing for speech, acoustic intervals that were potentially damaging to the finding of relative timing were simply eliminated (Weismer & Fennell, 1985). That is, when speech utterances were segmented into seven acoustically defined intervals, Intervals 6 and 7 were excluded with the explicit justification that previous research suggested that relative timing would *not* be demonstrated for these data! In the same study, data from certain subjects likewise were discarded. If relative timing of acoustic segments is a pervasive and robust feature of motor programming for speech, such questionable data elimination procedures should be unnecessary. To this point, other investigators' data also appear to indicate that so-called relative timing operates only for "selected" elements of a speech action sequence or for selected subjects. Tuller et al. (1981, 1982) conducted a large set of correlational analyses on 9 pairs of speech segment duration sets; of these, only one segment duration pair (11% of the tests) evidenced consistently high positive correlations. Similarly, Munhall (1985) found indications of relative timing for a limited subset of data from one subject (i.e., patterns were not consistent even within that one subset), but not in data from a second individual.

Other recent work also indicates that proportional timing is not operating consistently for all phases of even a simple aiming movement (Gielen, van den Oosten, & Pull ter Gunne, 1985; Zelaznik et al., 1986). Kinematic components occurring early in the movement appear to be fixed, whereas those occurring later are more affected by changes in movement time and appear to scale proportionately (Zelaznik et al., 1986). It thus appears that proportional durations, or relative timing, may be present for some movement intervals, in the data of some subjects, but are certainly not universal invariants in human movement. Obviously, results such as these cast doubt upon the concept of generalized motor programs, especially if so-called time rescalability (proportional duration) is alleged to be a major component of the program (Zelaznik et al., 1986).

In addition to procedural problems, in previous studies of relative timing there were, as noted, several questionable statistical procedures. The first of these is the practice of averaging data across trials, conditions, and subjects (cf. comments by Gentner, 1987). Most studies have used data averaged across trials within different rate conditions (Carter & Shapiro, 1984; Summers, 1975; Tuller et al., 1981, 1982; Weismer & Fennell, 1985). Although statistically convenient, this practice violates a necessary premise underlying the operation of a stereotyped motor program; for invariance to be indicative of a "regular" and unchanging motor program, it must be assumed that it will be manifest in each movement sequence produced by each individual subject (Gentner, 1987). Obviously, when data are averaged, variability, which could contradict the model, is obscured. If proportional durations are demonstrated for individual trials, within all subjects, then these durations will also be reflected in mean data; however, the reverse situation may not be true (Gentner, 1987). To analyze individual events and sequences, Gentner developed the "constant proportion test," whereby if there is no change in the interval's proportion as a function of movement speed (or total duration), a derived regression is close to zero. However, in data with large amounts of variability, Gentner (1987) suggests an ANOVA among intervals and rate conditions; significant interactions demonstrate that intervals are not proportional across rate conditions. When data from some of the studies claiming to demonstrate relative timing (e.g., Carter & Shapiro, 1984; Munhall, 1985; Summers, 1975; Terzuolo & Viviani, 1979) were reanalyzed using these methods, individual movement intervals did not, as claimed, maintain relative durations for the majority of the data (Gentner, 1987).

With regard to the part-whole correlation artifact present in some of the studies purporting to find evidence for relative timing, the reader is referred to Gentner's (1987) critique of Armstrong (1970) and Tuller et al. (1981). Briefly, the argument is that two intervals with a shared component (i.e, overall duration and the duration of an element) will be positively correlated, even if the two intervals are completely independent (see also Barry, 1983; Munhall, 1985). In the series of studies by Tuller et al. (1981, 1982), electromyographic patterns for tongue, lip, and jaw movements were examined in utterances produced at different rates and with different stress patterns. In the sequence of vowel-consonant-vowel, the vowel time was found to be highly correlated with the entire sequence. This finding, thought to support the proportional duration model, thus was apparently artifactual. Interestingly, Gentner (1987), using the constant proportion test, reanalyzed these data and determined that the ratio of vowel time to consonant time did not change significantly in 10/12 data sets, thus supporting the proportional duration model for this select data set. Recall, however, that the data reported by Tuller et al. (1981, 1982) were selected from a larger data set in which the significant positive correlations were manifest only in 11% of the total; that is, 89% of the results were contrary to relative timing! A further caveat might be that only 4 data points were present in each of Tuller et al.'s (1981, 1982) data sets, and statistical power in rejecting the proportional duration model was therefore very limited (Gentner, 1987).

In an apparent attempt to overcome the part-whole correlation artifact, a more recent study of relative timing in speech (Weismer & Fennell, 1985) used a difference-testing statistical technique. Unfortunately, there were serious statistical flaws in that study as well. As noted by Corcos, Agarwall, and Gottlieb (1985), in studies such as these, setting significance levels in hypothesis testing is complicated by the fact that the goal is to accept a hypothesis of no difference rather than to reject it. Weismer and Fennell's (1985) work provides an example of such misuse of hypothesis testing. Analyses consisted of 170 *t*-tests with significance levels ranging from 0.0014 to 0.00555. These small significance levels were apparently set to reduce the likelihood of producing Type I errors with multiple *t*-tests. Under most circumstances this is, of course, a careful statistical control. However, when the hypothesized model is supported by acceptance rather than rejection of the null hypothesis of no difference, to *support* the hypothesized model, setting a small significance level biases the experiment in that desired direction. In other words, this practice ensures support for the experimental hypothesis that differences would *not* be found in the mean ratios tested; finding significant differences among ratios from different speech rates was highly unlikely.

A second and related problem in this particular speech study was that Weismer and Fennel (1985) failed to demonstrate that each of the absolute sequence durations varied to a *statistically significant degree*. Without such a demonstration, the observation of no difference in "relative" timing may simply have reflected an absence of variations in absolute timing. Indeed, in attempting to deal with this issue, these authors noted that the total utterance may be subject to less compressibility than individual segments because the total utterance contained a high content of consonant sounds, which were alleged to be inherently less compressible than vowels. Interestingly, this line of reasoning violates a basic premise of relative timing as a major factor in speech motor programming; if an entire class of elements generated during speech production (e.g., consonants) must be excluded from the model, then the model clearly cannot account for the complex coordinative behaviors observed during speech.

A final problem in several of the studies under discussion relates to the measures on which the conclusions about relative timing and hence "motor programs" were based. Although muscle activity or movements were in some cases examined directly, certain investigators used measures of motor output that were only *indirect* reflections of the actual movement patterns. For example, Terzuolo and Viviani's (1979; 1980) often cited work examined the timing of movement end points for typing movements (e.g., interkeystroke intervals) and found these end points to be highly regular in relative time. In typewriting, it appears that the use of such remote motor output measures yields inappropriate interpretations of underlying motor regularity. That is, examination of timing for *initiation and termination* of typing movements reveals that the overall movements were regular only in their termination times (Gentner, Grudin, & Conway, 1980). As shown in Figure 5.2, although keys were always pressed in the correct order (end point timing), the initiation of movement was variable (i.e., the order of initiation for the word *epic* was *i, e, p, c* in one example). As such, because regularity of finger movement termination was not paralleled by regularities in movement initiation, measures of the former were misleading as to the invariance of an underlying motor control process. These data thus indicate that relative timing was far from invariant when the most direct measures were made. If the goal of such investigations is to determine the operation of a "motor execution program," the final arbitration of the relative timing issue depends upon the patterns of the multiple muscle actions and movements involved. In this same vein, the durational measures of the speech acoustic signal used by some are only very indirect measures of the underlying movement and muscle activity. For example, like Terzuolo and Viviani (1979) before them, Weismer

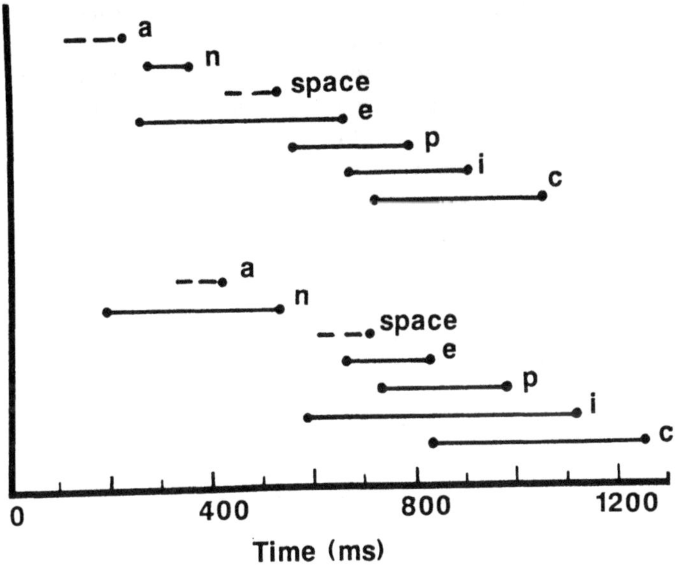

Figure 5.2. Illustration of the initiation and termination of finger movements for two typing trials of "an epic." Although the timing of the right-most point on each line (the key press and movement termination) is similar across the repetitions, the initiation of those movements (the left-most point on each line) shows no such regularity. *Note:* From "Finger Movements in Transcription Typing" (Tech. Rep. No. 8001) by D. R. Gentner, J. Grudin, and E. Conway, 1980, San Diego: University of California, Center for Human Information Processing. Adapted by Permission.

and Fennell (1985) appear to make the assumption that "relative timing" of the segments of the speech acoustic signal directly represents the underlying output of speech motor execution. This assumption, of course, does not stand up to empirical observation.

In summary, analyses of the relative timing of motor output have not provided support for any particular mechanisms underlying motor coordination. Ironically, most of the experimental results cited in this discussion actually cast doubt upon the viability of any predetermined motor program as a major factor in motor coordination. This is not to say, of course, that movement timing is not organized in some manner. However, in future experiments on this issue, it would be useful to have some a priori hypotheses based upon the underlying neurobiology as well as more stringent procedures for data measurement and analysis. Another issue raised by these observations is whether unfocused searches for invariance as a means to discern underlying control processes are fruitful at all. Without more specific hypotheses regarding underlying neurobiological processes and/or more thoughtful approaches to these issues, one must question the significance of such eagerly sought regularities; some aspects of a motor output may be "roughly" the same across a variety of conditions because they lack functional significance! As one of our colleagues recently argued, "Because the body temperature of these subjects presumably was unchanged during variations in speech rate, does this imply that this variable is a preprogrammed aspect of speech motor output?" (B. B. Edin, personal communication, September, 1988).

Sensorimotor Mechanisms and Coordination

An alternate view of the means by which the multiple concurrent and sequential actions for speech are coordinated is via on-line influence on motor output by sensory processes. The evidence for sensory-based coordination among multiple speech movements and muscles comes largely from a series of recent studies in which unanticipated perturbations were introduced into one of several constituent elements of complex motor gestures. Similar results have been obtained with speech and hand movements, although the results of particular interest here are those in speech.

The basic set of studies addressing afferent contributions to multiple movement coordination involved upper lip, lower lip, and jaw actions for oral closure during speech (cf. Abbs et al., 1984; Abbs & Gracco, 1984; Gracco & Abbs, 1985a; 1986; 1988). Unanticipated load perturbations were delivered approximately 15% of the time on either the lower lip (Abbs & Gracco, 1984; Gracco & Abbs, 1985a), the jaw (Folkins & Abbs, 1975, 1976), or the upper lip (Abbs, Shaiman, Gracco, & Cole, 1985) during these various speech tasks. In these studies, the perturbed movement was a component of a larger speech gesture; for example, the closing movements of the lower lip were perturbed prior to and during oral closure for a *p*, involving upper lip, lower lip, and jaw movements. As illustrated in Figure 5.3, striking spatial reorganization was observed in both the perturbed and nonperturbed muscle groups; this pattern was seen the very first time a load was introduced in naive subjects.

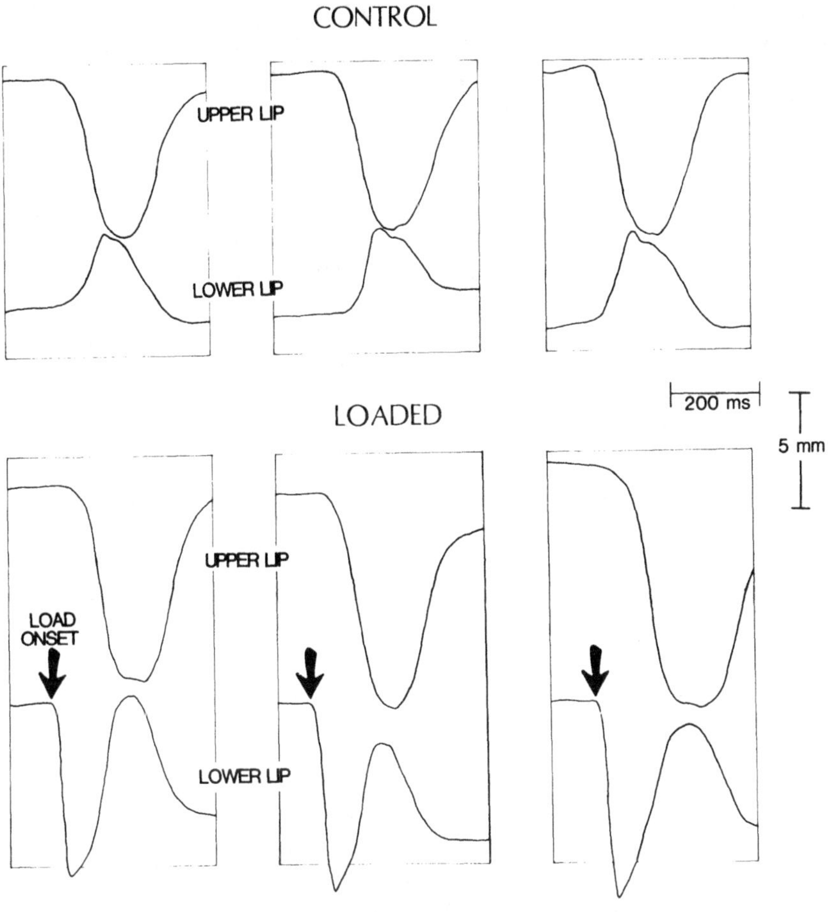

Figure 5.3. Illustration of movement compensations (upper and lower lip) that occur in response to unanticipated perturbations of the lower lip. These responses, described in detail in Abbs and Gracco (1984), were consistently accompanied by corresponding adjustments in muscle activity. *Note*: From "Control of Complex Motor Gestures: Orofacial Muscle Responses to Load Perturbations of Lip during Speech" by J. H. Abbs and V. L. Gracco, 1984, *Journal of Neurophysiology, 51*, p. 710. Copyright 1984 by The American Physiological Society. Adapted by permission.

Unexpected perturbations caused motor compensations in the disturbed movement and in other movements cocontributing to the total action, even those physically remote from the site of perturbation (Shaiman, Abbs, & Gracco, 1985). Systematic variations in movement amplitude, movement duration, and muscle activity were observed as part of the compensatory responses to unanticipated loads (Abbs & Gracco, 1984; Gracco & Abbs, 1985a). These durational variations, in and of themselves, argue against predetermined patterns of timing, as proposed by the so-called relative timing model. Additionally, these load-related alterations in magnitude and timing were seen even when loads were introduced 30 ms after muscle activity onset in the perturbed muscle groups. Interestingly, the compensations for these unanticipated perturbations were sufficiently effective that despite the significant kinematic and muscle activity changes, listeners were not able to discern differences in speech patterns for load versus control trials.

Results from these perturbation studies thus provide direct support for sensorimotor contributions to coordination among these movements. Obviously, strict models of central motor programming would not be compatible with these findings. Instead, these models would predict alterations (i.e., adjustments in magnitude or timing) in only the perturbed muscle group, and other constituent actions would unfold according to the so-called central plan. In an additional experiment, the issue of sequence control was addressed directly by introducing unanticipated perturbations for the initial portion of a speech motor sequence and then in some conditions eliminating that load (cf. Gracco & Abbs, 1985b). Subjects generated two successive lip and jaw closing movements associated with the first and second *p*'s in *sapapple*. By selectively perturbing the lower lip during the first or both of these *p* elements in this sequence, it was possible to discern characteristics of the presumed sequential movement programming. In the Load (LN) condition, the perturbation remained on for both sequential movements. In the Load On/Off (LNF) condition, the perturbation was removed at variable times prior to the second closing movement.

As illustrated in Figure 5.4, analyses indicated that the muscle activity and resulting kinematics for the second *p* closure were differentially affected by the load conditions. In the LN condition, substantial increases were observed in lower lip muscle activity and

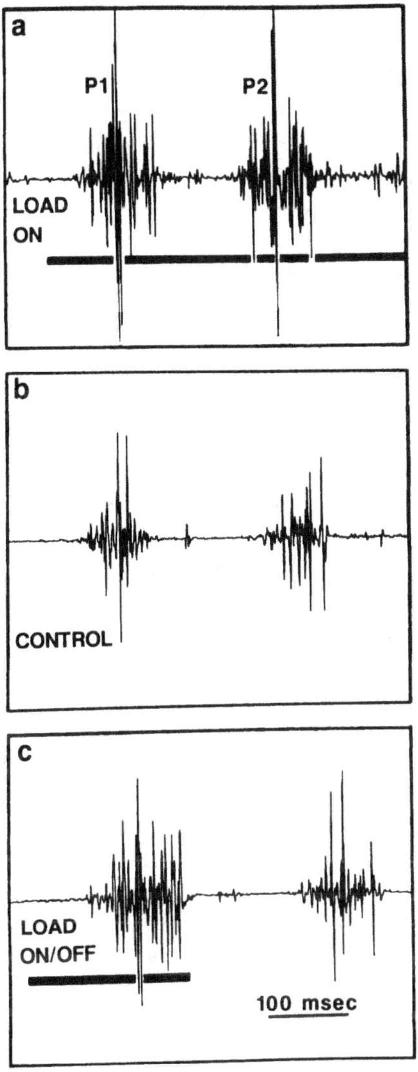

Figure 5.4. Pattern of Orbicularis Oris muscle activity for the two *p* oral closures in *sapapple*, for two experimental load conditions and for the control condition. (Adapted from Gracco & Abbs, 1985b).

movement associated with the second *p* closure. However, in the LNF condition, the lower lip muscle activity and movement were significantly reduced compared to the LN condition, and slightly increased from the control condition. This latter result indicates that when the load is removed, the system readjusts. As such, each of these movements is executed separately and under the influence of sensory information even well after a sequence is initiated.

Another finding of particular interest with regard to control of inter-element timing was also seen in this study. The interval between the muscle activity and movement onsets for the first and second *p* movements also was found to be differentially influenced by the two load conditions. For the LN condition, the interburst interval was reduced (re: no load), whereas for the LNF condition, the interburst interval was increased (re: no load). These load-dependent variations in timing between the two sequential movements suggest that the timing of sequential elements also is modifiable with variations in sensory input.

The latter results thus indicate the use of sensorimotor mechanisms to update and adjust *individual serial actions on a movement-to-movement basis*. The consequence of this neuromotor scheme is that specification of movement parameters for sequential motor acts is a dynamic process under continual modification via central and peripheral inputs. As suggested by Abbs et al. (1984), motor programs for the sequences of speech may be considered to reflect certain generalized movement actions (e.g., oral opening, oral closing) with adjustments of timing and magnitude and details of motor execution being determined on-line by sensory processes. In this context, a motor program is obviously not simply a set of efferent commands but a dynamic representation of a movement goal, the multiple constituents which contribute to that goal, and sensorimotor processes to guide the detailed composition of those contributions.

REFERENCES

Abbs, J. H., & Gracco, V. L. (1984). Control of complex motor gestures: Orofacial muscle responses to load perturbations of the lip during speech. *Journal of Neurophysiology*, *51*, 705-723.

Abbs, J. H., Gracco, V. L., & Cole, K. J. (1984). Control of multimovement coordination: Sensorimotor mechanisms in speech motor programming. *Journal of Motor Behavior, 16,* 195-231.

Abbs, J. H., Shaiman, S., Gracco, V. L., & Cole, K. J. (1985, October). *Task-dependent sensorimotor actions are inherent in speech motor programs.* Paper presented at the meeting of the Society for Neuroscience, Dallas.

Armstrong, T. R. (1970). *Training for the production of memorized motor patterns* (Tech. Rep. No. 26). Ann Arbor: University of Michigan, Human Performance Center.

Atkeson, C. G., & Hollerbach, J. M. (1985). Kinematic features of unrestrained vertical arm movements. *Journal of Neuroscience, 5,* 2318-2330.

Barry, W. J. (1983). Some problems of interarticulator phasing as an index of temporal regularity in speech. *Journal of Experimental Psychology: Human Perception and Performance, 9,* 826-828.

Bernstein, N. A. (1967). *The co-ordination and regulation of movements.* New York: Pergamon.

Brooks, V. B., & Thach, W. T. (1981). Cerebellar control of posture and movement. In V. B. Brooks (Ed.), *Handbook of Physiology: Sec. 1. The Nervous system: Vol. II. Motor control: Part 2* (pp. 877-946). Bethesda, MD: American Physiological Society.

Carter, M. C., & Shapiro, D. C. (1984). Control of sequential movements: Evidence for generalized motor programs. *Journal of Neurophysiology, 52,* 787-796.

Corcos, D. M., Agarwall, G. C., & Gottlieb, G. L. (1985). A note on accepting the null hypothesis: Problems with respect to the mass-spring and pulse-step models of movement control. *Journal of Motor Behavior, 17,* 481-487.

Fentress, J. C. (1984). The development of coordination. *Journal of Motor Behavior, 16,* 99-134.

Folkins, J. W., & Abbs, J. H. (1975). Lip and jaw motor control during speech: Responses to resistive loading of the jaw. *Journal of Speech & Hearing Research, 18*, 207-220.

Folkins, J. W., & Abbs, J. H. (1976). Additional observations on responses to resistive loading of the jaw. *Journal of Speech & Hearing Research, 19*, 820-821.

Gentner, D. R. (1982). Evidence against a central control model of timing in typing. *Journal of Experimental Psychology: Human Perception and Performance, 8*, 793-810.

Gentner, D. R. (1987). Timing of skilled motor performance: Tests of the proportional duration model. *Psychological Review, 94*, 255-276.

Gentner, D. R., Grudin, J., & Conway, E. (1980). Finger movements in transcription typing (Tech. Rep. No. 8001). San Diego: University of California, Center for Human Information Processing.

Gielen, S. C. A. M., van den Oosten, K., & Pull ter Gunne, F. (1985). Relation between EMG activation patterns and kinematic properties of aimed arm movements. *Journal of Motor Behavior, 17*, 421-442.

Glencross, D. J. (1973). Temporal organization in a repetitive speed skill. *Ergonomics, 16*, 765-776.

Gracco, V. L., & Abbs, J. H. (1985a). Dynamic control of the perioral system during speech: Kinematic analyses of autogenic and nonautogenic sensorimotor processes. *Journal of Neurophysiology, 54*, 418-432.

Gracco, V. L., & Abbs, J. H. (1985b, October). *Programming of serial multiarticulate movements: Data from speech movement sequences.* Paper presented at the meeting of the Society for Neuroscience, Dallas.

Gracco, V. L., & Abbs, J. H. (1986). Variant and invariant characteristics of speech movements. *Experimental Brain Research, 65*, 156-166.

Gracco, V. L., & Abbs, J. H. (1988). Central patterning of speech movements. *Experimental Brain Research, 71*, 515-526.

Hogan, N. (1985). The mechanics of multi-joint posture and movement control. *Biological Cybernetics, 52*, 315-331.

Holmes, G. (1939). The cerebellum of man. *Brain, 62*, 1-30.

Hollerbach, J. M. (1981). An oscillation theory of handwriting. *Biological Cybernetics, 39*, 139-156.

Kent, R. D. (1976). Models of speech production. In N. J. Lass (Ed.), *Contemporary issues in experimental phonetics* (pp. 79-104). New York: Academic Press.

Kent, R. D., & Moll, K. L. (1972). Tongue body articulation during vowel and diphthong gestures. *Folia Phoniatrica, 24*, 286-300.

Lashley, K. S. (1951). The problem of serial order in behavior. In L. A. Jeffress (Ed.), *Cerebral mechanisms in behavior* (pp. 112-136). New York: Wiley.

Lubker, J. F., Fritzell, B., & Lindqvist, J. (1970). Velopharyngeal function: An EMG study. *Speech Transmission Laboratory Quarterly Progress & Status Report, 4* , 9-20.

Lubker, J. F., & Parris, P. J. (1970). Simultaneous measurement of intraoral pressure, force of labial contact, and labial electromyographic activity during production of the stop consonant cognates /p/ and /b/. *Journal of the Acoustical Society of America, 47*, 625-633.

Munhall, K. G. (1985). An examination of intro-articulator relative timing. *Journal of the Acoustical Society of America, 78*, 1548-1553.

Pew, R. W. (1974). Human perceptual-motor performance. In B. H. Kantowitz (Ed.), *Human information processing: Tutorials in performance and cognition* (pp. 1-39). Hillsdale, NJ: Erlbaum.

Schmidt, R. A. (1975). A schema theory of discrete motor skill learning. *Psychological Review, 82*, 225-260.

Schmidt, R. A. (1985). The 1984 C. H. McCloy Research Lecture: The search for invariance in skilled motor behavior. *Research Quarterly for Exercise and Sport, 56,* 188-200.

Schmidt, R. A. (1988). *Motor behavior: Programming, control and acquisition* (2nd ed.). Berlin: Springer.

Shaiman, S., Abbs, J. H., & Gracco, V. L. (1985). Sensorimotor contributions to oral-laryngeal coordination for speech. *Society for Neuroscience Abstracts,* Vol. I, Part 1, p. 76.

Sternberg, S., Monsell, S., Knoll, R. L., & Wright, C. E. (1979). The latency and duration of rapid movement sequences: Comparison of speech and typewriting. In G. E. Stelmach (Ed.), *Information processing in motor control and learning* (pp. 118-150). New York: Academic Press.

Summers, J. J. (1975). The role of timing in motor program representation. *Journal of Motor Behavior, 7,* 229-241.

Terzuolo, C. A., & Viviani, P. (1979). The central representation of learned motor patterns. In R. E. Talbott & D. R. Humphrey (Eds.), *Posture and movement* (pp. 113-121). New York: Raven Press.

Terzuolo, C. A., & Viviani, P. (1980). Determinants and characteristics of motor patterns used for typing. *Neuroscience, 5,* 1085-1103.

Tuller, B., & Kelso, J. A. S. (1984). The timing of articulatory gestures: Evidence for relational invariants. *Journal of the Acoustical Society of America, 76,* 1030-1036.

Tuller, B., Kelso, J. A. S., & Harris, K. S. (1981). Phase relationships among articulator muscles as a function of speaking rate and stress. *Status Report on Speech Research SR-65* (pp. 63-90). New Haven, CT: Haskins Laboratories.

Tuller, B., Kelso, J. A. S., & Harris, K. S. (1982). Interarticulator phasing as an index of temporal regularity in speech. *Journal of Experimental Psychology: Human Perception and Performance, 8,* 460-472.

Tuller, B., Kelso, J. A. S., & Harris, K. S. (1983). Converging evidence for the role of relative timing in speech. *Journal of Experimental Psychology: Human Perception and Performance, 9*, 829-833.

Weismer, G., & Fennel, A. M. (1985). Constancy of (acoustic) relative timing measures in phrase-level utterances. *Journal of the Acoustical Society of America, 78*, 49-57.

Zelaznik, H. N., Schmidt, R. A., & Gielen, S. C. A. M. (1986). Kinematic properties of rapid aimed hand movements. *Journal of Motor Behavior, 18*, 353-372.

Perspectives on the Coordination of Movement
S.A. Wallace (Editor)

COMPARATIVE COORDINATION (A STORY OF THREE LITTLE P'S IN BEHAVIOR)§

John C. FENTRESS*

Departments of Psychology and Biology
Dalhousie University

ABSTRACT

Studies of comparative coordination rest upon a proper understanding of what I here call the "three little p's" of behavior: *process*, *pattern*, and *phenotype*. How do we understand the dynamic flowing qualities of behavior, the rules of separation and combination of abstracted behavioral events, and the deeper structures from which they are derived? Here I focus upon recent research on the organization of rodent grooming sequences and, to a lesser extent, on the social patterns of behavior in wolves.

The problems raised by these multilevel investigations offer many challenges not only to future comparative research but also to the establishment of links between comparative research and human performance plus neuroscience. In spite of species diversity, as well as the diversity that necessarily occurs when one crosses levels of analysis or abstracted behavioral events, there may be deeper commonalities among a

---

*Address correspondence to: John C. Fentress, Department of Psychology, Dalhousie University, Halifax, Nova Scotia B3H 4J1, Canada.

§Preparation of this chapter was supported in part by research grants from the Medical Research Council, the Natural Sciences and Engineering Research Council, and the Faculty of Graduate Studies at Dalhousie University.

wide range of movement phenomena. For such commonalities to be discerned, however, we need better tools for conceptualizing and analyzing dynamic balances among the interactive and self-organizational tendencies that are observed in coordinated action throughout the animal kingdom. Comparative analyses provide essential tools for the achievement of these goals.

> Ethologists characteristically study natural as opposed to contrived behavior and try also to discern the *structure* of natural behavior, that is to say, to discern a functionally coherent, quasi-purposive performance in what to an inexperienced observer would present itself as a sequence of isolated and teleologically unconnected performances.
>
> (Medawar & Medawar[1])

The study of coordinated action patterns has long been a basic foundation of ethological research (Lorenz, 1950; Tinbergen, 1963). These studies often begin with observations of unrestricted movements in diverse species. A major challenge in recent years has been to refine observations of naturally occurring movement sequences so that underlying processes may be more clearly understood.

In this chapter I borrow from the pre-ethological folklore of the bad wolf and the three pigs and offer the tripartite perspective of coordinated movement as resting upon "three little p's": *process, pattern,* and *phenotype.*

Process (the straw house) refers to the essentially dynamic nature of behavioral expression. Indeed, movement is change in the relations of body parts in reference to one another or the external world, or both. I argue that underlying processes must also be pursued from an explicitly dynamic perspective. It is often much easier to construct static representations of process, but these representations can obscure the fundamental nature of process itself.

Pattern (the stick house) emphasizes that the processes in behavior cannot be entirely seamless, meaning that they have a division. A degree of modularity is essential for organization of behavior to become established and maintained. Animal actions can often be classified into more or less discrete events that are then ordered sequentially and hierarchically. Analyses of patterns involve the fundamentally dual task of finding both rules of separation and interconnection. Underlying systems must represent a dynamic balance between tendencies towards self-order and interaction (Fentress, 1976), and careful analyses of coordinated patterns can help clarify this essential duality.

Phenotype (the brick house) reflects the deeper stabilizing features in behavior that result from a still poorly understood interplay between genetic and experiential factors. The notion of "species-characteristic" behavior, the hallmark of ethological research, reflects phenotypic stability in behavior. Phenotypic stability is, of course, a relative concept, and ethologists have also devoted much attention to understanding the processes of evolution and development in behavior, as well as the functional consequences of any given performance (Tinbergen, 1963).

The division of coordinated movement into the three little p's is obviously something of a conceptual fairy tale. It does, however, highlight two deeper polarities in the ways investigators measure and explain behavior (Fentress, 1984; Fentress & McLeod, 1986). The first polarity is that of continuity-discontinuity: the degree to which nature appears to be seamless or to occur in jumps. The second polarity is that of change-stability: the degree to which dynamic and more static properties of nature are emphasized. In the study of coordinated movement, each of these polarities can be represented over a number of time frames and levels of organization. A major concern is how our perceptions of these polarities change as we cross time frames and levels.

The notion of co-ordering can imply that otherwise distinguishable rules of order are not fully separable. Perhaps the most difficult task of all is to conceptualize systems that are necessarily both distinct and interconnected (Fentress, 1989; in press). My basic argument is that we must focus properly upon *dynamic balances* among the interactive and self-organizing systems in process, pattern, and phenotype if we are to achieve coherent models of comparative coordination. I suspect that more powerful links between animal and human literature may occur if workers in each area design their efforts within this framework.

TAXONOMIES OF MOVEMENT PROCESS

The initial and difficult task in the study of comparative coordination is the translation of observation into movement taxonomy (cf. Golani, 1976; Fentress, 1984; Jacobs et al., 1988). Properties of action must necessarily be abstracted for subsequent analysis. Decisions made at this point can to a large extent dictate the data collected and the conclusions drawn from them. A fundamental distinction is that between descriptions of movement form and movement function (e.g., Hinde, 1982). Movement form refers to changes in the relations among body parts and between them and the environment. Such descriptions provide the investigator with the morphology of movement without in themselves conveying the purpose to which the animal applies the movement. Descriptions of movement function indicate what the consequences of movement are without necessarily implying any particular properties of form.

The complementary nature of these two perspectives in description can be appreciated as soon as one recognizes that movement patterns differing greatly in their form may serve a common function, just as movement patterns that are similar in form may serve quite different functions. Thus, animals might fight their social partners through a variety of movement forms that range from biting to kicking. Conversely, animals might employ certain biting movements in attacking a conspecific or eating food. Because of the hierarchical nature of movement organization, this distinction between form and function can at times blur. Thus the term *bite* at a more refined level is an abstraction about function and does not in itself say anything specific about the form of movement used (cf. the more formal distinction between movement types and tokens suggested by Jacobs et al., 1988).

By focusing early upon relatively simple properties of coordinated movement that are species-characteristic, ethologists often are able to simplify their initial descriptive task. Thus the classical notion of a fixed action pattern can be used to summarize both the form that a given movement takes and the target towards which it is directed. Obviously there is linkage between form and function. However, in more detailed analyses it is often useful to separate these movement properties.

Another fundamental perspective emphasized by early ethologists is that rules of linkage among abstracted movement properties are as important as individual movement properties. This emphasis allows comparative investigators to examine the organization of movement at a number of complementary levels. To use a simple musical metaphor, both the note structure and the melody line of animal movements can in this way be evaluated. Such multilevel perspectives also allow the investigator to examine relative flexibilities and consistencies in movement across different frames of organization. One might thus in this way evaluate the range of movement properties animals apply toward a given goal (cf. Bernstein, 1967; Lashley, 1951).

Even at the descriptive level, coordinated movement patterns can often be understood most fully when appropriate attention is paid to the broader expressive contexts within which the movement patterns occur. This is true both for (a) rules of where or when particular action sequences occur and (b) possibly changing rules of action form or function in different contexts. These considerations in the description of basic movement can have important ramifications for subsequent questions of movement control.

Ethological analyses of communicative courtship displays in birds are illustrative (e.g. Hinde, 1982). Many courtship displays in birds contain motor elements that are also seen in attack, fleeing, and copulatory encounters. Behavioral evidence suggests that such displays may be the result of conflicts among more global ("motivational") action tendencies. Such conflicts are often observed when animals are undergoing broad changes in their behavioral profiles, such as during early phases of a reproductive season (Figure 6.1). Once the early ambivalences in behavior are replaced by stronger tendencies to behave in one way or another, the displays of early courtship disappear. Simultaneous expression of both cooperative and conflicting action tendencies is a common observation in ethological research. This finding suggests that operational systems in comparative movement may overlap to varying degrees as an overall function of system dynamics (Fentress, 1976, 1984, 1989, in press).

Rodent grooming patterns are frequently elicited when the animals give evidence of being in conflict or uncertainty over different possible courses of action (Fentress, 1968-a, 1968-b). One can thus often account for the timing and form of rodent grooming most satisfactorily by considering transitions that occur between

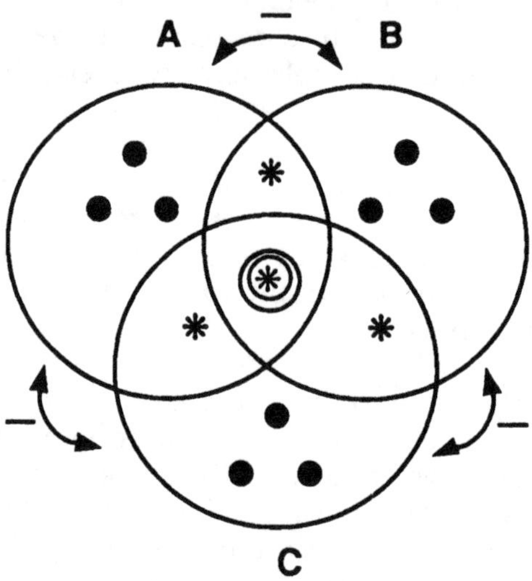

Figure 6.1. Schematic representation of three higher order behavioral systems (A,B,C), each of which contains a number of expressive elements (dots). The higher order (e.g., motivational) systems are normally mutually antagonistic, but they may also share elements (stars). Many ethological data suggest that animal displays often occur during motivational conflict. The animals may then generate motor actions from several systems in rapid sequence and also generate novel movements (represented by star within the double circle at center of figure).

exploration and rest, or avoidance responses and subsequent exploration. An important implication of this brief reference to bird courtship and rodent grooming is that individually defined motor sequences are embedded among relations that exist among different higher-order states. Further, these higher-order states may both compete and share with one another even though they support the same particular motor actions.

Rodent Grooming Sequences: An Example of Individually Coordinated Action

Rodent grooming sequences provide a convenient illustration of the complementary frameworks from which comparative actions can be described. The potential importance of contextual associations has already been noted. The term *grooming* illustrates the convenience of short-hand functional labels. Grooming movements are interpreted as attention to various parts of the body surface. This definition, however, does not in itself clarify the component action or the underlying dimensions of action that may be involved in grooming (cf. Fentress, 1984; Golani, 1976; Jacobs et al., 1988).

By a combination of several criteria, rodent grooming sequences can be divided into various types of strokes (Fentress & Stilwell, 1973). Once these types are discerned, their rules of sequential ordering can also be determined. These rules are hierarchically defined in the interesting descriptive sense that the same abstracted action components can be part of several more broadly blocked sequences—rather like the individual notes that contribute to several melodies in music; (Fentress, 1978; cf. Hoyle, 1985). More subtly, the individual action components may change in some of their descriptive characteristics (amplitude, duration, etc.) as a function of the particular sequential relations they have with other abstracted action components. This interdependency implies a certain lack of separation among these abstracted action components.

Detailed descriptions of grooming action (Golani & Fentress, 1985) indicate that some of these more perceptually hidden relations can be clarified by separation of the dimensions of grooming into kinematic properties of individual limb segments, limb trajectories in space, and contact pathways between limb segments (e.g., forepaws and face). Once pursued to this level of analysis, two fundamental points become clear. First, the unitary separation among movement "acts" can become blurred. Second, there are several simultaneously occurring dimensions of movement that have their own "chordal" structure.

Golani and Fentress's (1985) descriptions of rodent grooming indicate frequent constancies or invariances among otherwise separate action components or dimensions. These movement properties could not be anticipated from measures of variation for the isolated action properties. This principle of "relational invariance" has many parallels in the movement literature (cf. Bernstein, 1967). To return to the music analogy, the details of individual notes in a melody are not so important as long as the notes maintain the appropriate relations to one another in time (e.g., as seen in cases of musical transposition). Analogies such as this one have force only when they take both temporally discrete and overlapping events into account (cf. Eshkol & Wachmann, 1958). In summary, the picture of grooming that emerges is one of sequentially and hierarchically ordered movement events; and the sequences also contain partially overlapping properties (see Fentress, 1984; Fogel & Thelen, 1987, for related comparative discussions of movement dynamics).

Berridge, Fentress, and Parr (1987) quantified sequences of abstracted grooming acts in rats by examining both overall sequential order (through the use of information statistics) and particular sequencing rules. They found that grooming sequences contain bouts of action perseveration and action alternation. Perseveration here refers to repetitions of a defined action class. Alternation refers to reciprocal, numerical transitions between actions that may have quite different individual probabilities (Figure 6.2).

Certain phases of grooming sequences may be relatively variable; others are much more stereotyped (Berridge & Fentress, 1986; Berridge et al., 1987; cf. Fentress, 1972; see Figure 6.3). Introduction to these more invariant sequence phases of grooming are signaled by highly rhythmic licking movements whose temporal properties are then extended to subsequent sequences of facial grooming strokes. Clearly, movement properties may be shared by a number of otherwise separately defined action components (for elaboration, see Fentress, 1986, 1989, in press). The point is especially important if these action components are otherwise viewed as totally isolated events. Movement boundaries are often more subtle than this. Further, different classificatory criteria may highlight different

Figure 6.2. Sample three-way transition matrix for spontaneous postprandial grooming. Preceding pairs of actions are shown in left column with third action across the top row. Boxes indicate simple perseveration, hexagons indicate simple alternations (ABA), and circles indicate reciprocal transitions of an "alternating perseveration" form (AAB, BAA). Note transitional symmetries that occur in spite of different individual probabilities. (Action code: TP = tongue protrusion, LTP = lateral tongue protrusion, FW = face wash stroke, FF = forelimb flail, HS = head shake, MM = mouth movements, PL = paw or body licking.) *Note.* From "Natural Syntax Rules Control Action Sequence of Rats" by K. C. Berridge, J. C. Fentress, and H. Parr, 1987, *Behavioural Brain Research, 23*, p. 65. Copyright 1987 by Elsevier Science Publishers. Reprinted by permission.

3-way transition matrix; postprandial; intact

	Subsequent												
	TP	LTP	G	CR	FW	FF	HS	PT	PL	MM	PD	LO	Total
TP/ TP	122	2	-	-	1	-	-	-	-	21	-	-	146
TP/LTP	-	1	-	-	-	-	-	-	-	3	-	-	4
TP/ FW	-	-	-	-	2	-	-	-	-	-	-	-	2
TP/ MM	14	4	-	-	-	-	-	-	1	19	-	-	38
LTP/ TP	1	1	-	-	-	-	-	-	-	-	-	-	2
LTP/LTP	21	3	-	-	-	-	-	-	1	18	-	-	23
LTP/ FW	-	-	-	-	-	-	-	-	-	2	-	-	2
LTP/ HS	-	1	-	-	-	-	-	-	-	-	-	-	1
LTP/ PL	-	2	-	-	-	-	-	-	-	-	-	-	2
LTP/ MM	2	43	-	-	-	-	-	-	1	49	-	-	95
FW/ TP	-	-	-	-	-	-	-	-	-	1	-	-	1
FW/LTP	-	1	-	-	-	-	-	-	-	-	-	-	1
FW/ FW	1	-	-	-	58	3	-	-	20	2	-	-	84
FW/ FF	-	-	-	-	2	-	-	-	1	-	-	-	3
FW/ PL	-	-	-	-	24	-	-	-	14	-	-	-	38
FW/ MM	-	1	-	-	-	-	-	-	-	3	-	-	4
FF/LTP	-	-	-	-	-	-	-	-	-	1	-	-	1
FF/ FW	-	-	-	-	4	-	-	-	1	-	-	-	5
FF/ FF	-	-	-	-	1	-	-	-	-	-	-	-	1
FF/ MM	-	-	-	-	-	-	-	-	-	1	-	-	1
HS/LTP	-	-	-	-	-	-	-	-	-	1	-	-	1
PL/LTP	-	1	-	-	-	-	-	-	1	1	-	-	3
PL/ FW	-	-	-	-	19	-	-	-	22	-	-	-	41
PL/ FF	-	-	-	-	-	1	-	-	-	-	-	-	1
PL/ PL	-	1	-	-	13	-	-	-	9	-	-	-	23
PL/ MM	-	1	-	-	-	-	-	-		1	-	-	2
MM/ TP	17	1	-	-	1	-	-	-	-	14	-	-	33
MM/LTP	1	17	-	-	2	-	1	-	-	75	-	-	96
MM/ FW	-	-	-	-	-	1	-	-	-	-	-	-	1
MM/ FF	-	1	-	-	2	-	-	-	-	1	-	-	4
MM/ PL	-	-	-	-	-	-	-	-	-	1	-	-	1
MM/ MM	14	53	-	-	1	3	-	-	-	167	-	-	238
Totals:	173	134	-	-	130	8	1	-	71	381	-	-	

Grand total: 898

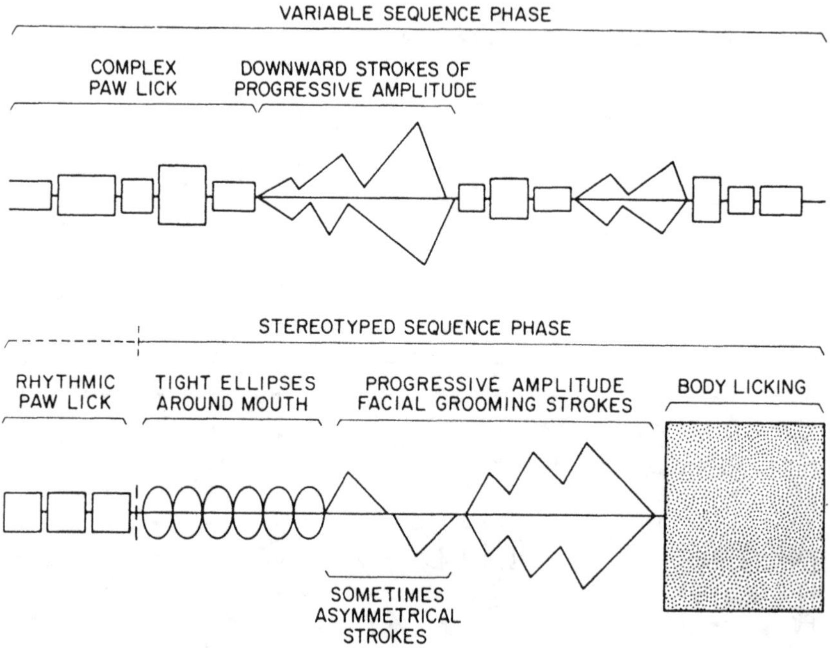

Figure 6.3. Schematic representation of variable sequence phase and stereotyped sequence phase of rat grooming. Horizontal axis represents midline of the animal, with deviations from the line symbolizing different classes of movement. *Note.* From "Natural Syntax Rules Control Action Sequence of Rats" by K. C. Berridge, J. C. Fentress, and H. Parr, 1987, *Behavioural Brain Research, 23*, p. 66. Copyright 1987 by Elsevier Science Publishers; from "Contextual Control of Trigeminal Sensorimotor Function" by K. C. Berridge and J. C. Fentress, 1986, *Journal of Neuroscience, 6*, p. 329. Copyright 1986 by the Society for Neuroscience. Adapted by permission.

organizational properties (cf. Keele & Ivry, 1987, for useful evaluation of modular and shared timing properties in human skills).

Wolf Movement Patterns: An Example of Socially Coordinated Action

Many animal movement patterns must be coordinated with the actions of a moving social partner (Golani, 1976). A fundamental task in the analysis of such socially coordinated movements is to evaluate the extent and nature of mutual constraints in the movements of these social partners. As illustration, Moran (1978; see Moran, Fentress, & Golani, 1981) examined ritualized movement sequences in pairs of wolves by combining three measures: (a) the distance between two animals, (b) the orientation of one animal's longitudinal axis to the other's, and c) the closest point of opposition of one animal along the body axis of the other. These three measures were summarized in an interaction cube (Figure 6.4). The first observation is that the animals often maintained relatively invariant relations to one another in time even though each of the animals was in motion as measured by an environmental referent. Second, these socially "fixed" nodes of movement occupied restricted regions within the interactional cube. Third, when the animals changed their relative position, the trajectories between the fixed nodes followed predictable courses. Finally, it is worth noting that the relational measures often provided a simpler description of important movement features than did evaluations of either of the participating wolves in isolation. This means that the movements were mutually constrained (i.e., co-ordered).

Note that Moran's (1978) diagram (Figure 6.4) emphasizes the mutual stabilities and more or less continuous flows of these abstracted properties in the social behavior of the wolves. These nodes and flows could be produced through a variety of particular movements of each of the participants. The result is a social "motor equivalence" (cf. Lashley, 1951).

Havkin (1977; cf. Havkin & Fentress, 1985) extended Moran's (1978) analysis by quantifying the degree of symmetry in the behavior of two interacting wolves. Havkin's study emphasized mutual anatomical points of nearest proximity or actual contact. In this way Havkin was able to articulate issues relevant to the differentiation of social roles. In the Havkin and Fentress paper (1985), analysis of data indicated that the animals had a progressive ability to adopt multiple strategies in the accomplishment of singly defined social acts (such as one animal toppling the other animal over on its back or side).

McLeod (1987) has subsequently used a variety of information theory, clustering, and multidimensional scaling techniques to evaluate the simultaneous rules of social display as defined within and between animals. As did Havkin (1977; Havkin & Fentress,

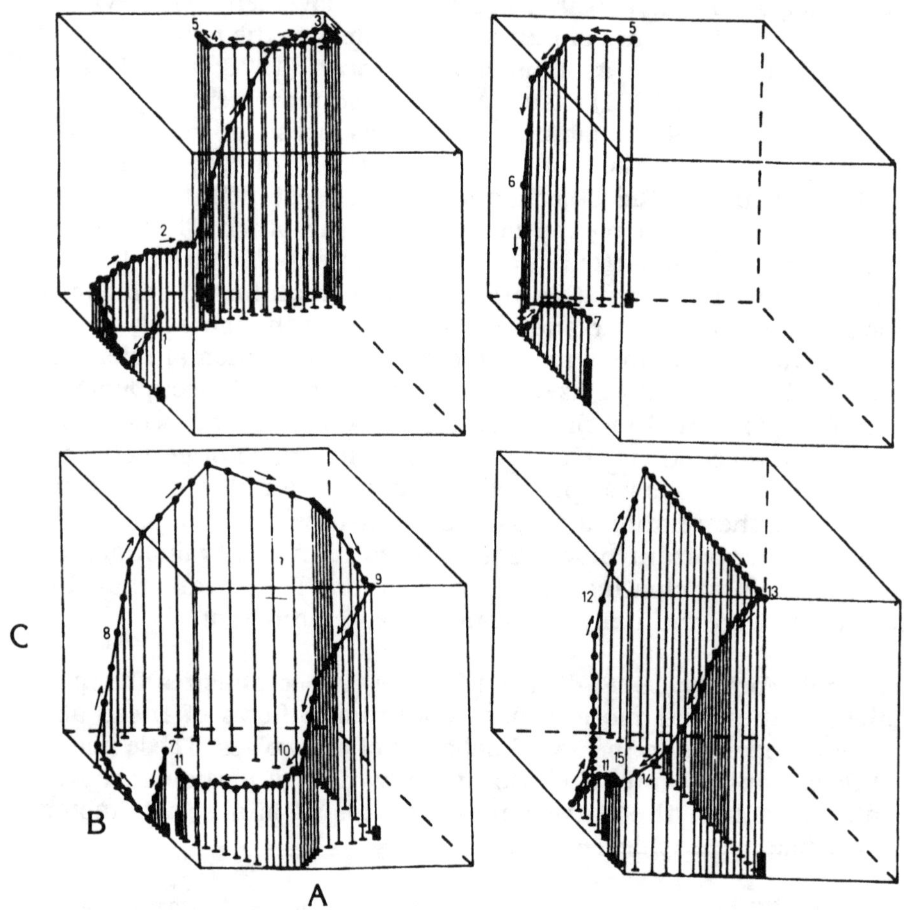

Figure 6.4. Four examples of "social interaction cubes" based upon three relational measures defined for two wolves. Front axis (A) is relative distance between wolves, rear projecting axis (B) is the relative orientation of the two wolves, and the vertical axis (C) is the point of nearest opposition defined along the body surface. Both fixed locations and trajectories through the cube are rule-defined and can be maintained through a variety of individual movements defined by fixed environmental referents. This indicates that the animals "co-order" their actions with each other as each moves through external space (cf. Moran, Fentress & Golani, 1981). *Note.* From *The Structure of Movement in Supplanting Interactions in the Wolf* by G. Moran, 1978, unpublished doctoral dissertation, p. 163. Copyright 1978 by G. Moran. Adapted by permission.

1985), McLeod emphasized the developmental emergence and coordination of social behavior. McLeod showed that the intraanimal constraints defined by informational measures can continue to increase over a 3-month period whereas the interanimal constraints show an increase and then decrease (Figure 6.5). This analysis illustrates the value of looking at relative movement regularities defined both within and between animals. For the comparative behavioral scientist, distinctions revealed by alternative descriptive frames can often provide especially important insights (cf. Fentress & McLeod, 1986, 1988).

In a recent study, Tooze (1988) examined in detail the formal properties of long-distance vocalization patterns in wolves (Figure 6.6). She measured variability within and between animals through a number of multivariate techniques. Tooze was therefore able to evaluate these variability measures against more globally defined characteristics of individual phenotype and social context. Tooze's initial playback experiments enabled her to begin an evaluation of linkages between behavioral form, expressive context, and function.

This brief mention of social behavior in wolves illustrates the multiple layers over which coordinated behavior must often be controlled. The pictures provided by such complementary perspectives are not mutually exclusive. Animal nervous systems must obviously be able to order flexibilities and constraints in their total action at many different levels. For each of these levels, the relative flexibility and constraint can also depend upon the particular measures that an investigator uses (cf. Fentress, 1984, 1989, in press).

UNRAVELING INTEGRATIVE PATTERNS

The rich multidimensional and multilevel flow of activities within and between animals inspires obvious questions, including those pertaining to the underlying network of integrative events. In brief, how do the various simultaneously and sequentially occurring properties of behavior become both appropriately separated from one another and combined into functionally coherent units? The complexity of such questions often makes it useful to concentrate on relatively simple forms of behavioral expression. Therefore, I return to analysis of rodent grooming.

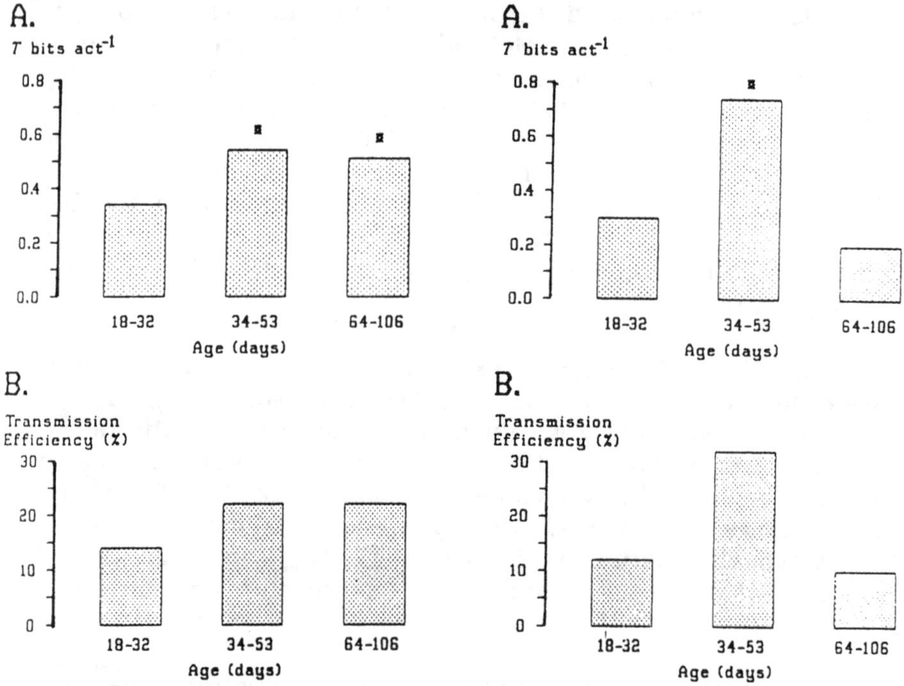

Figure 6.5. Information measures of shared information (A) and transmission efficiency (B) during early social interactions in wolves. Seven behavioral items were used. Left side of figure represents within-animal transitions; right side contains between-animal measures. Differences between the left- and right-hand graphs indicate different rules of behavioral order development for within-animal and between-animal profiles. *Note.* From *Aspects of the Early Social Development of Timber Wolves (Canis Lupus)* by P. J. McLeod, 1987, unpublished doctoral dissertation, p. 79. Copyright 1987 by P. J. McLeod. Adapted by permission.

Interactive, Self-Organizing Systems in Rodent Grooming

The basic approach of many ethologists and other comparative workers is to provide detailed descriptions of the ongoing flow of behavior and then to perturb this ongoing system at particular phases. The perturbations are then arranged along appropriate quantitative, qualitative, and temporal dimensions (cf. Fentress, 1976, 1984, 1989, in press, for elaboration). Finally, various com-

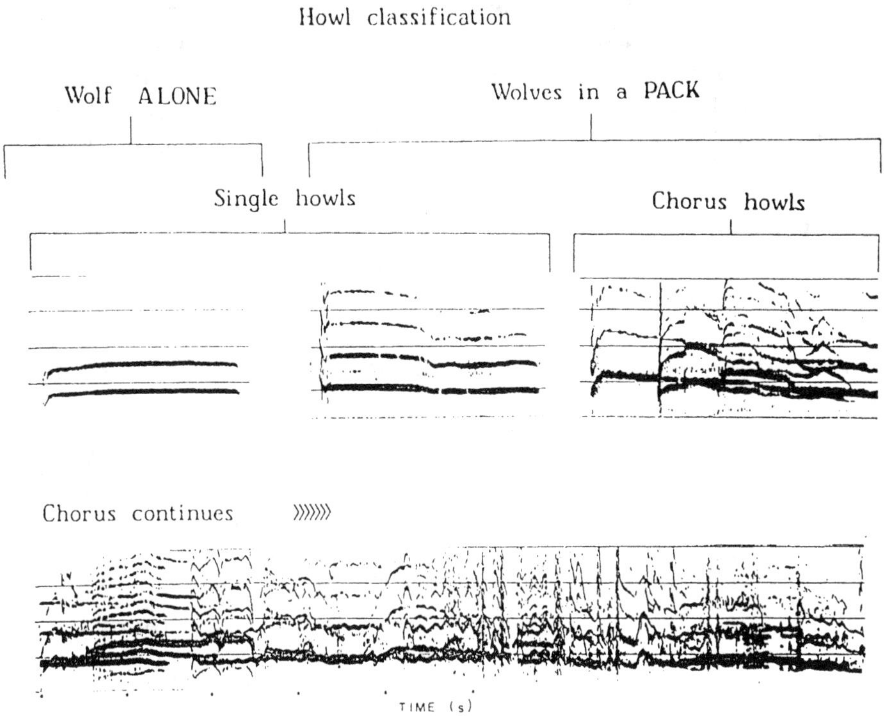

Figure 6.6. Wolf howl sonograms for single and chorus howls for a wolf alone and wolves in a pack. Individual and sex differences, age changes, and role of social context in the coordination of these long distance vocalizations can be derived from quantitative analyses of measures that include fundamental frequency, side bands, duration, and frequency modulation. *Note.* From *Some Aspects of the Structure and Function of Long-Distance Vocalizations of Timber Wolves (Canis Lupus)* by Z. Tooze, 1988, unpublished master's thesis, p. 18. Copyright 1988 by Z. Tooze. Reprinted by permission.

binations of perturbation are applied, and the range of consequences resulting from these perturbations singly and in combination are related to one another to provide a picture of *converging* and *diverging* networks of integrative function.

The consequences of any given perturbation can depend importantly upon the current state of the system under investigation.

That is, there can be multiple and state-dependent variations in response to a given manipulation. These variable responses are also rule given. As a simple illustration, Grillner (1985) summarized a number of data that demonstrate the fundamentally dynamic nature of reflex responses in various vertebrate species (cf. reviews by Forssberg, 1985, and Rossignol & Drew, 1985, in this same volume [Barnes & Gladden, 1985]). "Phase dependency" can be seen at essentially all levels of integrated motor movement control. Ethologists normally do not work at the level of spinal integration, but the same basic principles apply.

As illustration, Woolridge (cited in Fentress, 1984) demonstrated that momentary proprioceptive loads applied to the forelimbs of mice during different phases of grooming had strikingly different consequences. During slow and variable phases of grooming, these proprioceptive loads terminated the grooming sequence with a short latency. However, during rapid and stereotyped phases of facial grooming, these same phasic loads on the forelimbs (e.g., pulling the limbs outward from the face via flexible threads) did not terminate the grooming behavior. This was true even though the ongoing grooming movements were no longer functional in the sense that the normal paw-to-face contacts were prevented.

Berridge and Fentress (1986, 1987a) have shown that trigeminal deafferentation also produces a differential effect upon variable and stereotyped phases of rodent grooming. The stereotyped phases are much more immune to disruptions in both form and sequence. Most interesting from the perspective of this discussion, when changes in form were produced during the stereotyped phases of the grooming sequence and when the trigeminal deafferentation procedures even produced abnormal "forelimb flails," the sequence continued to its normal endpoint (Figure 6.7). This means that the

Figure 6.7. Grooming sequences in the stereotyped sequence phase for intact and trigeminally deafferented rats. The stereotyped face grooming phase is not disrupted and continues into body grooming even when excess forelimb flails are introduced. Earlier studies (Berridge & Fentress, 1986) had shown that the variable sequence phase of grooming was very sensitive to disruption by these same trigeminal lesions. *Note.* From "Deafferentation Does Not Disrupt Natural Rules of Action Syntax" by K. C. Berridge and J. C. Fentress, 1987, *Behavioural Brain Research, 23*, p. 74. Copyright 1987 by Elsevier Science Publishers. Reprinted by permission.

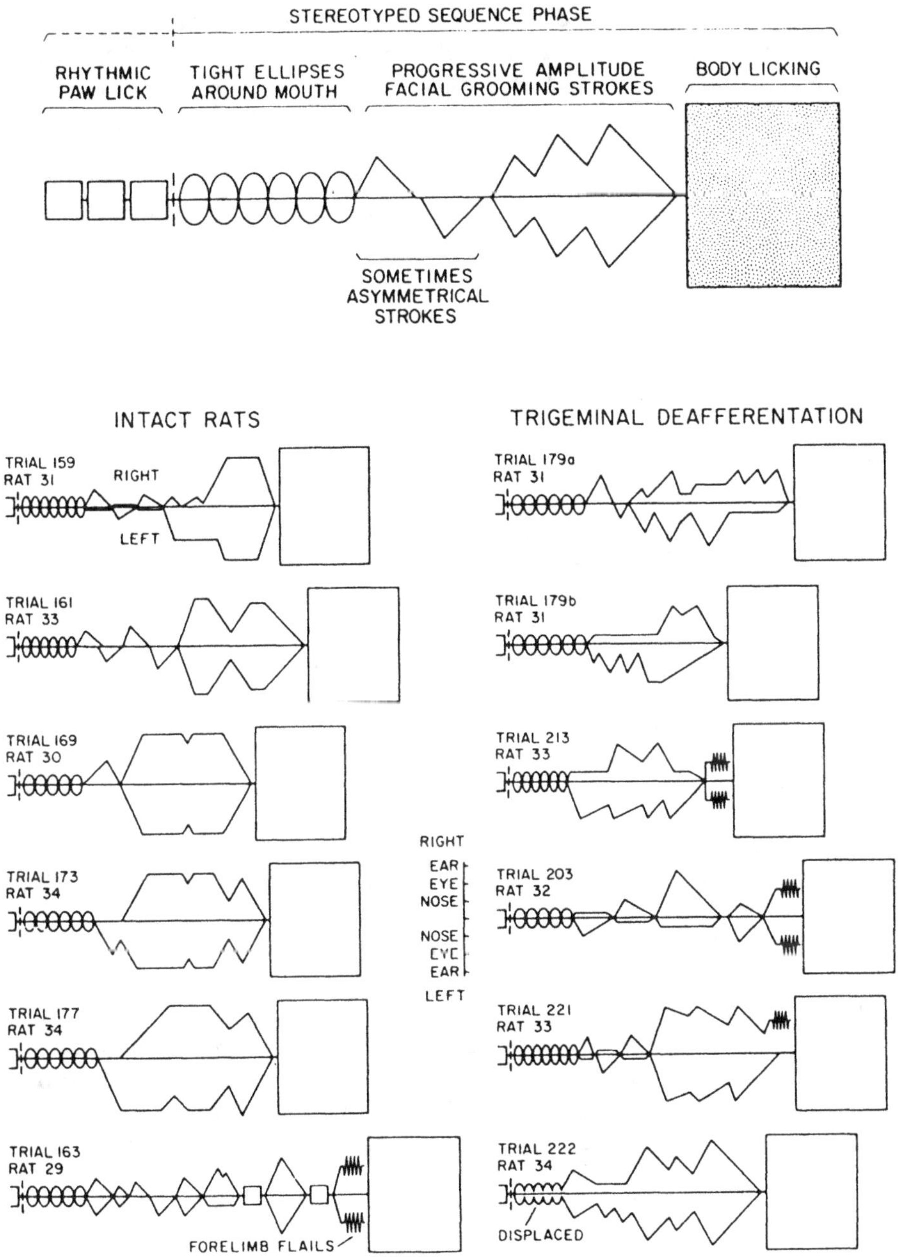
STEREOTYPED SEQUENCE PHASE
RHYTHMIC PAW LICK
TIGHT ELLIPSES AROUND MOUTH
PROGRESSIVE AMPLITUDE FACIAL GROOMING STROKES
BODY LICKING
SOMETIMES ASYMMETRICAL STROKES
INTACT RATS
TRIGEMINAL DEAFFERENTATION
TRIAL 159 RAT 31
RIGHT
LEFT
TRIAL 161 RAT 33
TRIAL 169 RAT 30
TRIAL 173 RAT 34
TRIAL 177 RAT 34
TRIAL 163 RAT 29
FORELIMB FLAILS
TRIAL 179a RAT 31
TRIAL 179b RAT 31
TRIAL 213 RAT 33
TRIAL 203 RAT 32
TRIAL 221 RAT 33
TRIAL 222 RAT 34
DISPLACED
RIGHT
EAR
EYE
NOSE
NOSE
EYE
EAR
LEFT

grooming sequence cannot be explained as chain reflexes or by simple models of sensory feedback. Rather, central patterning mechanisms must be involved. From their behavioral observations, early ethologists had postulated the importance of central modulatory and patterning mechanisms (Lorenz, 1950; cf. Ewert, 1987; Hoyle, 1985). The relative importance or salience of the central versus peripheral contributions to grooming patterns varies as a function of the overall sequence even for particular activities that are shared by these two overall sequence phases.

The Striatum and Beyond

One set of regions in the central nervous system that are important to the sequencing of movement and its modulation is the basal ganglia (striatum, pallidum, and related structures). These complexly interconnected areas (Early, Posner, & Reiman, 1988) have been interestingly implicated in the organization of many species-characteristic sequences (cf. MacLean, 1978; Murphy, MacLean, & Hamilton, 1981). Motor deficits following lesions of the striatum and associated structures have been recognized for many years, although more sophisticated and multilevel analyses of these deficits are needed (e.g., Brooks, 1986; Cools, 1985).

Berridge and Fentress (1987b) found that kainic acid lesions of portions of the striatal circuitry (a subset of structures within the basal ganglia) in rats prevented the completion of the stereotyped phases of grooming in most cases (Figure 6.8). Thus, the striatal lesions provide a complementary picture to data previously obtained with the trigeminal deafferentation procedures. In such ways one can begin to put together a picture of the dynamic balance between various defined central and peripheral events.

It often becomes useful to conceptualize various systems underlying the control of integrated movement as if they were ordered with a central core of excitation and a periphery of inhibition (see Fentress, 1984). Depending upon the dynamics of the individual systems and their rules of interaction, the relative salience or size

Figure 6.8. Grooming chain disruptions following striatopallidal lesions (cf. Figure 6.7). *Note.* From "Disruption of Natural Grooming Chains after Striatopallidal Lesions" by K. C. Berridge and J. C. Fentress, 1987, *Psychobiology, 15,* p. 340. Copyright 1987 by the Psychonomic Society, Inc. Reprinted by permission.

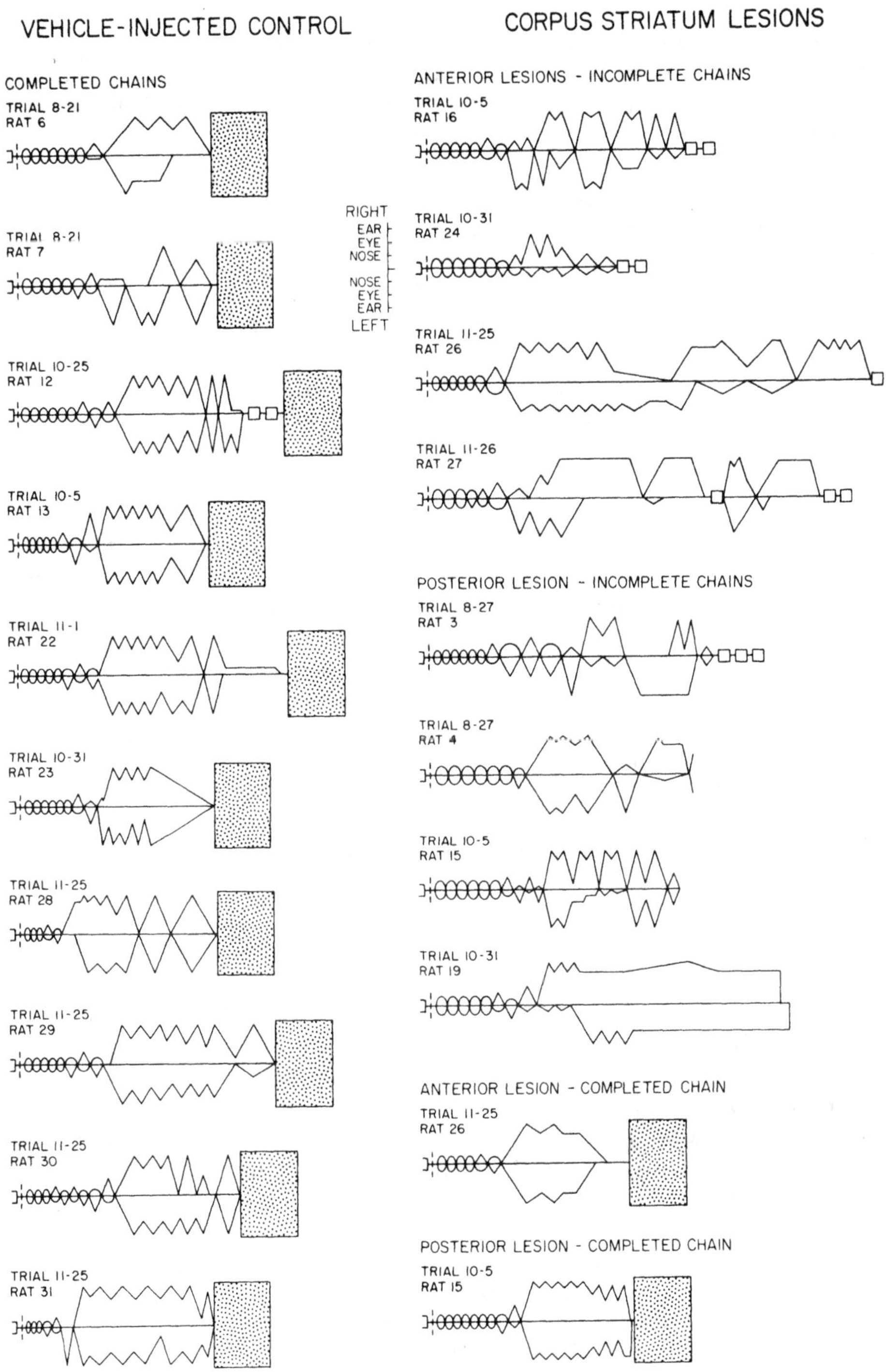
VEHICLE-INJECTED CONTROL
COMPLETED CHAINS
TRIAL 8-21
RAT 6
TRIAL 8-21
RAT 7
TRIAL 10-25
RAT 12
TRIAL 10-5
RAT 13
TRIAL 11-1
RAT 22
TRIAL 10-31
RAT 23
TRIAL 11-25
RAT 28
TRIAL 11-25
RAT 29
TRIAL 11-25
RAT 30
TRIAL 11-25
RAT 31
RIGHT
EAR
EYE
NOSE
NOSE
EYE
EAR
LEFT
CORPUS STRIATUM LESIONS
ANTERIOR LESIONS - INCOMPLETE CHAINS
TRIAL 10-5
RAT 16
TRIAL 10-31
RAT 24
TRIAL 11-25
RAT 26
TRIAL 11-26
RAT 27
POSTERIOR LESION - INCOMPLETE CHAINS
TRIAL 8-27
RAT 3
TRIAL 8-27
RAT 4
TRIAL 10-5
RAT 15
TRIAL 10-31
RAT 19
ANTERIOR LESION - COMPLETED CHAIN
TRIAL 11-25
RAT 26
POSTERIOR LESION - COMPLETED CHAIN
TRIAL 10-5
RAT 15

of these central excitatory cores and peripheral inhibitory regions may vary. By attempting to work out the rules of this variation, one can begin to piece together a better representation of the dynamic network of underlying integrative events.

The basal ganglia (striatum and associated structures) appear to be good candidates for ordering the dynamic rules of separation and connection necessary for effective coordinated movement. As summarized recently by Gerfen (1987), the mammalian striatum operates along both the principle of segregated pathways and more globally defined integrative functions (cf. Fentress, 1989, in press). In humans, damage to the striatum leads to a set of still poorly understood fragmentations and perseverant connections among particular dimensions of performance.

Recent models of striatal function emphasize its dual role in the "programming" of sequential motor events and in the gaiting and amplification of various sensory events (e.g., Cools, 1985; Early et al., 1988). Further, the striatum in humans has been implicated in a variety of higher-order attentional, motivational, and cognitive processes (Posner & Presti, 1987). The striatum has also been implicated in various multilevel human disorders such as adult schizophrenia (Early et al., 1988) and childhood autism (Maurer & Damasio, 1982). Full understanding of striatal function will depend in part upon careful dissections of behavior at multiple levels. Comparative approaches to coordinated movement in different contexts can be expected to make valuable contributions to this goal.

The complexity of striatal circuitry is further emphasized by the variety of excitatory and inhibitory neurotransmitters that are involved in its function. Further, distinguishable reentrant loops connect striatal function to a wide (but rigidly distributed) range of cortical areas, thalamic nuclei, and so on. As illustrated by the work of Cools and his colleagues (Cools, 1985), careful pharmacological dissections of striatal function combined with equally careful and multilevel behavioral analyses can add importantly to our currently very incomplete picture of striatal function.

A recent study by E. Buckle (1988) in my laboratory illustrates the value of the rodent grooming model in this context. Buckle found that neonatal applications of 6-hydroxydopamine in mice greatly reduced the ability of these animals to complete the stereotyped phase of face grooming (Figure 6.9). The animals initiated early phases of this sequence more frequently than did the controls, and later elements in the sequence occurred less often in the experimental animals than they did in the controls.

It is important to note that disruptions of striatal function not only can produce fragmentation of behavior but also under other circumstances can lead to the performance of perseverant motor stereotypes. As illustration, Berridge, Fentress, and Treit (1988) found that rats with kainic acid lesions of the striatum would perform protracted hyperkinetic paw treading sequences upon receiving intraoral infusions of relatively nonpalatable taste substances (Figure 6.10). The hyperkinetic paw treading movements were specifically linked to oral stimuli. They rarely occurred spontaneously, and a variety of other sensory inputs failed to elicit them reliably. The ethological tradition of isolating specific links between particular sensory events and particular classes of motor expression, in combination with the search for more global modulatory (e.g., motivational) processes, is important to both behavioral and neural science.

I anticipate that the study of normal flexibilities in movement as contrasted to these perseverant stereotypes (cf. Fentress, 1976) will provide a particularly valuable assay in future research. Interestingly, perseverant motor stereotypes can also be produced in animals through protracted confinement to restricted environments (e.g., the "zoo animal syndrome," exposure to a variety of stressors, etc.). Thus, as in the case of rodent grooming, future investigators should be able to assess central and environmental contributions to motor stereotypies against one another. This work could lead to more precise evaluations of the center-surround form of conceptualization previously alluded to (cf. Fentress, 1984).

Such studies on movement stereotypy and flexibility also indicate the potential importance of conducting more work on comparative coordination devoted to ontogenetic issues. Unfortunately, the scope of this chapter does not allow me to discuss in detail how such ontogenetic and integrative perspectives might be brought into closer register with one another (see Fentress, 1989, in press, for greater detail). However, a brief review of some of the issues and emerging principles in the construction of movement phenotypes may prove useful.

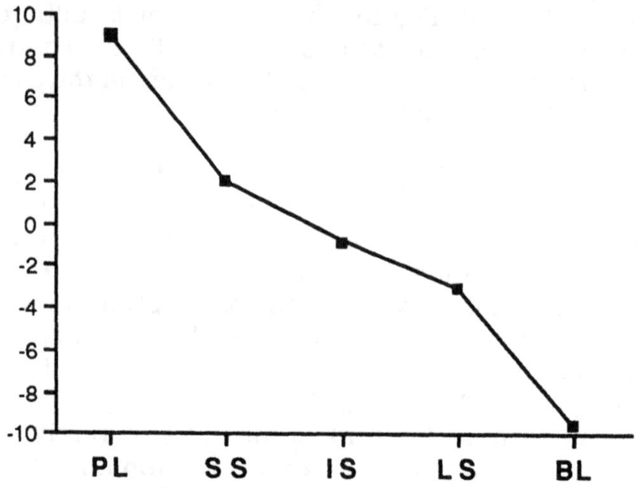

Figure 6.9. Proportional changes in comparison to controls for grooming actions in early and late phases for mice given neonatal treatment with 6-hydroxydopamine (n = 8 per group). Grooming sequences tend to begin with paw licks (PL), followed by small strokes of the face (SS), intermediate strokes (IS), large strokes (LS), and then body grooming (BL). Treated animals were less likely to complete the sequence. (Based on data from Buckle, 1988.)

CODING MOVEMENT PHENOTYPES

Ethological Profiles

The early ethological emphasis upon "instinctive behavior" has provided an important focus for the study of comparative coordination. Although the simple dichotomy between nature and nurture frequently implied in early ethological writings is no longer tenable, it is obvious that all forms of coordinated movement rest ultimately upon the organism's genetic machinery (as released, modulated, amplified, suppressed, or destroyed by a wide range of experiences during ontogeny). In recent years there have been a number of extensive reviews of motor development (e.g., Blass, 1986, 1988). One of the fundamental recurring issues is the interplay between processes of movement differentiation (i.e., separation) and integration (i.e., combination). It is now generally accepted that these two developmental trends occur together with

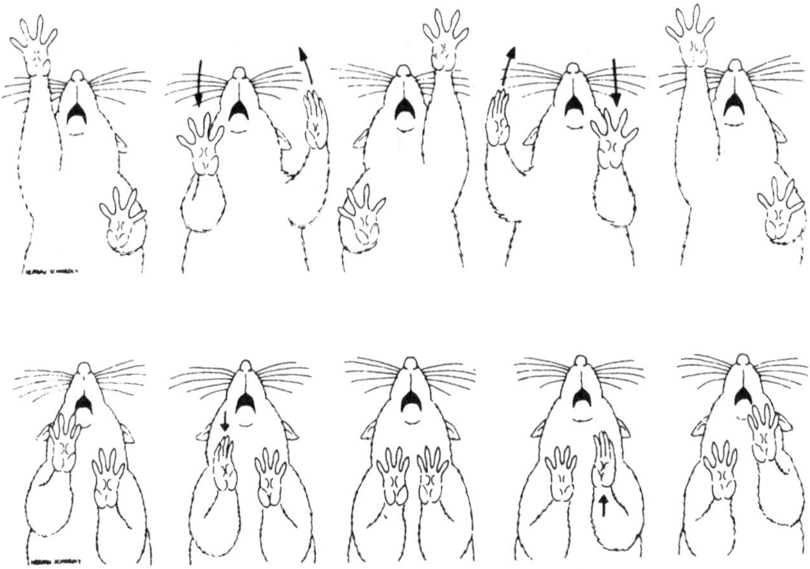

Figure 6.10. Large amplitude (top row) and small amplitude (bottom row) hyperkinetic paw treading movements produced in rats by a combination of kainic acid lesions of the corpus striatum and oral sensory stimulation. *Note.* From "A Triggered Hyperkinesia Induced in Rats by Lesions of the Corpus Striatum" by K. C. Berridge, J. C. Fentress, and D. Treit, 1988, *Experimental Neurology, 199*, p. 263. Copyright 1988 by Academic Press, Inc. Reprinted by permission.

different emphases depending in part upon one's own particular measures and level of organization. To cite a single illustration from my laboratory, Golani and Fentress (1985) found that in the ontogeny of facial grooming patterns in mice, early rich but loosely coordinated movements became refined and also coherently directed to the accomplishment by the forepaws of effective treatment of the body surface. Indeed, apparently simple questions such as whether the movements become more or less variable with age do not have any single answer. When mice are approximately 10 days old, they become very effective at guiding their movements to a particular (invariant) subregion of the face and at the same time express this precision of contact pathway through an increasing number of forepaw trajectories and kinematic details of the limb segments (cf. the concept of motor equivalence).

Interestingly, this issue of motor equivalence turns out to be an extremely valuable assay in a wide range of invertebrate and vertebrate species, and for vertebrates at neural organizational levels down to the spinal cord. For example, Berkinblit, Feldman, and Fukson (1986) have demonstrated in an elegant series of experiments that spinal frogs are able to maintain effective total wiping movements following perturbations of particular movement details. This ability is made possible by a complex network of mutual compensatory actions, a case of *co-ordering* in the most explicit sense! Recent research on the balance between central "pattern generators" and feedback in a number of invertebrate as well as vertebrate species has shown that such compensatory actions are widespread (e.g., reviews in Barnes & Gladden, 1985).

The Golani and Fentress study (1985) also emphasized a common finding that developmental trends in coordinated movement do not necessarily move in a single direction. There are often reversals (i.e., apparent regressions) during the development of movement as the animal consolidates earlier developmental phases. As illustration, mice between the ages of postnatal 0 to 100 hr show a rich variety of limb segment kinematics and forepaw trajectories that are aimed rather loosely at facial targets. Between 100 and 200 hr postnatally, the targets on the face become restricted and more precise, yet limited. Beyond approximately 200 hr postnatally, the contact pathways become re-elaborated as do the potential kinematic contributions to the execution of any one of these pathways.

In a recent study of early exploratory behavior in the house rat, Eilam and Golani (in press) demonstrated an orderly progression of movements from lateral to forward to vertical with age. Each dimension decreases systematically in amplitude and incorporates more body segments in a rostral-caudal direction. Across ontogeny, rats show an initial immobility upon being placed in a normal environment and then "warm up" in a manner that recapitulates earlier developmental stages. From these observations, Eilam and Golani were able to abstract "programming rules" by which "the spread of activity and build-up manifested in less advanced sequences is contained within the more advanced sequences." An earlier study by Szechtman, Ornstein, Teitelbaum, and Golani (1985) demonstrated that apomorphine (a direct dopamine agonist) produces a decrease in mobility that reverses the observed ontogenetic sequence. This principle of "last in–first out" occurs for a number of neurological (e.g., striatal) disorders, and its discovery shows how careful comparative analyses can help clarify underlying principles that may have broad importance. To obtain this link between developmental and integrative time frames, precise descriptions along explicitly defined movement dimensions are essential.

Toward a Developmental Neuroethology

Starting with the early ethological research on bird song (Thorpe, 1961), imprinting, and related topics in behavioral ontogeny, ethologists have progressively teased apart the network of developmental events that provide the foundations for many forms of coordinated action. As I suspect is well known to all readers of this volume, one of the fundamental contributions from this literature is the finding that during development, there are phases of particular sensitivity to particular sources of sensory input (i.e., "sensitive periods") and other phases in which the developmental trajectory appears to be much more immune to these same influences. Today there is an important convergence between comparative approaches to sensitive periods and research at the neurobiological (e.g., cellular) level. (For excellent recent reviews, including recent data on bird song, see Rauschecker & Marler, 1987).

Recent comparative perspectives on the development of coordinated action have emphasized that young animals are not simply imperfect adult animals. They can often accomplish remarkably sophisticated forms of motor coordination early during ontogeny. In most cases, these early coordinated movements serve critical ontogenetic functions (e.g., suckling movements in mammals; cf. Alberts & Cramer, 1988, for a recent comprehensive review of these so-called "ontogenetic adaptations").

Other circuits underlying movement may be prefunctional. Nervous systems of essentially all species studied to date assemble complex neural networks that the organism utilizes in movement only some time later. The assembly of such circuits thus cannot depend upon feedback from the movements themselves. The primacy of motor circuits during ontogeny is often referred to as *retrograde development* to emphasize the reverse direction to sensory and then motor activation in many forms of integrated performance. This term evokes a long historical debate on the spontaneity versus reflex nature of early movement patterns (cf. review in Fentress & McLeod, 1986).

Depending on the species and the particular movement in question, some early movements appear to occur without any eliciting stimulus. Others appear to be driven, so to speak, by particular sensory signals. In mammals, including humans, both spontaneity and sensory sensitivity during early ontogeny are intimately linked. Much future research is needed to separate the relative contributions of these "spontaneous" and "receptive" modes of expression. Circuits for motor performance are often latent in early ontogeny and become apparent only after appropriate contextual supports are provided to the organism (Fentress & McLeod, 1986).

An issue still not understood at all well is how *integrated networks* of behavioral expression become established (see, e.g., the recent review by Hogan, 1988). It is clear that the consequences of experience during development may generalize along channels that we simply have not investigated properly. The issue becomes still more complex once it is recognized that the same underlying circuit may contribute to different actions at different stages.

Roles of Experience

We also do not have good working models of precisely what we expect experience to do during development. The earlier literature stressed the concept of learning. However, it is clear that experience is a much more multifaceted domain than this term implies. For example, in much of the embryological research, experience is viewed more as a selective mechanism than as an instructional mechanism (e.g., Edelman, 1987; see Fentress, 1989, in press). *Selection* in this context has two senses. First, the experiences may serve to activate or trigger otherwise latent developmental potentialities. In such cases, there is no logical need for the final behavioral outcome to depend in its detail upon the organism's having received all detail from experiential sources. In animal play, for example, distinctions between early and later ontogenetic movement forms preclude any simple mapping of the details of ontogenetic experience and the details of later performance (e.g., Pellis & Pellis, 1987).

Even at the level of human language, investigators are beginning to emphasize inherent developmental capacities (e.g., Chomsky's

[1980] poverty of the stimulus). Petitto (1988) and Petitto and Bellugi (1988) have provided valuable recent reviews of the ontogeny and production of linguistic skills (speech and signing). Comparative analyses of constraints and flexibilities in other forms of coordinated action may eventually help clarify the boundary conditions at the linguistic level. Here the obvious point of comparison is the delineation of both similarities and constraints among coordinated skills.

The second use of the term *selection* is more strictly Darwinian (Edelman, 1987). This perspective emphasizes competition among alternative developmental pathways. At many levels, investigators are obtaining evidence for attrition of certain developmental capacities as others grow. One neurobiological example with which I have been associated is that considerable neuronal cell death occurs across precisely defined trajectories in the rodent striatum during early embryogenesis (Fentress, Stanfield, & Cowan, 1981). The relation between such anatomical profiles and profiles of behavioral expression, however, remains an elusive problem.

This note on embryogenesis serves to remind us that studies of motor development at all levels will remain incomplete until the prenatal as well as postnatal contributions of various factors are better understood (Fentress & McLeod, 1988). These comparative considerations will almost certainly prove to be important to our understanding of both normal human motor development and various clinical misdirections in this development. I suspect that future comparative studies may offer their most significant impact in this area.

THE THREE P'S IN RETROSPECT: FROM FAIRY TALES TO SCIENCE IN PROSPECT

I began this essay with the suggestion that comparative coordination deserves our consideration from three perspectives of process, pattern, and phenotype. There is, of course, nothing magical about this particular set of emphases. However, as a group I do believe that the "three little p's" serve well in reminding us that we wish to understand coordinated movement in dynamic, relational, and multilevel terms while also taking into account the deeper structural stabilities from which these ephemeral events are derived. I have made a few brief allusions to literature on human performance up to the level of language as well as down to cellular neurobiology. I did so because I think that the three p's are relevant to each of these enterprises as well as to more traditionally defined comparative research.

Indeed, it is encouraging to see that investigations of comparative coordination in recent years have begun to reach out to these sister disciplines. The recent advent of dynamic self-organizational models of pattern formation in biophysics (Kugler & Turvey, 1987; Yates, 1987) suggests one direction in which future conceptual links might be forged (cf. Fogel & Thelen, 1987; Kelso & Jeka, this volume). Such explorations must obviously be made critically and with caution, for the essence of the comparative approach is to highlight differences as well as similarities in organization across different domains of inquiry. By starting with natural history and a simple fascination with the diversity of behavior in nature, ethologists have now begun to bring their often very rich insights into a framework that should allow many, and coordinated, avenues of future research.

ACKNOWLEDGMENTS

My thanks to S. Wallace for his invitation to contribute to this volume as well as for his patience in my preparation of the final product, and to W. Danilchuk for her assistance in completing the manuscript from often imperfectly coordinated verbal tapes and written notes. To T., in memory of her joy for fable in the search of truth, and for her love of diversity in the search of principle.

FOOTNOTES

[1]From *Aristotle to Zoos* (p. 83) by P. B. Medawar and J. S. Medawar, 1983, Cambridge, MA: Harvard University Press. Copyright 1983 by P. B. Medawar and J. S. Medawar.

REFERENCES

Alberts, J. R., & Cramer, C. P. (1988). Ecology and experience: Sources of means and meaning of developmental change. In E. M. Blass (Ed.), *Handbook of behavioral neurobiology: Vol. 9. De-*

velopmental psychobiology and behavioral ecology (pp. 1-39). New York: Plenum Press.

Barnes, W. J. P., & Gladden, M. H. (Eds.). (1985). *Feedback and motor control in invertebrates and vertebrates*. London: Croom Helm.

Berkinblit, M. B., Feldman, A. G., & Fukson, O. I. (1986). Adaptability of innate motor patterns and motor control mechanisms. *Behavioral and Brain Sciences, 9*, 585-599.

Bernstein, N. (1967). *The co-ordination and regulation of movements*. New York: Pergamon Press.

Berridge, K. C., & Fentress, J. C. (1986). Contextual control of trigeminal sensorimotor function. *Journal of Neuroscience, 6*, 325-330.

Berridge, K. C., & Fentress, J. C. (1987a). Deafferentation does not disrupt natural rules of action syntax. *Behavioural Brain Research, 23*, 69-76.

Berridge, K. C., & Fentress, J. C. (1987b). Disruption of natural grooming chains after striatopallidal lesions. *Psychobiology, 15*, 336-342.

Berridge, K. C., Fentress, J. C., & Parr, H. (1987). Natural syntax rules control action sequence of rats. *Behavioural Brain Research, 23*, 59-68.

Berridge, K. C., Fentress, J. C., & Treit, D. (1988). A triggered hyperkinesia induced in rats by lesions of the corpus striatum. *Experimental Neurology, 99*, 259-268.

Blass, E. M. (Ed.). (1986). *Handbook of behavioral neurobiology: Vol. 8. Developmental processes in psychobiology and neurobiology*. New York: Plenum Press.

Blass, E. M. (Ed.). (1988). *Handbook of behavioral neurobiology: Vol. 9. Developmental psychobiology and behavioral ecology*. New York: Plenum Press.

Brooks, V. B. (1986). *The neural basis of motor control*. New York: Oxford University Press.

Buckle, E. A. (1988). *Effects of dopamine depletion on the form and sequencing of grooming in mice*. Unpublished honors thesis, Dalhousie University, Halifax, Nova Scotia.

Chomsky, N. (1980). *Rules and representations*. New York: Columbia University Press.

Cools, A. R. (1985). Brain and behavior: Hierarchy of feedback systems and control of its input. In P. H. Klopfer & P. Bateson (Eds.), *Perspectives in ethology* (Vol. 6, pp. 109-168). New York: Plenum Press.

Early, T. S., Posner, M. I., & Reiman, E. M. (1988). *Hyperactivity of the left striato-pallidal projection: An integrated model of multilevel pathology in schizophrenia*. (Manuscript submitted for publication).

Edelman, G. M. (1987). *Neural Darwinism: The theory of neuronal group selection*. New York: Basic Books.

Eilam, D., & Golani, I. (in press). The ontogeny of exploratory behavior in the house rat (*Rattus rattus*): The immobility-mobility gradient. *Developmental Psychobiology*.

Eshkol, N., & Wachmann, A. (1958). *Movement notation*. London: Weidenfeld & Nicholson.

Ewert, J.-P. (1987). Neuroethology of releasing mechanisms: Prey-catching in toads. *Behavioral and Brain Sciences, 10*, 337-368.

Fentress, J. C. (1968a). Interrupted ongoing behaviour in voles (*Microtus agrestis* and *Clethrionomys britannicus*): I. Response as a function of preceding activity and the context of an apparently 'irrelevant' motor pattern. *Animal Behaviour, 16*, 135-153.

Fentress, J. C. (1968b). Interrupted ongoing behaviour in voles (*Microtus agrestis* and *Clethrionomys britannicus*): II. Extended analysis of intervening motivational variables underlying fleeing and grooming activities. *Animal Behaviour, 16*, 154-167.

Fentress, J. C. (1972). Development and patterning of movement sequences in inbred mice. In J. Kiger (Ed.), *The biology of behavior* (pp. 83-132). Corvallis, OR: Oregon State University Press.

Fentress, J. C. (1976). Dynamic boundaries of patterned behavior: Interaction and self-organization. In P. P. G. Bateson & R. A. Hinde (Eds.), *Growing points in ethology* (pp. 135-169). Cambridge: Cambridge University Press.

Fentress, J. C. (1978). *Mus musicus:* The developmental orchestration of selected movement patterns in mice. In M. Bekoff & G. Burghardt (Eds.), *The development of behavior: Comparative and evolutionary aspects* (pp. 321-342). New York: Garland.

Fentress, J. C. (1984). The development of coordination. *Journal of Motor Behavior, 16*, 99-134.

Fentress, J. C. (1986). Development of coordinated movement: Dynamic, relational and multilevel perspectives. In H. T. A. Whiting & M. C. Wade (Eds.), *Motor development in children: Aspects of coordination and control* (pp. 77-105). Dordrecht: Martinus Nijhoff.

Fentress, J. C. (1989). Developmental roots of behavioral order: Systemic approaches to the examination of core developmental issues. In M. R. Gunnar & E. Thelen (Eds.), *The Minnesota symposia on child psychology: Vol. 22. Systems and development* (pp. 35-76). Hillsdale, NJ: Erlbaum.

Fentress, J. C. (in press). Organizational patterns in action: Local and global issues in action pattern formation. In G. M. Edelman, W. E. Gall, & W. M. Cowan (Eds.), *Signal and sense: Local and global order in perceptual maps*. New York: Wiley.

Fentress, J. C., & McLeod, P. (1986). Motor patterns in development. In E. M. Blass (Ed.), *Handbook of behavioral neurobiology: Vol. 8. Developmental processes in psychobiology and neurobiology* (pp. 35-97). New York: Plenum Press.

Fentress, J. C., & McLeod, P. J. (1988). Pattern construction in behavior. In W. P. Smotherman & S. R. Robinson (Eds.), *Behavior of the fetus* (pp. 63-76). Caldwell, NJ: Telford Press.

Fentress, J. C., & Stilwell, F. P. (1973). Grammar of a movement sequence in inbred mice. *Nature, 244*, 52-53.

Fentress, J. C., Stanfield, B. B., & Cowan, W. M. (1981). Observations on the development of the striatum in mice and rats. *Anatomy and Embryology, 163*, 275-298.

Fogel, A., & Thelen, E. (1987). Development of early expressive and communicative action: Reinterpreting the evidence from a dynamic systems perspective. *Developmental Psychology, 23*, 747-761.

Forssberg, H. (1985). Phase dependent step adaptations during human locomotion. In W. J. P. Barnes & M. H. Gladden (Eds.), *Feedback and motor control in invertebrates and vertebrates* (pp. 451-475). London: Croom Helm.

Gerfen, C. R. (1987). The neostriatal mosaic: The reiterated processing unit. In M. Sandler, C. Feuertein, & B. Scatton (Eds.), *Neurotransmitter interactions in the basal ganglia* (pp. 19-29). New York: Raven Press.

Golani, I. (1976). Homeostatic motor processes in mammalian interactions: A choreography of display. In P. P. G. Bateson & P. H. Klopfer (Eds.), *Perspectives in ethology* (Vol. 2, pp. 69-134). New York: Plenum Press.

Golani, I., & Fentress, J. C. (1985). Early ontogeny of face grooming in mice. *Developmental Psychobiology, 18*, 529-544.

Grillner, S. (1985). Neural control of vertebrate locomotion-central mechanisms and reflex interaction with special reference to the cat. In W. J. P. Barnes & M. H. Gladden (Eds.), *Feedback and motor control in invertebrates and vertebrates* (pp. 35-36). London: Croom Helm.

Havkin, G. Z. (1977). *Symmetry shifts in the development of interactive behaviour of two wolf pups (Canis lupus)*. Unpublished master's thesis, Dalhousie University, Halifax, Nova Scotia.

Havkin, G. Z., & Fentress, J. C. (1985). The form of combative strategy in interactions among wolf pups (*Canis lupus*). *Zeitshrift fur Tierpsychologie, 68*, 117-200.

Hinde, R. A. (1982). *Ethology: Its nature and relations with other sciences*. Oxford: Oxford University Press.

Hogan, J. A. (1988). Cause and function in the development of behavior systems. In E. M. Blass (Ed.), *Handbook of behavioral neurobiology: Vol. 9. Developmental psychobiology and behavioral ecology* (pp. 63-106). New York: Plenum Press.

Hoyle, G. (1985). Generation of behaviour: The orchestration hypothesis. In W. J. P. Barnes & M. H. Gladden (Eds.), *Feedback and motor control in invertebrates and vertebrates* (pp. 57-75). London: Croom Helm.

Jacobs, W. J., Blackburn, J. R., Buttrick, M., Harpur, T. J., Kennedy, D., Mana, M. J., MacDonald, M. A., McPherson, L. M., Paul, D., & Pfaus, J. G. (1988). Observations. *Psychobiology, 16*, 3-19.

Keele, S. W., & Ivry, R. I. (1987). Modular analysis of timing in motor skill. In G. H. Bower (Ed.), *The psychology of learning and motivation* (Vol. 21, pp. 183-228). New York: Academic Press.

Kugler, P. N., & Turvey, M. T. (1987). *Information, natural law, and the self-assembly of rhythmic movement*. Hillsdale, NJ: Erlbaum.

Lashley, K. S. (1951). The problem of serial order in behavior. In L. A. Jeffries (Ed.), *Cerebral mechanisms in behavior* (pp. 112-136). New York: Wiley.

Lorenz, K. (1950). The comparative method in studying innate behaviour patterns. *Symposia of the Society for Experimental Biology, 4*, 221-268.

MacLean, P. D. (1978). Effects of lesions of globus pallidus on species-typical display behavior of squirrel monkeys. *Brain Research, 149*, 175-196.

Maurer, R. G., & Damasio, A. R. (1982). Childhood autism from the point of view of behavioral neurology. *Journal of Autism and Developmental Disorders, 12*, 195-205.

McLeod, P. J. (1987). *Aspects of the early social development of timber wolves (Canis lupus).* Unpublished doctoral dissertation, Dalhousie University, Halifax, Nova Scotia.

Medawar, P. B., & Medawar, J. S. (1983). *Aristotle to zoos.* Cambridge, MA: Harvard University Press.

Moran, G. (1978). *The structure of movement in supplanting interactions in the wolf.* Unpublished doctoral dissertation, Dalhousie University, Halifax, Nova Scotia.

Moran, G., Fentress, J. C., & Golani, I. (1981). A description of relational patterns during 'ritualized fighting' in wolves. *Animal Behaviour, 29,* 1146-1165.

Murphy, M. R., MacLean, P. D., & Hamilton, S. C. (1981). Species-typical behavior of hamsters deprived from birth of the neocortex. *Science, 213,* 459-461.

Pellis, S. M., & Pellis, V. C. (1987). Play-fighting differs from serious fighting in both target of attack and tactics of fighting in the laboratory rat *Rattus norvegicus. Aggressive Behavior, 13,* 227-242.

Petitto, L. A. (1988). "Language" in the prelinguistic child. In F. S. Kessel (Ed.), *The development of language and language researchers: Essays in honor of Roger Brown* (pp. 187-221). Hillsdale, NJ: Erlbaum.

Petitto, L. A., & Bellugi, U. (1988). Spatial cognition and brain organization: Clues from the acquisition of a language in space. In J. Stiles-Davis, M. Kritchevsky, & U. Bellugi (Eds.), *Spatial cognition: Brain bases and development* (pp. 299-326). Hillsdale, NJ: Erlbaum.

Posner, M. I., & Presti, D. E. (1987). Selective attention and cognitive control. *Trends in Neurosciences, 10,* 13-17.

Rauschecker, J. P., & Marler, P. (Eds.). (1987). *Imprinting and cortical plasticity: Comparative aspects of sensitive periods.* New York: Wiley.

Rossignol, S., & Drew, T. (1985). Interactions of segmental and suprasegmental inputs with the spinal pattern generator of locomotion. In W. J. P. Barnes & M. H. Gladden (Eds.), *Feedback and motor control in invertebrates and vertebrates* (pp. 355-377). London: Croom Helm.

Szechtman, H., Ornstein, K., Teitelbaum, P., & Golani, I. (1985). The morphogenesis of stereotyped behavior induced by the dopamine receptor agonist apomorphine in the laboratory rat. *Neuroscience, 14*, 783-798.

Thorpe, W. H. (1961). *Bird song*. Cambridge: Cambridge University Press.

Tinbergen, N. (1963). On aims and methods of ethology. *Zeitschrift fur Tierpsychology, 20*, 410-433.

Tooze, Z. (1988). *Some aspects of the structure and function of long-distance vocalizations of timber wolves (Canis lupus)*. Unpublished master's thesis, Dalhousie University, Halifax, Nova Scotia.

Yates, F. E. (Ed.). (1987). *Self-organizing systems: The emergence of order*. New York: Plenum Press.

SECTION 2

DEVELOPMENTAL ISSUES

Perspectives on the Coordination of Movement
S.A. Wallace (Editor)

MASTERING REACHING AND GRASPING: THE DEVELOPMENT OF MANUAL SKILLS IN INFANCY

Claes von HOFSTEN*

Department of Psychology
Umeå University

ABSTRACT

This chapter deals with the development of reaching and grasping during the 1st year of life. As the maturation of the nervous system imposes the most severe constraints on this development, an attempt is first made to summarize what is known about the motor pathways and the sensorimotor systems involved in manual development. The chapter then sketches the presumed major steps in the manual development of the human infant. Visual control of arm movements is already present in newborn infants, but the arm and the hand are synergistically coupled in these movements. As the arm extends, the hand opens and vice versa. This synergy is broken up at about 2 months as infants start to fist the hand vigorously as the arm extends. That pattern gives way to a more functional one in which the hand opens up during arm extension in preparation for grasping an object. Around 4 months of age, infants start to reach and grasp objects successfully. Major changes in the organization of the manual movements accompany this event. Vision takes a more prominent role in the control of hand movements as the infant gains ability to monitor the hand relative to a target. The major change in manual control during the second half of the 1st year of life is the appearance of differentiated finger movements and the

*Address correspondence to: Claes von Hofsten, Department of Psychology, Umeå University, S-90187 Umeå, Sweden.

mastering of the pincer grasp. When successful reaching emerges, the movements are already characterized in several ways by anticipatory control, and the hand aligns itself with the object to be grasped and starts closing before the target is encountered. However, the most impressive element of anticipation is the infant's ability to catch fast moving objects.

FUNDAMENTALS OF THE CONTROL OF REACHING AND GRASPING

The development of manual skills in infancy is both dramatic and intriguing. In a matter of months, the crudely coordinated limb movements of the neonate are transformed into elegant and precise reaching and grasping acts. By the end of the 1st year of life, the infant can pick up most kinds of objects, including very tiny ones, and examine and manipulate them. How is this rapid development achieved? The purpose of this chapter is to examine this problem.

Motor Pathways

At least three distinct motor pathways control the movements of the upper limbs, the pyramidal system projecting from the motor cortex and the lateral and ventromedial subcortical systems named after their termination in the spinal grey matter. Knowledge of these structures has come mainly from elaborate studies of the rhesus monkey by Kuypers and his associates (Kuypers, 1962, 1964, 1973; Lawrence & Kuypers, 1968a, 1968b; Lawrence & Hopkins, 1972, 1976). The manual system of the rhesus monkey is in several respects similar to the human one. It has an opposing thumb and an index finger that can be moved independently of the other fingers.

The pyramidal system projects directly to the motoneurons of the distal extremity muscles. It seems to be responsible for independent finger movements. Lawrence and Kuypers (1968a) showed that the interruption of both pyramidal tracts initially severely affected independent movements of the hand and that a permanent loss of individual finger movements occurred. The monkey could no longer move the index finger without also moving the rest of the fingers and could no longer pick up a pellet from a small depression.

Lawrence and Kuypers (1968b) found it appropriate to distinguish between two subcortical systems, one lateral and one ventromedial, grouped according to their termination in the spinal grey matter. Lawrence and Kuypers studied how interruption of these two systems affected the motor control of the limb in monkeys with prior bilateral interruption of the pyramidal tracts. Lesions of the lateral brainstem system produced an impairment of independent hand movements and an impaired capacity to flex the extended thumb. Initially animals with such lesions could not reach out to pick up pieces of food by closure of the hand. However, movements involving the whole limb and the body, as in walking and climbing, were only minimally affected. After some recovery, these animals regained the capacity to close the hand but only as a part of a total arm movement.

Interruption of the ventromedial pathways produced severe impairment of axial and proximal extremity movements and of the maintenance of body posture (Lawrence & Kuypers, 1968b). When the monkeys could finally sit and walk, they were unsteady. The head and trunk would slump forward, and when approached with food, the animals showed immobility of the head, trunk, and limbs. Despite these impairments, however, they could pick up pieces of food with their hands if the limbs were appropriately supported and brought to the food.

In summary, the ventromedial pathways seem to be especially related to the maintenance of erect posture and the integration of movements of the trunk and limbs. The lateral brainstem pathways superimpose upon this control the capacity for independent use of the extremities, particularly the hands. Finally, the cortical system exerts control over the distal part of the arm, hand, and the individual finger movements.

In the newborn rhesus monkey, these pathways seem rather undifferentiated. There are no connections made yet through the pyramidal tract. In the earliest reaching attempts of the rhesus monkeys studied by Lawrence and Hopkins (1972, 1976), the aim was rather inaccurate and the arm movements unsteady so that the hand frequently missed the food. If the food was grasped at all, the movement was accomplished by closure of all the fingers together, and the subsequent releasing of the food in the mouth when the arm was flexed was reported to be difficult. In other words, the movements of the hand seemed synergistically constrained by the movements of the arm. At around 3 months of age, these difficulties had disappeared. Reaching was smooth and accurate, and there were no longer any difficulties in releasing the grip.

Effective removal of food from a small container with the index finger was observed in monkeys between 3 and 4 months of age, and an adult level of performance was judged to be present at around 8 months. The establishment of relatively independent finger movements was observed to occur in parallel to the emergence of direct cortico-motoneuronal connections. The number of connections increased markedly up to 8 months of age. Complete interruption of both pyramidal tracts at 4 weeks of age did not affect the early development of reaching in the rhesus monkey; however, independent finger movements never appeared. Apparently, the direct cortico-motoneuronal connections are crucial for such movements.

Information Used for the Control of Reaching and Grasping

Three kinds of information are required for the solution of manual tasks: first, information about the object to be manipulated, its position and orientation in space, its size and form, and its substance and texture; second, information about the positions and movements of the arms and hands relative to the subject's own body; and finally, information about the positions and movements of the arms and hands relative to the object to be manipulated. Vision, proprioception, and haptics all contribute to the efficiency of manual action. They seem to collaborate and supplement one another in an optimal way to provide integrated informational support for manual action. Vision gives detailed information about the position of the target and its orientation in space before it is encountered and about the target's form and size. Vision gives equally good information about the positions of the hands in space. Vision is therefore superior in the close guidance of the hands relative to a target. The shortcoming of vision is that information is available only within the visual field, more detailed in the center and less detailed in the periphery. This means that if the positions and movements of a body part are to be monitored in any precise way, vision has to be directed toward that part of the body. In reaching, however, the subject needs to look at the target toward which the action is directed. If the target and the body part used for the action are much separated, the subject will encounter problems in visually monitoring both.

Proprioception provides a means for defining the position and movements of the hands and arms relative to the body. Thus, proprioception makes it possible to detach the acting hand from the visual field while directing it toward a visual target. During the first phase of reaching, vision typically defines the position of the target; and proprioception, the position of the reaching limb.

The mature approach movement is fast and continuous and carries the hand to the vicinity of the target. Therefore, it is often called ballistic. Because visual information about the body does not seem to be needed to update the movement, the approach movement is sometimes described as an open loop, that is, controlled by feedforward (Arbib, 1981). However, this description is of doubtful utility because it is not at all clear whether feedback from the proprioceptors enters into the control of the approach movement.

When the hand comes close to the target, proprioception is not precise enough to ensure a smooth grasp of the target. Therefore, the reach passes into a more visually driven mode. The visible position of the hand relative to the visible position of the target is used to control the final adjustments before grasping. In contrast to the first part of the reach, this part has often been called *guided*. Surely, visual information about the position of the hand guides the movement during this part of the reach. However, even when vision of the reaching hand is prevented, two distinct phases may still be seen (Jeannerod & Biguer, 1982).

REACHING AND GRASPING IN THE NEWBORN INFANT

Birth does not in any way define the starting point of development, but it is our first real opportunity for studying organized limb movements. It is also an important transition point in development, both biomechanically and psychologically. Obviously, the different sensorimotor systems involved in the control of limb movements need to be calibrated and stabilized in their new extra utero environment, which imposes very different biomechanical constraints on the system. Calibration is especially crucial for visually linked sensorimotor systems, which for obvious reasons have not been functioning before birth.

Reaching

Contrary to traditional belief (see, e.g., Piaget, 1953, 1954), neonates have the ability to visually control their arm movements. Hofsten (1982) studied the arm movements of 5-day-old infants placed in semireclining seats that supported each infant's waist and trunk but allowed free movement of the arms. A spherical tuft made of bright red, blue, and yellow yarn was moved slowly and irregularly in front of the neonate along a horizontal, circular path of 140 cm diameter. A slowly moving rather than a stationary object was chosen because of the low resolution of the neonate's visual system, which Dobson and Teller (1978) and Held (1979) estimated has less than 1/20th the acuity of the adult's visual system. Although neonates might therefore fail to notice a stationary object, they are apt to notice a moving target because motion adds powerful information separating the target from the background. All neonates tested in the experiment detected the moving target, were attracted by it, and followed it for shorter or longer periods with their eyes and head. This tracking also made it easier to detect whether or not the infant observed the target.

The movements were recorded with two video cameras, placed at a 90° angle to each other so that the three-dimensional trajectories of the movements could be reconstructed. For calculation of aiming, the movements were subdivided into functional units, each of which consisted of one acceleration and one deceleration phase (see Brooks, Cooke, & Thomas, 1973). The unit that carried the hand closest to the trajectory of the object was examined further. This part of the movement should be aimed at the object if there is visual control but not otherwise.

Each forward extension performed during a 7-min recording was measured in the way just described. In addition, two independent observers scored the infant's looking behavior for each forward extended movement. The experiment showed that when the infants fixated the target, the movement of the hand was off the target by an average of 32°. In comparison, the hand was off the target by an average of 52° when the infant did not fixate it and by 54° when the infant's eyes were closed. Table 7.1 shows the results from the 5 most frequent reachers in the study. The effect of fixation is substantial in each of these subjects.

Table 7.1.

Mean Approach Angles for Different Individuals and Different Looking Behaviors

Subject	Nonfixation and closed eyes	Fixation
G.H.	51.4° (13)	21.9° (14)
G.K.	52.4° (16)	32.8° (14)
F.L.	62.0° (5)	28.4° (10)
F.S.	50.0° (13)	21.9° (8)
F.N.	53.5° (12)	28.8° (7)

Note. The number of reaches on which each mean is based is shown in parentheses. From "Eye-Hand Coordination in Newborns" by C. von Hofsten, 1982, *Developmental Psychology, 18*, p. 457. Copyright 1982 by the American Psychological Association. Adapted by permission.

The neonate's ability to visually control the movements of the arms also demonstrates that visual space and proprioceptive space are connected at this age. As the arm is not placed in a stereotyped position before the initiation of the movement, both the position of the target and the position of the arm need to be defined for the production of an aimed movement. The infant fixates the target, and the starting position of the arm therefore needs to be defined by some other means, that is, proprioceptively.

The neonate can also direct arm movements toward the mouth, that is, reach toward a proprioceptively defined target (Butterworth, 1986; Rochat, Blass, & Hoffmeyer 1987). Butterworth (1986) reported that the mouth was significantly more likely to be open throughout the arm movement when the hand went directly to the mouth than when the hand first contacted other parts of the face. Butterworth also found that the hand could be guided to the mouth after it had contacted other parts of the face. He found "no evidence of rooting after contact;" the head was held still and the hand moved "immediately in the direction of the mouth" (p. 28).

Although Hofsten (1982) showed that vision had a substantial effect on the clustering of the neonates' forward directed movements relative to a target, the visually directed movements were still not very precise. On the average, they were off the target by 32°. Everyday observations support this finding. The reaching movements of the newborn infant are generally not strikingly well coordinated. Sometimes, when all conditions are optimal, the movements produced appear amazingly mature, but in most instances they are not.

There are two reasons for this lack of coordination. A specific coordination may require a certain amount of postural control that the neonate does not have, and coordination will therefore be demonstrable only if the neonate is supported in certain ways. Neonate walking is a good example. To be able to walk freely, the infant needs to master the balancing of upright posture and integrate balance with the production of walking movements. Another example comes from studies by Fentress (1984) of neonatal mice. When neonatal mice were supported in an upright position, they produced rich but poorly coordinated grooming-like movements from the 1st day after birth.

A second important reason that a sensorimotor system may not function appropriately at birth is that the neurostructures involved are not sufficiently differentiated. In the neonate, the motor mechanisms controlling the arm are in certain ways coupled to the motor mechanisms controlling the neck. The asymmetric tonic neck response, which can be most clearly seen during the 2nd and 3rd months postnatally, is a classical example of this coupling (Touwen, 1976). Grenier (1980, 1981) suggested that the neck impotence of the neonate would also hamper the other part of the synergy, the movements of the arm. Grenier held the neonate's neck in position for a certain time and found much more coordinated arm movement than when the infant's neck was not supported.

The motor system of the upper limb is undifferentiated in yet another way that makes coordinated reaching and grasping difficult. An appropriately coordinated reach and grasp require the arm to extend and the hand to flex around the object without any interruption. In the neonate, the arm and hand instead extend and flex in a synergy (Hofsten,1982). The synergistic properties of neonatal reaching do not seem to be influenced by vision. In 70% of the cases (Hofsten, 1984), the reaching hand opened before or during the extension of the arm, regardless of whether the infant looked at the target. In the extended phase of the arm, the subjects never flexed the hand to grasp the target, not even when the hand ended up with the object on its palm.

Another expression of the synergistic coupling between the motor system of the arm and hand is the so-called "traction response" (Twitchell, 1965). The neonate's hand is pulled by the wrist stretching shoulder adductors and flexors. The response is a synergistic flexing of all joints, including the joints of the hand.

The synergy between the hand and arm does not imply absence of independent hand movements. On the contrary, the neonate makes many hand movements in the absence of arm movements. Thus, it seems to be the movement of the arm that locks the hand into synergistic movement patterns, not vice versa. One of the most interesting aspects of neonatal hand movements is that the movements are in no way confined to the whole hand. Hofsten and Rőnnqvist (in preparation) recorded finger movements on video during a 5-min period of alertness in 20 neonates, placed either in front of a slowly moving object or in front of their mothers. Altogether, 2,530 movements were scored, or one movement approximately every 2.5 s. As many as 550 of these movements (21.7%) involved only the thumb and the index finger. This is remarkable because the use of these movements for manipulation requires cortico-motoneuronal connections that are absent in the neonate. Differentiated finger movements in the neonate therefore seem to be the expression of spinal motor loops constituting a manual vocabulary for fine motor skill. The cortico-motoneuronal connections do not act directly on the muscles of the limb but indirectly through these spinal loops (Kandel & Schwartz, 1985).

DEVELOPMENT DURING THE PREREACHING PERIOD

Hofsten (1984) studied infants longitudinally during the prereaching period from 1 to 19 weeks of age. The infants were seen every 3rd week. The recording arrangement was identical to the one used in the neonatal studies (Hofsten, 1982). For each infant and session, every forward extended arm movement was scored from the videotape except those originating from Moro responses or startles and those associated with yawning, sneezing, and sudden forward head movements. Each forward extension was scored with respect to the infant's looking behavior and the movement of the hand. Hofsten found a dramatic change in the reaching pattern around 2 months of age. The extension synergy between arm and hand movements broke up. Instead of opening the hand as the arm extended forward, infants at this age clenched the hand into a fist in the extended position. The occurrence of this latter behavior increased from less than 10% in the neonate to around 70% in the 2-month-olds (see Figure 7.1). Visual fixation of the target did not affect this tendency. Apparently, the hand is gaining independent status at this age.

Another indication of increasing independence of the hand around 2 months of age is the grasp reflex observed by Twitchell (1965). According to Twitchell, the tactually elicited grasp reflex is not fully developed in the newborn. To elicit grasping in the newborn, pressure needs to be applied to the palm or fingers to stimulate the proprioceptors, and the grasping is accompanied by a synergistic flexion of the arm. By 4 to 8 weeks of age, a stimulus applied between the thumb and the index finger will produce an adduction of the fingers (later also flexion), and this movement is followed by a synergistic flexion of all the digits of the limb (facilitation of the traction response). Two to 4 weeks later, contact stimulus applied to the palm will result in flexion of all fingers but not flexion of the arm.

Differentiation between the motor systems of the arm and the hand (reflected in the ventro-medial and lateral spinal pathways) is necessary for development of functional reaching and grasping patterns. At around 3 months of age, the infants in my study (Hofsten, 1984) started to open the hand again when extending the arm but this time only when fixating the target. At the same time, the number of reaching attempts increased greatly.

THE EMERGENCE OF SUCCESSFUL REACHING AND GRASPING

Around the age of 4 months, infants start to be able to reach out for an object and eventually grasp it. This major accomplishment is the result of several parallel developments in the child. The differentiation between the motor systems of the arm and the hand is only one of the determinants.

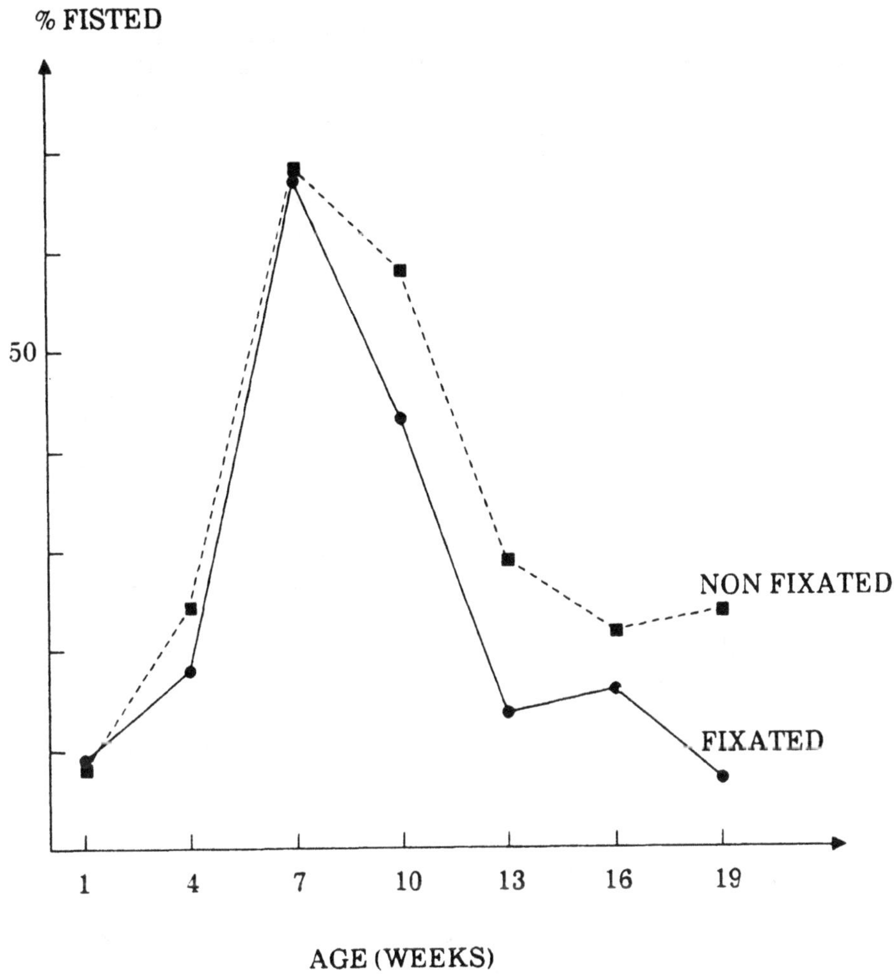

Figure 7.1. Percentage of fixated movements and nonfixated movements at which the hand is fisted for different prereaching age levels. *Note.* From "Developmental Changes in the Organization of Prereaching Movements" by C. von Hofsten, 1984, *Developmental Psychology, 20,* p. 385. Copyright 1984 by the American Psychological Association. Reprinted by permission.

Parallel to this development, better means of obtaining precise visual information about spatial relations in reaching space emerge. The sensitivity to binocular disparity develops very rapidly between 3 and 5 months of age. Fox, Aslin, Shea, and Dumais (1980) found that 3-1/2-month-olds but not 2-1/2-month-olds would track a moving virtual object specified by binocular disparity in a dynamic random-dot stereogram. Held and colleagues (Birch, Gwiazda, & Held, 1982; Held, Birch, & Gwiazda, 1980), using a modified preferential looking technique, found the same developmental trend. They observed a rapid rise in detection of fine disparities from 3 months of age to an adult-like level between 5 and 7 months of age.

Two other factors that contribute to the development of successful reaching deserve mention. One is the uncoupling of head and arm movements. This coupling seems to be strongest around 2 months of age when the infant has gained control over the neck muscles. The asymmetric tonic neck reflex is easily demonstrated at this age (Touwen, 1976). However, in the next 2 months, arm movements become increasingly independent of neck movements, and the asymmetric tonic neck reflex becomes more and more difficult to elicit. This uncoupling allows for more flexible integration between eye-head movements and manual coordination. The other factor that contributes to the emergence of successful reaching is the appearance of postural stability of the upper trunk at around 4 months of age (Gallahue,1982). Postural stability enables the infant to sit with support and is an appropriate base for the construction of reaching movements.

Differentiation of the sensorimotor systems for approaching and grasping, sensitivity for binocular disparity, decoupling of arm and neck movements, and postural stabilization of the upper trunk all appear around the same age. They are essential for the emergence of successful reaching. Therefore, the emergence of successful reaching can be described in a sense as an emergent property of several converging developments. Thelen (1985) has discussed the development of coordinated leg movements from this perspective. She proposed the existence of an early coordinative structure for leg movements that leads to highly patterned output. The elaboration and differentiation this coordinative structure would undergo during development would allow for more flexible movements. At any point in development, however, movements are not specified by this pattern generator alone but by the systems outcome of a number of interacting components, each with its own developmental course and each acting within definite constraints and opportunities defined by the context. In the case of locomotion, Thelen enumerated seven components apart from pattern generation itself: tonus control, articulator differentiation, extensor strength, postural control, visual flow sensitivity, body constraints, and motivation.

THE REFINEMENT OF APPROACHING

The mature approach, as mentioned earlier, consists of two distinct phases. The first is a transport phase carrying the hand to the vicinity of the target. After about 70-80% of the total duration of the approach, a discontinuity in the velocity profile marks the beginning of the second phase. During this phase, the hand assimilates the target. The approach is not divided in this way just to allow visual correction for undershooting the target: Jeannerod (1984) showed that the same phasing occurs in the absence of visual feedback. Rather, the phasing seems to be a basic organizational property of the approach, allowing visual feedback, when present, to be incorporated within the preexisting structure of the movement (Jeannerod, 1986).

Visual guidance of the later part of the approach begins to be a prominent feature of infant reaching toward the 4th month of life. I have found that at this age, when the target was stationary or moved slowly, a common behavior was to leave the limb extended at the target for a considerable time while minor adjustments were made. This behavior was previously described by Piaget (1953) and White, Castle, and Held (1964). White et al. (1964) wrote: "Occasionally, one hand will be raised, looked at, and brought slowly to the stimulus while the glance shifts from hand to object repeatedly (p. 356)." These findings show that shortly before 4 months of age, the infant starts to be able to use the purely visual mode of control whereby the visible position of the hand is related to the visible position of the object. McDonnell (1975) has shown that a 4-month-old infant will correct a reach for a target seen through horizontally displacing prisms.

In terms of the number of phases or steps, the approach of the target in early reaching looks rather different from that of the adult. When the infant first starts to reach for targets successfully, the movement often consists of several steps or phases, and the approach is awkward and crooked (Hofsten, 1979). However, this awkwardness changes very quickly with age. The movement path straightens up, and the number of units decreases dramatically during the first cou-

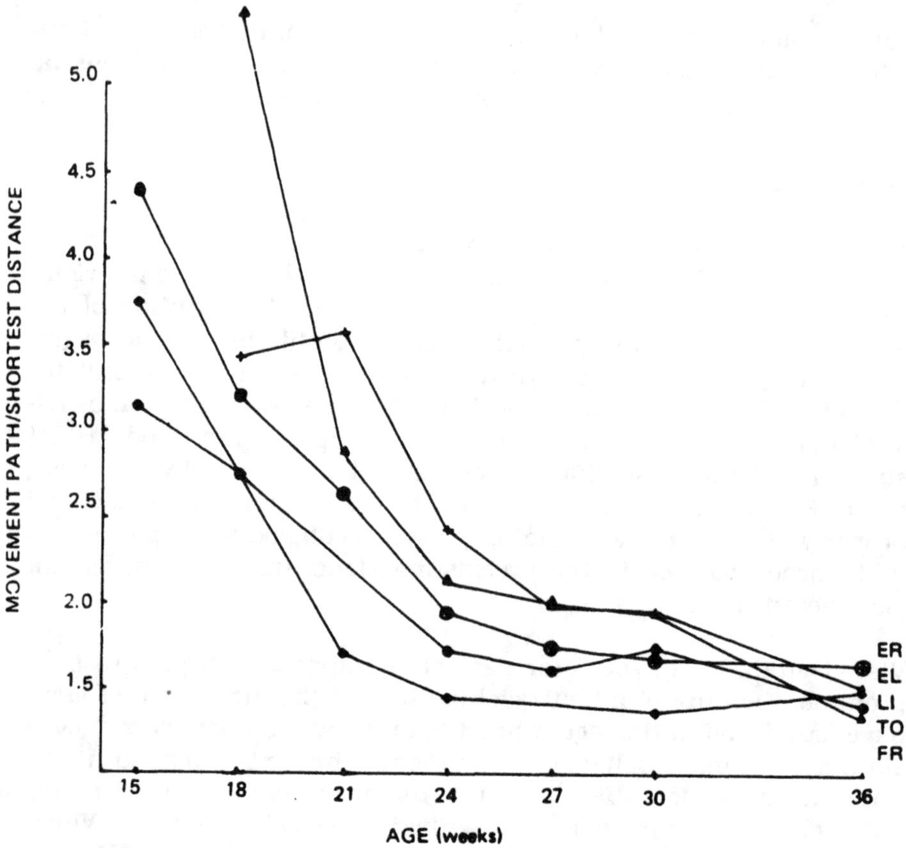

Figure 7.2. Relative length of movement path as a function of age for 5 subjects. *Note.* From "Development of Visually Guided Reaching: The Approach Phase" by C. von Hofsten, 1979, *Journal of Human Movement Studies*, *5*, p. 166. Copyright 1979 by Teviot-Kimpton Publications. Reprinted by permission.

ple of months of successful reaching. Figure 7.2 shows, for 5 subjects individually, how the approach path straightened up between 15 and 24 weeks of age. The figure shows that individual differences in performance were stable during the period studied but that the developmental function was similar for all 5 infants.

Hofsten (1979) divided movements according to their velocity profile into movement steps, each step consisting of one acceleration and one deceleration phase. Early reaches consist of several steps relatively similar in duration. With age, the first step grows in importance. More and more of the approach and power of the reach become concentrated in this step. Subsequent steps become increasingly subordinate to the first one and have less to do with the approach and more with increasing the precision of the reach.

As age increases, the number of units in the approach decreases. At around 6 months of age, the adult-like reaching pattern starts to dominate. From that age on, most approaches consists of, at most, two steps: presumably, one visual-proprioceptive "transport" step and one visually guided "correction" step.

That visual guidance becomes more prominent as the approach differentiates and refines with age is further supported by Lasky (1977). Lasky had infants reach for an object seen through a horizontally placed mirror. At the proper position underneath the mirror, an object identical to the reflected one was placed. In the control condition, a panel of clear plastic replaced the mirror. Lasky found that being able to see the reaching hand had very little effect on the performance of the 4-1/2-month-old infants. They retrieved the object in 23% of the attempts performed in the mirror condition and in 28% of the attempts performed in the control condition. The rate of retrieval in the 5-1/2- and 6-1/2-month-olds did not improve in the mirror condition, but in the clear-plastic condition, the improvement was substantial. In this condition, the 5-1/2-month-olds retrieved the object in 48% of the reaches, and the 6-1/2-month-olds, in 68% of the reaches. The rate of contact with the target in the mirror condition actually decreased in the two older age groups from 0.39 to 0.12 and 0.11/s respectively.

When infants grow still older, they seem to become less dependent on seeing the reaching hand. Reaching has become more precise and more automatized. Older infants may even look away while reaching and do not seem to be as bothered as the 6-month-olds are when sight of the reaching hand is disrupted (Bushnell, 1985). However, in reaching for small targets, when control of individual finger movements is essential, visual guidance is always prominent in controlling the movements. The fine pincer grasp appearing around 9 months of age presupposes delicate visual guidance during the approach as well.

THE REFINEMENT OF GRASPING

When infants first start to encounter objects successfully, they do not yet grasp them very well. Contact is often made with the back of the hand, and grasping, if any, is slow and awkward. Hofsten and Lindhagen (1979) had infants reach for moving objects and found that at 15 weeks of age (3-1/2 months), although the target was frequently contacted, it always slipped out of the hand and was lost during or before the infant's attempts to grasp it (see Figure 7.3). However, development is fast after this age. At 18 weeks of age, infants grasped the target in a majority of reaches.

One of the most important aspects of grasping skill is the degree to which the grasp is prepared for. Such anticipatory adjustments are visually controlled and are of two kinds. First, there are spatial adjustments of the reaching hand to the orientation, form, and size of the object. Second, the securing of the target is timed in such a way that the hand starts to close around the target in anticipation of and not as a reaction to encountering the object.

Gearing the Grasp to Object Properties

Young infants adjust the orientation of the hand to the orientation of the object before it is encountered (Lockman, Ashmead, & Bushnell, 1984; Hofsten & Fazel-Zandy, 1984; Morrongiello & Rocca, 1986). Such adjustments are of great advantage to the child. They need to be performed for the hand to close around the object in an adequate way. Hofsten and Fazel-Zandy (1984) and Morrongiello and Rocca (1986) found that 5-month-old infants already made preparatory adjustments of hand orientation. At this age, the preparation is rather crude and occurs mainly in the early phase of the reach or before the start of the approach. After 6 months of age, these adjustments improve greatly and are often made during the whole approach. However, strategies vary among individuals, and some 5-month-olds also showed a tendency to adjust the orientation of the hand during the approach.

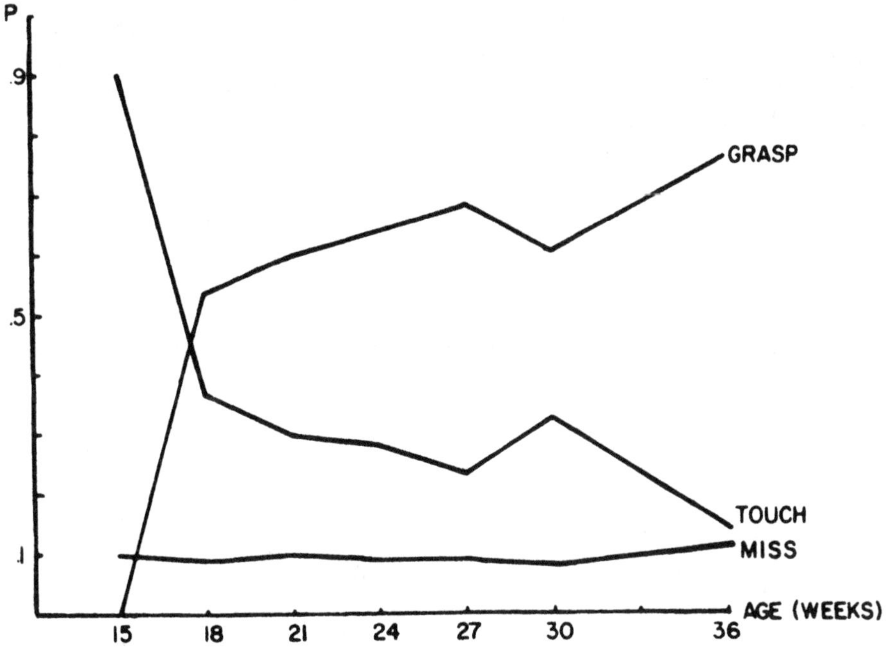

Figure 7.3. Type of contact with moving and stationary objects. Proportion of total number of reaches resulting in grasping the object, touching the object, and missing the object at different ages. *Note.* From "Observations on the Development of Reaching for Moving Objects" by C. von Hofsten and K. Lindhagen, 1979, *Journal of Experimental Child Psychology, 28*, p. 167. Copyright 1979 by Academic Press, Inc. Reprinted by permission.

Adults also alter the opening of the hand as a function of the size of the object they are reaching for (Jeannerod, 1981). However, preparatory adjustments to object size are less crucial than preparatory adjustments to object orientation. Instead of adjusting the opening of the hand to the size of the object, it is also possible to always open the hand fully during the approach. The disadvantage of this strategy is that it takes more time to close a fully opened hand than to close a partly opened hand. The advantage of opening the hand more fully is that it implies decreased demands for end-point accuracy. Therefore, adults use this strategy when they have to make a fast reach (Wing, Turton, & Fraser, 1986).

Hofsten and Rönnqvist (1988) monitored the distance between thumb and index finger with an optoelectronic technique (Selspot) in reaches performed by 5- to 6-month-olds, 9-month-olds, and 13-month-olds. Spherical targets 15, 25, and 35 mm in diameter were used. The infants in the two older age groups adjusted the opening of the hand to the size of the target reached for, but infants in the youngest age group did not. This finding is displayed in Figure 7.4, which shows for the three age groups the maximum opening of the hand during the approach and the opening at touch as a function of target size.

Why was the opening of the hand not systematically related to target size at 5-6 months? One possibility is that the infants were not yet able to perceive the differences in target size (15, 25, and 35 mm) at the distances at which they were presented, perhaps because of immature binocular perception. Granrud (1986) found that infants insensitive to disparity information did not discriminate between two different-sized objects. The timing results argue against this explanation, however, and suggest a rather fine ability to perceive distances in space at the youngest age studied. However, even if the 5-month-old infant perceives relative size correctly, that does not mean that an infant of that age can perceive absolute size as well: Size perception may not be in a form suitable for the control of manual action. Another and more plausible interpretation of the result is that at 5-6 months of age, infants do not use predominantly the thumb and index finger in grasping objects (see, e.g., Halverson, 1931). Instead, they use the medial part of the hand and the palm. It may be that monitoring the distance between the thumb and index finger does not reflect the most central properties of the spatial adjustments of grasping at this age.

Although older infants adjusted the opening of the hand to the size of the target, their adjustments were less differentiated than those of adults, and the hand was more fully opened during the approach for all the different target sizes used. There may be at least two reasons for this behavior. It may reflect incomplete differentiation between the motor systems of the arm and the hand. As the arm extends, extension of the digits may be facilitated, too. From a neurological perspective, Twitchell (1970) observed abduction or dorsiflexion of the fingers during the approach and discussed it in terms of "approach and avoidance" responses constraining the movements of the infant's upper limbs. Another interpretation of this behavior is

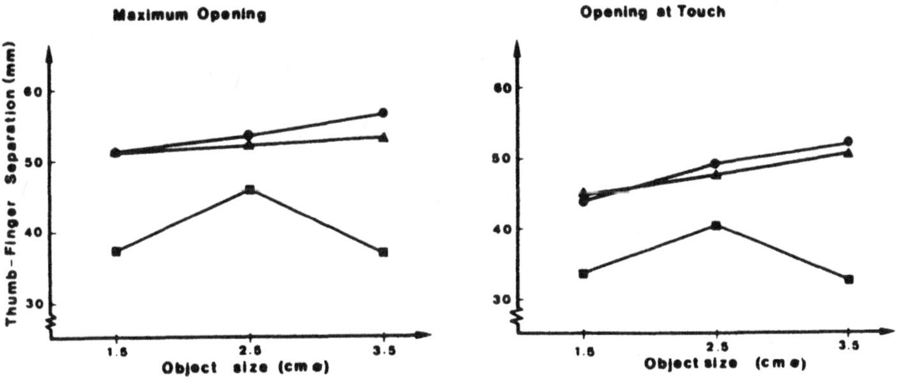

Figure 7.4. Mean maximum opening of the hand during the approach and opening at touch as a function of object size and age. *Note.* From "Preparation for Grasping an Object: A Developmental Study" by C. von Hofsten and L. Rönnqvist, 1988, *Journal of Experimental Psychology: Human Perception and Performance, 14,* p. 617. Copyright 1988 by the American Psychological Association. Reprinted by permission.

that it is adaptive. A fully opened hand maximizes the probability of capturing the object if the movement is spatially less precise. Both interpretations could be valid, of course.

The relation of grasping strategy to object size is also reflected in the way the various digits of the hand are used. In mature grasping, the whole hand is used to grasp large objects, and only the thumb and index finger are used to grasp tiny ones. Several factors indicate that this ability to allocate an optimal number of digits in grasping an object of a specific size is not yet developed in early reaching. As mentioned previously, the medial part of the hand then plays a more prominent role, and the whole hand is typically used in grasping any object (Halverson, 1931). During the second 6 months of life, the focus of the grasp moves toward the radial part of the hand, and the infant starts to master the pincer grasp in picking up tiny objects. Parallel to this development is the establishment of the cortico-motorneuronal connections, which allow the child to control independent finger movements (see, e.g., Kuypers, 1973).

In a recent study, Hofsten and Siddiqui (in preparation) monitored the fingers used by 5-, 7-, and 9-month-old infants in grasping objects of various sizes: 0.5, 1.0, 3.5, 7.0, and 14.0 cm in diameter. Among other things, Hofsten and Siddiqui found that the number of grasps involving only the two or three most radial digits (thumb, index finger, and long finger) increased greatly over this age span. At 9 months of age, this kind of grasp was 9 times more frequent than at 5 months of age. However, at each age level, when only the two or three most radial digits were used, the reaches were typically directed at the two smallest objects. Finally, the number of reaches for the two smallest objects more than doubled during the period studied, whereas the number of reaches for the larger objects increased by only a third. This result indicates that the problem for the infant is neither a perceptual one nor a problem of knowing when to use different kinds of grasps. Rather, it is a problem of the availability to the infant of different kinds of grasps. In older infants, the cortico-motorneuronal connections are better established than in younger infants, and the reaching patterns are more differentiated. Grasping patterns are therefore more efficient, and infants show a greater willingness to grasp the tiny objects that were difficult to grasp earlier.

The Timing of the Grasp

A smooth reaching action requires that the grasp be adequately timed relative to the encounter with the object. This is especially true if the object to be grasped is moving. If the hand closes too late, the object will just bounce on the palm and be lost. If the hand closes too early, the object will hit the knuckles. Alderson, Sully, and Sully (1974) found that the precision of timing in grasping a lightly thrown ball had to be around 14 ms. As most adults master this task with ease, it can be concluded that precise timing is a prominent feature of adult grasping.

Precise timing of the grasp requires the grasp to be planned for and initiated in anticipation of the encounter with the target. Such planning can occur only under visual control. Tactual control of the grasping act implies that the grasp is initiated after the target has been touched, too late for securing the target if it is moving. Because tactually controlled grasping is initiated after contact, it necessarily interrupts the reach-and-grasp act. Thus, it is obvious that the emergence of visually controlled, well-timed grasping is crucial for the development of manual skill. It is well known that grasping can be tactually controlled by the prereaching infant (Twitchell, 1970), but little is known about the emergence of effective visual control. When during development does the child start to use visual information for controlling grasping in an anticipatory way?

Hofsten and Rönnqvist (1988) studied the timing of the grasp in infant reaching. They monitored the distance between thumb and index finger during approach to the target and determined precisely when this distance started to diminish. They also monitored the position of the target. When the hand encountered the target, it was immediately displaced, and the time of this event could be precisely determined as well. An example is shown in Figure 7.5. At all three age levels (5-6 months, 9 months, and 13 months), the closing of the hand was well timed in relation to the encounter with the object. This finding is shown in Figure 7.6, which depicts the distribution of differences for individual reaches between the time when the hand started to close, and touch. In the two younger age groups, the distribution of timing centered around touch. For the 5- to 6-month-olds, 75% of the grasps started within ±0.1 s of the encounter with the target. For the 13-month-olds, the distribution of timing did not center around touch. At this age, the grasping action typically started well before touch. In a study of adult grasping, Hofsten and Rönnqvist (1988) found that subjects only partly adjusted the maximum opening of the hand to the size of the object. Instead, they opened the hand much more than needed and adjusted the timing of the grasp to the size of the target. In other words, they began to close the hand earlier for the smaller target than for the larger target. At no age were there any signs of this kind of adjustment in infant subjects. This refinement of grasping is yet to be developed after 13 months of age.

The capacity for planning and initiating the grasp in anticipation of the encounter with the object is a necessary but not sufficient requirement for the emergence of integrated reach-and-grasp action. Even if both the approach and the grasp are well controlled, there may very well be a discontinuity between these two subactions. Jeannerod and Biguer (1982) have shown that approach and grasp are well integrated and synchronized in the adult subject. The beginning of the second, visually guided part of the approach also marks the beginning of the closing of the hand.

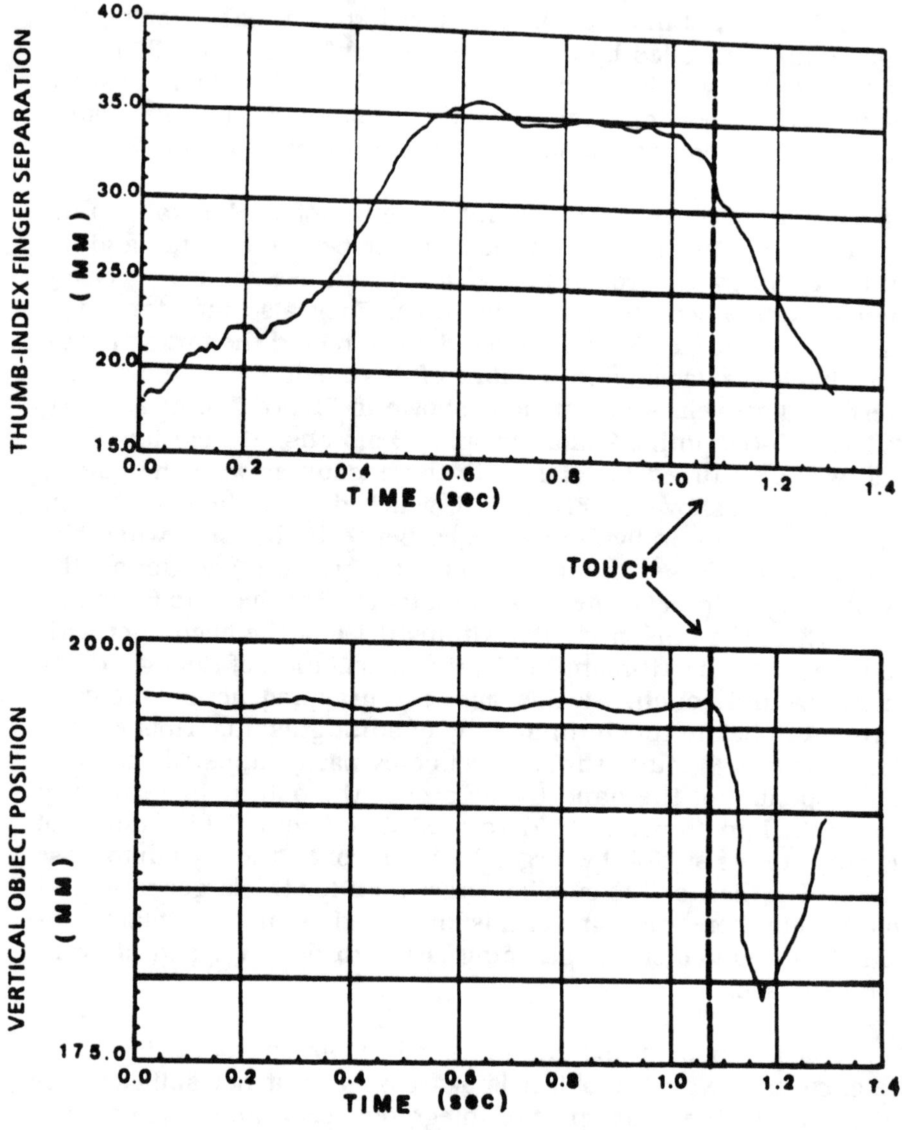

Figure 7.5. A reach performed by a 5-month-old infant. The target size was 15 mm in diameter. The upper graph shows the distance between the thumb and index finger as a function of time. The lower graph shows the position of the target as a function of time and its displacement when contacted. Note that the hand starts to close before contact.

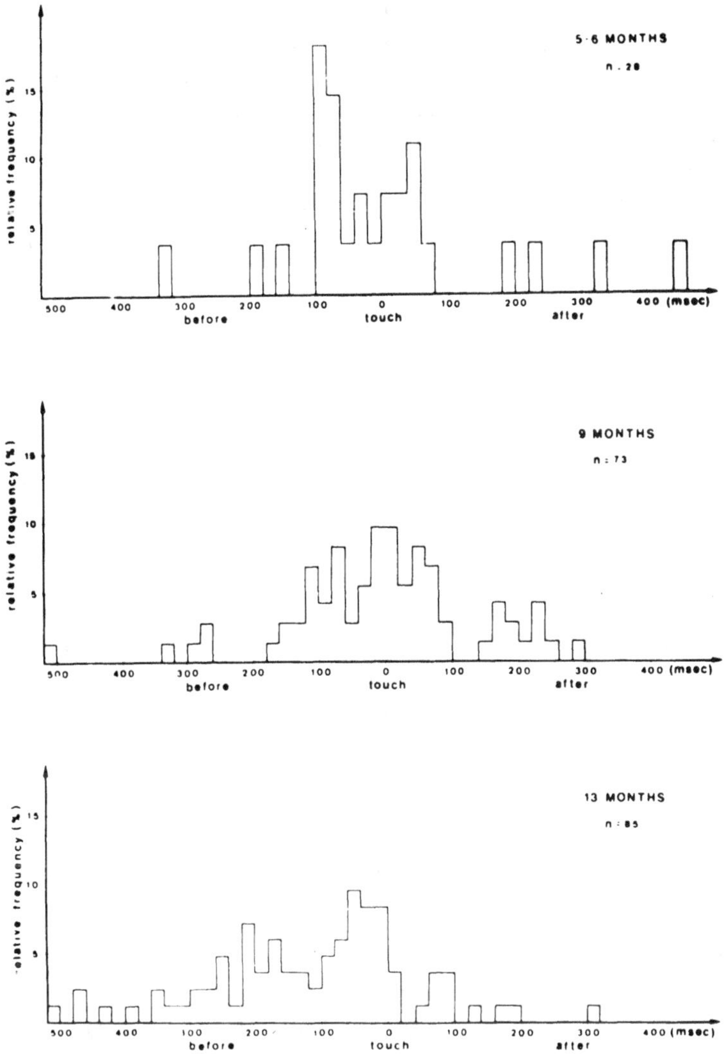

Figure 7.6. Relative frequency distributions of time intervals between starting to close the hand and touch for individual reaches performed by the 5- to 6-month-old, 9-month-old, and 13-month-old infants in Experiment 2. *Note.* From "Preparation for Grasping an Object: A Developmental Study" by C. von Hofsten and L. Rönnqvist, 1988, *Journal of Experimental Psychology: Human Perception and Performance, 14,* p. 618. Copyright 1988 by the American Psychological Association. Reprinted by permission.

How well integrated are approach and grasp when infants start to reach for and grasp objects successfully at 4-5 months of age? Is the target first approached and then grasped, or is the grasping action initiated during the approach, as in the adult? Hofsten and Rönnqvist (1988) found that their two younger age groups (5-6 and 9 months of age) most often started closing the hand in the close vicinity of the target. However, the 13-month-olds were more inclined to start closing the hand earlier in the approach, as shown in Figure 7.7.

CATCHING

The nicest and most striking example of anticipation in infants' manual action is definitely the ability to catch fast moving objects. To catch a ball, for example, the subject needs to perceive not only the instantaneous position of the ball but also its direction and velocity. The reach should not be directed to the point where the object is seen when the reach is initiated, or the hand will end up behind the object. A successful reach has to be aimed at some point ahead of the object where the hand and the object will meet; and as the hand gets there, it should close around the object at the right time—otherwise, the object will be lost. Obviously, timing has to be extremely precise.

Developmental psychologists who have been thinking about the ontogenesis of ball catching have mostly been struck by the complexity of the task. Kay (1970), for instance, suggested that catching ability would appear at the earliest around 5 years of age.

In a series of studies, I have found that young infants already possess a remarkable capacity to catch objects (Hofsten, 1980, 1983; Hofsten & Lindhagen, 1979). Hofsten and Lindhagen (1979) studied this problem longitudinally in a group of 11 infants, who were 12-24 weeks old at the first session. They were seen at 3-week intervals until they were 30 weeks old, and were seen for the last time at 36 weeks of age. Each infant was presented with an object moving at the height of the infant's nose in a horizontal, circular path 115 cm in diameter. The object passed the infant at a nearest distance of either 11 or 16 cm. It moved at 3.4, 15, or 30 cm/s and stopped moving when it was grasped. For each condition, the object was placed randomly to one side and was then moved back and forth from one side to the other until the infant grasped it. This procedure was repeated until

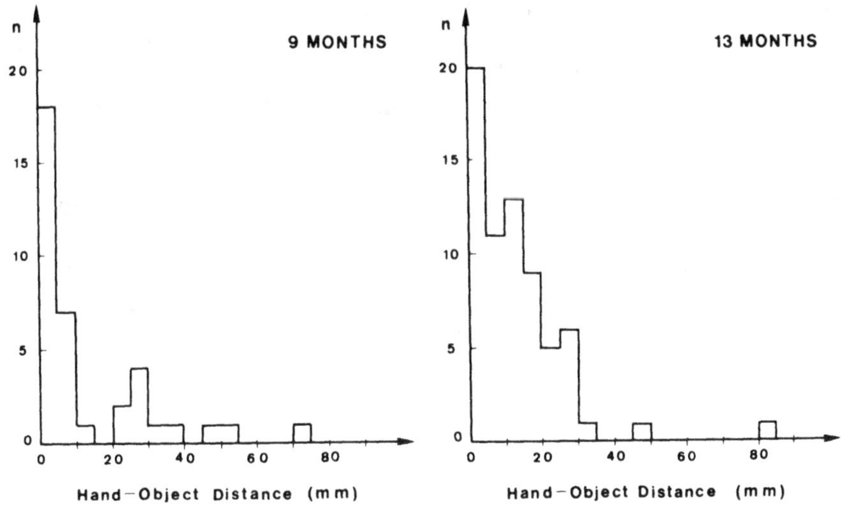

Figure 7.7. Distributions of distances between hand and target at the time when the hand started to close for 9- and 13-month-old infants. *Note.* From "Preparation for Grasping an Object: A Developmental Study" by C. von Hofsten and L. Rönnqvist, 1988, *Journal of Experimental Psychology: Human Perception and Performance, 14,* p. 620. Copyright 1988 by the American Psychological Association. Reprinted by permission.

three reaches were secured or the object had passed in front of the infant at least six times.

Hofsten and Lindhagen (1979) found that from the very age when infants start to master reaching for stationary objects, they also reach successfully for fast moving ones. Eighteen-week-old infants caught the object as it moved at 30 cm/s. To be able to catch an object moving at this velocity, the infant must have at least some predictive ability. As the length of the infant's arm at that age is less than 20 cm, the infant needs to start reaching for the target before it is actually within reach.

To evaluate the predictive skill reflected in these reaches, I performed a quantitative analysis of the three-dimensional trajectories (Hofsten, 1980). The movements were divided into units, each consisting of one acceleration and one deceleration phase. The aiming of each unit relative to the meeting point was then calculated. The analysis showed that for all age groups, most reaches were aimed at the meeting point right from the beginning; that is, a predictive strategy was employed. The predictive reaching was typically performed with the hand contralateral to the direction from which the object arrived.

Over the period of the study, it was mainly the mobility of reaching that improved. I also found a paradoxical dependency between motor performance and task demands. Reaching trajectories straightened out, and approach time decreased with increased target velocity. In other words, the subjects performed better when they needed to. Wade (1980) found a similar relationship in older children whose task was to strike a target moving from right to left by rolling an object down a track way. Especially the younger subjects (7-9 years old) showed much better performance with faster than with slower targets.

Hofsten and Lindhagen's (1979) longitudinal study of infants' catching behavior concerned aiming but not timing and left open many questions about the principles used by the infant to catch fast moving objects. To answer some of these questions, I performed a second study that took both aiming and timing into account (Hofsten,1983).

Fifteen healthy, full-term infants between 34 and 36 weeks of age took part in the experiment. As before, the target moved in a horizontal, circular path of approximately 153 cm in diameter. Velocity and starting position were systematically varied. Velocity was either 30, 45, or 60 cm/s. A subgroup was also tested with 90 and 120 cm/s targets. The results were clear and impressive. The infants caught the presented targets in all conditions, even the fastest ones. Figure 7.8 shows an example of an infant catching an object moving at 120 cm/s.

Aiming was calculated as before. The "best" angle ahead was expressed as the angle between the position of the object at the start of the movement unit (*A*), the position of the hand at the same time (*B*), and the position of the object at the end of the reach (*C*), as shown in Figure 7.9. The obtained angle ahead was expressed as the angle between *A*, *B*, and the end of the reach (*C*), as shown in Figure 7.9. The obtained angle ahead was expressed as the angle between *A*, *B*, and the position of the hand at the end of the movement unit (*D*). As an estimate of timing, the time at which the reach ended was compared

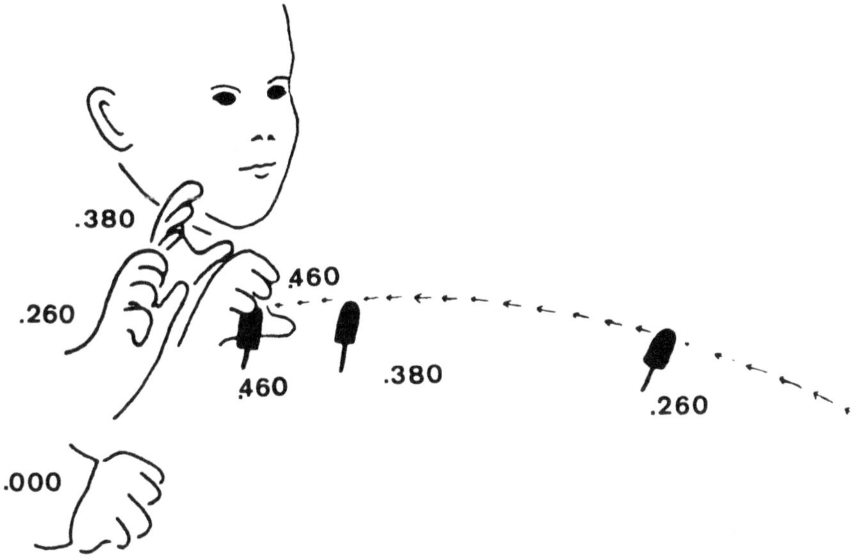

Figure 7.8. An infant catching an object moving at 120 cm/s in front of her. The duration of the reach was 460 ms. The object was encountered at 430 ms after its initiation. The figure shows the position of the hand and the object at different time intervals from the start of the reach. Note that during the movement, the hand first opened and then closed just in time to grasp the object. *Note.* From "Catching Skills in Infancy" by C. von Hofsten, 1983, *Journal of Experimental Psychology: Human Perception and Performance, 9*, p. 82. Copyright 1983 by the American Psychological Association. Reprinted by permission.

to the time at which the hand came closest to the target. The end of the reach was thereby defined as the time when the hand came to a standstill or when its deceleration had stopped.

The initial aiming of reaches in the different velocity conditions of the experiment is shown in Table 7.2. Although the required angle ahead increases as the velocity of the target increases, so does the obtained angle ahead. The reaches were at all instances directed close to the meeting point with the target. Table 7.2 also shows the systematic and variable timing errors. At no velocity was the systematic timing error greater than 17 ms. The variable timing error was found to be between 54 and 59 ms.

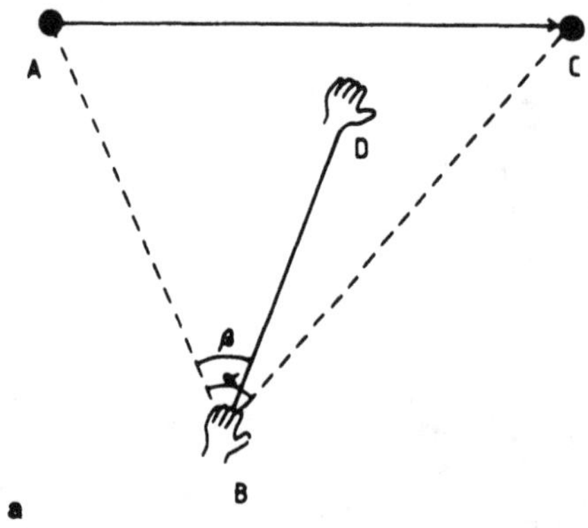

Figure 7.9. Calculation of the aiming of a movement unit. *A* is the position of the object at the beginning of the step; *B* is the hand at the same time; *C* is the position of the object at the meeting point; and *D* the position of the hand at the end of the step. α is the "best" direction ahead; β is the obtained direction ahead. *Note.* From "Catching Skills in Infancy" by C. von Hofsten, 1983, *Journal of Experimental Psychology: Human Perception and Performance, 9*, p. 78. Copyright 1983 by the American Psychological Association. Reprinted by permission.

MASTERING REACHING AND GRASPING

The reviewed research shows that the sensorimotor systems underlying manual action function early in development. Indeed, the newborn child is already able to direct arm movements both visually and proprioceptively. The movements may show various signs of immaturity, but the coordinative structures are there. By the end of the 1st year of life, reaching and grasping are more or less adult-like. The approach is straight and consists of one visual-proprioceptive transport phase and one visually guided correction phase, as it does in the adult. The grasp is initiated during the visually guided phase and is carried out smoothly and precisely. The infant is able to use the hand and its fingers in a differentiated way, and the grasp is well geared to the properties of the target. Small targets are grasped with the thumb and index finger in a pincer grasp, and large targets are grasped with the whole hand or with both hands.

Table 7.2.

Initial Aiming and Timing of Reaches for Objects Traveling at 30, 45, and 60 cm/s

		Initial aiming (°)		Timing (ms)	
Velocity (cm/s)	n	α	β	*M*	*SD*
30	47	30.2	33.9	9.4	54
60	46	39.4	43.4	4.4	57
90	45	48.3	48.9	-17.0	59

Note: α is the obtained angle ahead and β the deviation of the obtained angle ahead from the "best" angle ahead (see Figure 7.9). *n* = number of reaches analyzed; *M* = mean timing reflecting the systematic timing error; *SD* reflects the variable timing error.

It seems to be mainly sensory and neuromotor maturation that sets the limits of the manual system. As the brain matures, more and more possibilities for manual action emerge, and the child seems to be ready for them. When infants begin to reach for and grasp objects, they do it intelligently from the start. The hand is oriented to the orientation of the target, and the grasp is prepared beforehand and timed relative to the encounter. Even such a complex and highly specific skill as catching is mastered, in principle, as soon as the sensory and motor systems are ready for it.

In the transition periods, new manual skills do not always emerge in a badly coordinated state and then slowly improve. Rather, in some instances the skill seems almost perfect, and in others it is hardly evident at all. Reaching in the newborn is typically expressed in this way. The movements may look more or less randomly distributed in space for a while, and then suddenly the mist seems to clear and a beautiful reach is elicited. In this case, the determinants of variance in performance appear to be internal. In other instances, however, the determinants are definitely external. The improvement in reaching performance with increasing speed of the target is an example of the latter.

The fact that sensory and neuromotor development sets the limits of the manual system in development does not mean that experience plays an unimportant role. On the contrary, experience is probably a crucial factor in development at all stages. However, the environment does not set the limits, and experience in itself does not constrain development. The opportunities are always there when the child is ready for them. Therefore, it is hard to trace the impact of experience on an age curve. Only in cases when the availability of experience is manipulated is this possible, as in the work of Held, Hein, and their associates (see, e.g., Bauer & Held, 1975; Hein, 1974; Hein & Held, 1967; Held & Hein, 1963). They showed that deprivation of sight of the forelimbs in cats and monkeys during earliest development produces deficits in visually guided control of these members. More specifically, Held and Hein (1963) found that visual feedback from self-produced movements was necessary for normal development of visually guided reaching. However, they also found that after only, at most, a few days of free sight of the limbs, the performance of the experimental animals approached that of normal ones, even if restoration of sight occurred after as much as 6 months of deprivation. Bauer and Held (1975) therefore concluded that it was more appropriate to describe this kind of learning as a form of calibration of the metrical relation between space of vision and the motor space than to characterize it as motor learning in an ordinary sense.

REFERENCES

Alderson, G. J. K., Sully, D. J., & Sully, H. G. (1974). An operational analysis of a one-handed catching task using high speed photography. *Journal of Motor Behavior, 6*, 217-226.

Arbib, M. A. (1981). Perceptual structures and distributed motor control. In V. B. Brooks (Ed.), *Handbook of physiology: Sec. 1. The nervous system: Vol. 2. Motor control* (pp. 1449-1480). Bethesda, MD: American Physiological Society.

Bauer, J., & Held, R. (1975). Comparison of visually guided reaching in normal and deprived infant monkeys. *Journal of Experimental Psychology: Animal Behavior Processes, 1*, 298-308.

Birch, E. E., Gwiazda, J., & Held, R. (1982). Stereoacuity development for crossed and uncrossed disparities in human infants. *Vision Research, 22*, 507-513.

Brooks, V. B., Cooke, J. C., & Thomas, J. S. (1973). The continuity of movements. In R. B. Stein, K. G. Pearson, R. S. Smith, & J. B. Redford (Eds.), *Control of posture and locomotion*. New York: Plenum Press.

Bushnell, E. (1985). The decline of visually guided reaching during infancy. *Infant Behavior and Development, 8*, 139-156.

Butterworth, G. (1986). Some problems in explaining the origins of movement control. In M. G. Wade & H. T. A. Whiting (Eds.), *Motor development in children: Aspects of coordination and control* (pp. 23-32). Dordrecht: Martinus Nijhoff.

Dobson, V., & Teller, D. Y. (1978). Visual acuity in human infants: A review and comparison of behavioral and electrophysiological studies. *Vision Research, 18*, 1469-1483.

Fentress, J. C. (1984). The development of coordination. *Journal of Motor Behavior, 16*, 99-134.

Fox, R., Aslin, R. N., Shea, S. L., & Dumais, S. T. (1980). Stereopsis in human infants. *Science, 207*, 323-324.

Gallahue, D. L. (1982). *Understanding motor development in children*. New York: Wiley.

Granrud, C. E. (1986). Binocular vision and spatial perception in 4- and 5-month-old infants. *Journal of Experimental Psychology: Human Perception and Performance, 12*, 36-49.

Grenier, A. (1980). Révélation d'une expression motorice différente par fixation manuelle de la nuque [Appearance of a different motor expression by manually fixating the neck]. In A. Grenier & C. Amiel-Tison (Eds.), *Evaluation neurologique du nouveau-né et du nourrison*. Paris: Masson.

Grenier, A. (1981). "Motricité libérée" par fixation manuelle de la nuque au cours des premières semaines de la vie ["Liberated movements" by manual fixation of the neck during the first weeks of life]. *Archives Françaises de Pédiatrie, 38*, 557-561.

Halverson, H. M. (1931). Study of prehension in infants. *Genetic Psychology Monographs, 10*, 107-285.

Hein, A. (1974). Prerequisite for development of visually guided reaching in the kitten. *Brain Research, 71*, 259-263.

Hein, A., & Held, R. (1967). Dissociation of the visual placing response into elicited and guided components. *Science, 158*, 390-391.

Held, R. (1979). Development of visual resolution. *Canadian Journal of Psychology, 33*, 213-221.

Held, R., Birch, E. E., & Gwiazda, J. (1980). Stereoacuity of human infants. *Proceedings of the National Academy of Science U.S.A., 77*, 5572-5574.

Held, R., & Hein, A. (1963). Movement-produced stimulation in the development of visually guided behavior. *Journal of Comparative and Physiological Psychology, 56*, 872-876.

Hofsten, C. von. (1979). Development of visually guided reaching: The approach phase. *Journal of Human Movement Studies, 5*, 160-178.

Hofsten, C. von. (1980). Predictive reaching for moving objects by human infants. *Journal of Experimental Child Psychology, 30*, 369-382.

Hofsten, C. von. (1982). Eye-hand coordination in newborns. *Developmental Psychology, 18*, 450-461.

Hofsten, C. von. (1983). Catching skills in infancy. *Journal of Experimental Psychology: Human Perception and Performance, 9*, 75-85.

Hofsten, C. von. (1984). Developmental changes in the organization of prereaching movements. *Developmental Psychology, 20*, 378-388.

Hofsten, C. von, & Fazel-Zandy, S. (1984). Development of visually guided hand orientation in reaching. *Journal of Experimental Child Psychology, 38*, 208-219.

Hofsten, C. von, & Lindhagen, K. (1979). Observations on the development of reaching for moving objects. *Journal of Experimental Child Psychology, 28*, 158-173.

Hofsten, C. von, & Rönnqvist, L. (1988). Preparation for grasping an object: A developmental study. *Journal of Experimental Psychology: Human Perception and Performance, 14*, 610-621.

Hofsten, C. von, & Rönnqvist, L. (in preparation). *Finger movements in the neonate.*

Hofsten, C. von, & Siddiqui, A. (in preparation). *Object size as a determinant of grasping in infancy.*

Hofsten, C. von, & Spelke, E. S. (1985). Object perception and object directed reaching in infancy. *Journal of Experimental Psychology: General, 114*, 198-212.

Jeannerod, M. (1981). Intersegmental coordination during reaching at natural visual objects. In J. Long & A. Baddeley (Eds.), *Attention and performance IX* (pp. 153-168). Hillsdale: Erlbaum.

Jeannerod, M. (1984). The timing of natural prehension movements. *Journal of Motor Behavior, 16*, 235-254.

Jeannerod, M. (1986). The formation of the finger grip during prehension: A cortically-mediated visuo-motor pattern. In H. T. A. Whiting & M. G. Wade (Eds.), *Themes in motor development* (pp. 183-205). Dordrecht: Martinus Nijhoff.

Jeannerod, M., & Biguer, B. (1982). Visuomotor mechanisms in reaching within extrapersonal space. In D. J. Ingle, M. A. Goodale, & R. J. W. Mansfield (Eds.), *Analysis of visual behavior* (pp. 387-409). Cambridge, MA: MIT Press.

Kandel, E. R., & Schwartz, J. H. (1985). *Principles of neural science* (2nd ed.). New York: Elsevier.

Kay, H. (1970). Analyzing motor skill performance. In K. Connolly (Ed.), *Mechanisms of motor skill development* (pp. 139-159). London: Academic Press.

Kuypers, H. G. J. M. (1962). Corticospinal connections: Postnatal development in the rhesus monkey. *Science, 138*, 678-680.

Kuypers, H. G. J. M. (1964). The descending pathways to the spinal cord, their anatomy and functions. In J. C. Eccles, and J. C. Shade (Eds.), *Organization of the spinal cord* (pp. 188-202). Amsterdam: Elsevier.

Kuypers, H. G. J. M. (1973). The anatomical organization of the descending pathways and their contribution to motor control especially in primates. In J. E. Desmedt (Ed.), *New developments in electromyography and clinical neurophysiology* (Vol. 3, pp. 38-68). New York: S. Krager

Lasky, R. E. (1977). The effect of visual feedback of the hand on reaching and retrieval behavior of young infants. *Child Development, 48*, 112-117.

Lawrence, D. G., & Kuypers, H. G. J. M. (1968a). The functional organization of the motor system in the monkey: I. The effects of bilateral pyramidal lesions. *Brain, 91*, 1-14.

Lawrence, D. G., & Kuypers, H. G. J. M. (1968b). The functional organization of the motor system: II. The effects of the descending brainstem pathways. *Brain, 91*, 15-36.

Lawrence, D. G., & Hopkins, D. A. (1972). Developmental aspects of pyramidal motor control in the rhesus monkey. *Brain Research, 40*, 117-118.

Lawrence, D. G., & Hopkins, D. A. (1976). The development of motor control in the rhesus monkey: Evidence concerning the role of cortimotorneuronal connections. *Brain, 99*, 235-254.

Lockman, J. J., Ashmead, D. H., & Bushnell, E. W. (1984). The development of anticipatory hand orientation during infancy. *Journal of Experimental Child Psychology, 37*, 176-186.

McDonnell, P. (1975). The development of visually guided reaching. *Perception & Psychophysics, 18*, 181-185.

Morrongiello, B., & Rocca, P. (1986, April). *Age-related changes in reaching behavior*. Paper presented at the meeting of the International Conference for Infant Studies, Los Angeles.

Piaget, J. (1953). *The origins of intelligence in the child.* New York: Routledge.

Piaget, J. (1954). *The construction of reality in the child.* New York: Basic Books.

Piaget, J. (1970). Piaget's theory. In P. H. Mussen (Ed.), *Carmichel's manual of child psychology* (3rd ed., vol. 1, pp. 703-732). New York: Wiley.

Rochat, P., Blass, E. M., & Hoffmeyer, L. B. (1987). *Oropharyngeal control of hand-mouth coordination in newborn infants.* Unpublished manuscript.

Spelke, E. S., Hofsten, C. von, & Kestenbaum, R. (1988). *Object perception and object-directed reaching in infancy: Interaction of spatial and kinetic information for object boundaries.* Manuscript submitted for publication.

Thelen, E. (1985). Developmental origins of motor coordination: Leg movements in human infants. *Developmental Psychobiology, 18*, 1-18.

Touwen, B. C. L. (1976). Neurological development in infancy. *Clinics in a developmental medicine*, Serial No. 58. London: Heinemann.

Twitchell, T. E. (1965). The automatic grasping responses in infants. *Neuropsychologia, 3*, 247-259.

Twitchell, T. E. (1970). Reflex mechanisms and the development of prehension. In K. Connolly (Ed.), *Mechanisms of motor skill development* (pp. 25-38). London: Academic Press.

Wade, M. G. (1980). Coincidence anticipation of young, normal and handicapped children. *Journal of Motor Behavior, 12,* 103-112.

White, B. L., Castle, P., & Held, R. (1964). Observations on the development of visually directed reaching. *Child Development, 35,* 349-364.

Wing, A. M., Turton, A., & Fraser, C. (1986). Grasp size and accuracy of approach in reaching. *Journal of Motor Behavior, 18,* 245-261.

Perspectives on the Coordination of Movement
S.A. Wallace (Editor)

EVOLVING AND DISSOLVING SYNERGIES IN THE DEVELOPMENT OF LEG COORDINATION§

Esther Thelen*

Department of Psychology
Indiana University

ABSTRACT

The development of leg coordination during the first year of infancy is described from a dynamical systems perspective. In this view, the patterns of coordination seen in infant leg movements are context-assembled synergies that are preferred, but not obligatory, movement configurations. Patterns of movement emerge and dissolve as a function of the maturational status of the neuromuscular system and the task context. Adaptive and effective coordinative patterns are selected from a larger universe of movement topologies by multimodal mapping of the multiple sensory consequences of natural movements. In this sense, mature patterns are carved out and built up through movement experience. Three infant-task situations are especially crucial in reorganizing leg movements for adaptive actions: the transition to extrauterine life, the transition to supporting the weight on the limbs, and the transition to dynamic balance.

---

*Address correspondence to: Esther Thelen, Department of Psychology, Indiana University, Bloomington, IN 47405, U.S.A.

§This work is supported by a grant from the National Institutes of Health(R01 HD 22830) and a Research Scientist Development Award from the National Institutes of Mental Health.

Human infants are born with rudimentary motor coordination, and they perform few, if any, voluntary movements. Because infants acquire motor skill very gradually over a span of several years, they offer an unparalleled opportunity to observe the construction of coordinated movements for intentional and adaptive actions. Developmental studies are important because they allow us to unpack the processes by which movement emerges in natural activities. In mature subjects, motor performance is smooth, efficient, and exquisitely integrated with the perceptual and cognitive demands of the task. It looks easy. We come to appreciate the complexity of even our most automatic actions like standing and walking when we see infants build these skills over many weeks and months by systematically solving problems that appear trivial, but are not.

By 2 years of age, children can stand, walk, run, jump, climb stairs, ride a tricycle, and perform many intricate manual actions from feeding themselves to scribbling on paper. Where does this coordination come from? Here, I follow the lead of Bernstein (1967) in suggesting that coordination is not so much imposed on action as carved out of myriad possibilities in a continual problem-solving dialogue between the infant, the environment, and the task. In traditional accounts of motor development, coordinated movement appears as the inevitable result of the maturation of the central nervous system, which marches in predictable, stage-like sequences toward adult functioning. Early motor development does assume this orderly and progressive character when we look at many infants over a long sweep of time. But when we begin to dissect the process to ask what is actually pushing the system forward at any particular time, another picture emerges. At closer range, we see variability, flexibility, exploration, and a process more of gradually sculpting out and building up than stamping from a precision mold. I show how the developing coordination of the legs in the first year belies mechanistic analogies of the motor system and supports a dynamic and emergent view.

This approach rests on the assumption that behavioral events unfolding in time share certain fundamental organizational principles common to pattern formation in a more general class of complex, open systems (Thelen, Kelso, & Fogel, 1987). Human infants are actors in "real time" as they move their limbs and body segments in continual interchange with their physical and social worlds. We can think of each movement trajectory as circumscribing an area in the space of all possible movement trajectories, a real-time state space. But as time passes, the whole movement repertoire undergoes change, spilling into new areas of the state space and retreating from other areas. The infant, therefore, has two concurrent dynamic states: the immediate topology of movement and the longer-term trajectory of ontogeny.

Over both time scales, we ask what brings forth, in any given action, this particular way of organizing the limbs and body segments over space and time. This is a specific instance of the more general question, how physical and biological systems composed of many, many separate elements produce ordered behavior, a question being asked by researchers and theorists in a wide variety of fields, including mathematics, physics, chemistry, embryology, physiology, molecular biology, neuroscience, and cognitive studies (see, for example, Kelso, Mandell, & Shlesinger, 1988; Yates, 1987). Insights from the study of such complex cooperative systems have proved to be especially fruitful for conceptualizing mature motor behavior. These principles may be equally useful for understanding how motor coordination develops during early life. Thus, I view emerging leg coordination as the product of dynamical systems evolving over these two interacting and interdependent time scales.

THE NATURE OF COORDINATION

We call movements coordinated when the activities of the joints and muscles are related to one another in an ordered and regular way in time and space. Here, I adopt a view of coordination that Kelso, Kugler, Turvey, and their colleagues, inspired by Bernstein and using principles from the fields of synergetics, thermodynamics, and nonlinear systems, have put forth over the last decade (Kelso, Holt, Kugler, & Turvey, 1980; Kugler, Kelso, & Turvey, 1980). Because these views are described in detail elsewhere in this volume, I extract and summarize the key concepts for the developmental account that follows (see also Schöner & Kelso, 1988).

1. In the human perception-action system, a coordinated movement is a stable solution assembled from the many available anatomical elements (neural, muscular, skeletal) in a specific task context. When these elements, which offer potentially many degrees of free-

dom, cooperate, they produce spatially and temporally ordered movement that has a unitary and cohesive character.

2. Perception-action tasks do not have predetermined or prewired solutions that are muscle or even muscle synergy specific. The same task may be accomplished by a variety of different coordinated movements. This flexibility to accomplish tasks by different means and from different initial postural, dynamic, and attentional states means that all solutions cannot be instantiated a priori.

3. Some movement solutions, however, may be preferred under certain conditions, either because they are energetically efficient or because they are special learned coordinations. It is a general characteristic of complex dynamical systems to settle into a preferred behavioral configuration within particular boundary conditions, known in dynamic terminology as an attractor state. This configuration acts as a kind of dynamic magnet, such that when the system is perturbed, it tends to return to that state. The interaction among organism, task, and environment sets the boundary conditions for the coordinative outcome.

4. When the boundary conditions change because one or more elements in the system exceed certain values, the system may exhibit entirely new behavioral configurations and undergo a phase shift. These phase shifts occur because the internal cohesiveness of the system is disrupted under the new conditions, and the system seeks a new level of stability. Thus, a task may be accomplished with a particular muscle synergy within a range of organism or task metrics, but may require new configurations of movement beyond that range. Note that the system may be sensitive to small changes in only one or a few elements, known in dynamic terminology as the control parameters. Control parameters are not executives; they do not encode or represent change. Rather, they are crucial variables whose changes can cause system-wide reverberations. They may be rather unspecific and outside the organism, but they act to reorganize the system in specific ways.

5. Coordinated movement patterns are assembled and modulated by dynamic information, primarily from the visual and haptic perceptual systems. As in other dynamic interactions, the couplings between the perceptual fields and the action fields are nonlinear; some couplings are preferred, others are unstable.

THE NATURE OF MOTOR DEVELOPMENT

Given this very brief outline of a theory of coordination, I address the developmental questions. How does it happen that humans find coordinative solutions to adaptive tasks? What kind of a developmental process ensures this fit between the demands of the world and the constraints and opportunities of the neuromotor system? And how do humans develop such globally stable, yet individually flexible modes of coordination? I first provide some general assumptions and then apply these principles to developing leg coordination (see also Thelen, 1988, in press-b.)

Development as a Self-Organizing Process

My first premise extends the notion of self-organizing systems into the developmental time scale. That is, developing organisms are composed of many elements and subsystems, which cooperate over time to produce pattern and order as a compression of the many degrees of freedom. The appearance of new motor forms during development is like the assembly of coordination in real time: It is an emergent rather than a prescriptive process. There is no formula, timetable, schema, or clock in the genes or the nervous system that prescribes the accretion of new and more complex motor forms. Rather, these forms emerge from the nonlinearity of complex systems as phase shifts and bifurcations that are the natural consequence of change within the organism and between the organism and the task. Developmental phases are thus like coordinative patterns: They are soft wired or soft molded (Kugler & Turvey, 1987) and highly task and context dependent.

In particular, we can envision developing motor coordination as a series of such emergent attractor states, or preferred, but not obligatory, configurations. These attractor states evolve and dissolve as the component elements themselves change in nonlinear and asynchronous ways over time and as the context of the infant also changes. During development, the organism generates a series of stable solutions as a function of its developmental status and the task at hand.

One of Bernstein's (1967) important insights was that in the musculoskeletal system, there can be no privileged or one-to-one correspondence between the pattern of motor impulses and the resulting movement outcome. The reason is that the moving segments are subject to forces in addition to those generated by the contracting muscles: gravitational and inertial forces, and the forces generated by the other segments in the linked system. This means that the postural and dynamic context of the movement is all-important in determining what muscles are used. A particular arm movement uses different muscle synergies when the subject is standing up from those used when the subject is supine. At the same time, the same pattern of muscle innervation will produce entirely different movement outcomes in the two different postural conditions. The motor system is able to do this because the contributing elements are free to assemble or reassemble in response to the task and the current status of the moving segments.

A similar task-specific fluidity is apparent during development. That is, infants are free to assemble a variety of coordinative solutions to task requirements, within the constraints of their maturational status and current activities. Some solutions will be easier or preferred, others may be available but difficult, and still others may be sufficiently unstable as to never appear. With development, however, the preferred solutions will come to dominate the repertoire. They will become more frequently (although not exclusively) used and will be accurate, smooth and efficient—the hallmark of skilled activity.

COORDINATION EMERGES IN DIALOGUE WITH THE PERIPHERY

Following Edelman (1987), I suggest that the process by which functional coordinated activity emerges during early life is one of selection rather than imposition. A core assumption, to be substantiated in this chapter, is that the patterned output of the central nervous system is a reflection not only of its anatomical construction but also of the demands and constraints of the periphery. The central nervous system is capable of generating far more coordinative patterns than it eventually uses; these come to be grouped into functional categories, or coordinative structures, only as they meet the demands of the periphery. A second core assumption, therefore, is that even at a very early age, infants must be sensitive to the multiple sensory consequences of their own movements and must be using that information to further select and refine patterns of coordination.

Edelman (1987) provides an elegant model explaining how this process might happen. Very briefly, he postulates that every motor ensemble generates multiple and simultaneous sensory inputs (visual, kinesthetic, auditory), that are captured in the form of local network maps. These maps overlap to a large extent with each other and with the motor ensemble connections. This overlap allows features of the input-output array to be continuously and multiply correlated to produce a more global mapping of the motor gesture and its sensory consequences. As each slightly different variant of a movement combination is generated in presumably slightly different contextual conditions, the resulting sensory features are fed back into this global mapping so that they may become associated with their motor responses.

Current models show that this process of feature correlation can produce stable categories of action in response to repetition alone. The system learns and generalizes by this reentry procedure (Kuperstein, 1988), but there is no explicit instructor either existing as a schema within the organism beforehand, or using traditional reinforcement. Action categories are truly self-organizing by the dynamic interplay of all the system components. The crucial developmental questions, then, are first, what drives the system into new coordinative modes? How do the specific couplings between the perceptual maps and the motor maps that are preferred and stable at one age become disrupted and nonadaptive, leading the system to seek a new level of stability? Put in dynamic terminology, we seek the developmental control parameters, those few variables out of many elements that reorganize the system. Second, once a new qualitative mode is established, how does the system become smoother and more efficient? Again, in dynamic terminology, the question is not only what are the origins of new attractor regimes but also what are the processes by which established regimes become increasingly stabilized.

Toward a Dynamical Account of Development

The first requirement for such a dynamical account of coordination development is a description of the stable coordinative modes or attractor states over time. This descriptive level tells us what the organism prefers to do given a particular maturational state and a particular immediate context, and allows us to identify the developmentally interesting points of transition into new modes. At those points of transition, dynamical principles predict that the system has lost stability and can be experimentally probed to determine the control parameters, or those organismic or contextual factors that move the system forward.

Recall that I characterized developing humans as complex systems whose coordinative outcomes are a cooperative product of many component elements. Each of those component elements is itself a dynamical system—that is, it shows nonlinear changes over time. The rates of maturation and functional efficiency of the neural bases of perceptual and motor performance are asynchronous and asymmetrical: Some elements are comparatively accelerated and others quite retarded. Likewise, physical growth changes in skeletomuscular components occur in spurts and plateaus, not as steadily increasing functions. Organismic changes in the infant lead to dramatic changes in the social and physical context of the child; as infants mature, their entire task space evolves. The result is that the developmental process as a whole is highly nonlinear. The system may be sensitive to particular parameters at one point and to entirely different parameters when the system has reorganized into new modes.

The consequence of this nonlinearity is that no single mechanism of change may apply across the developmental time scale or across particular task domains. Thus, an analysis must include the specific capabilities of the infant in relation to the very particular demands of the task in a circumscribed environment. It is this continual assembly and reassembly of coordinative solutions in contexts that are in some ways similar and in some ways different that forms the raw material of mature, skilled motor performance. In the following sections, I trace the emerging patterns of coordination in the leg movements of infants from this perspective.

LEG COORDINATION IN THE FIRST YEAR

Although with training, legs can be used for highly articulated activities such as dancing or playing the organ, humans use their legs primarily for support and locomotion. It takes human infants about one year to master these two primary tasks. For several reasons, this developmental course takes a long time (Thelen, 1984). Most important is that moving bipeds are inherently unstable, and the neuromuscular requirements are severe for maintaining balance on a small base of support, especially during the weight shifts necessary for forward locomotion. The dual problem facing infants, therefore, is to generate the coordinative patterns needed to move the body forward simultaneously with those synergies needed to maintain dynamic balance. The challenge is to conquer the forces of gravity, which continually act to destabilize the infant. The infant accomplishes this through a continual dialogue in which coordinative patterns are established and dissolved as a system sensitive to these dynamic forces explores new regions of its body and movement space.

The challenge of gravity has three major developmental epochs: the transition to extrauterine life, the transition to supporting the weight on the limbs, and the transition to dynamic balance.

Newborn Synergies and the Transition to Extrauterine Life

The leg movements of newborn humans are remarkably well coordinated, especially in contrast to the seemingly random thrashings of the upper limbs. The predominant newborn leg movement is a staccato, nearly simultaneous flexion and extension of the hip, knee, and ankle joints, often occurring in rhythmic succession, and with frequent alternations between right and left legs (Thelen & Fisher, 1983b). These movements are seen both when infants are supine and when they are held upright, the latter giving the impression that they are "stepping." The frequency of these kicking movements appears to be a function of generalized behavioral activation. Infants move their legs directly in proportion to their overall state of excitement, although neither sleepy nor highly distressed infants perform these coordinated movements (Thelen, Fisher, Ridley-Johnson, & Griffin, 1982). Early kicking and stepping is distinct in being a largely flexor-activated synergy. That is, movements are initiated by a strong co-contraction in the large antagonist pairs of leg muscles, resulting in a strong flexion. The extension phase appears largely reactive, as little active muscle contraction can be detected, and the recoil results presumably from the inertial and elastic properties of the leg. It is important to note that the flexor dominance of movement in the newborn period is echoed in the flexor dominance of resting posture at this age.

What is the developmental significance of these very early motor synergies? One important clue is that these postnatal movements, which persist for a month or two, appear identical to leg movements that Prechtl (1986) and his colleagues detected prenatally by ultrasound: impulsive, flexor contractions that may alternate between the legs. My colleagues and I (Thelen, Kelso, & Fogel, 1987) have suggested that these newborn patterns are in fact the self-organized product of established neurological pathways and anatomical structures under certain energetic constraints, and that they are molded by the uterine environment. That is, although the movements look step-like, they are the result of converging multiple subsystems rather than a dedicated locomotor generator. The evolution of these subsystems into actual locomotion is thus not so much driven by changes in a locomotor executive as emergent from the confluence of the changing subsystems in a continually changing dynamic context.

Consider the first environment for the action of the developing human, the fluid-filled uterus. Two features are distinctive: the relative attenuation of the effects of gravity and the flexed, spherical position necessary as the fetus grows to fill the available space. This means that whatever the structure of the primary motor and sensory tracts and the neuromuscular junctions, fetal movements are constrained by the flexed posture but relatively immune from the sensory consequences of gravity. I suggest that leg movements in the newborn period reflect the confluence of the neuro-musculo-skeletal anatomy of the fetus and its ecological niche for 9 months (Thelen, in press-a). In dynamic terms, when the system is activated within a certain energy flux, it self-assembles into a preferred steady-state periodic attractor, that is, an alternating rhythmic mode of flexions and extensions at all three joints.

More precisely, it appears that when fetuses and newborns are in a "kicking mode," energy delivered to the muscles contracts flexor and extensor muscles in a periodic regime. Because of the relative flexor dominance, such co-contractions result in flexor movements. The cycle is completed by the passive extension. The movements look clock-like, but no central generator or clock need be invoked. Rather, these patterns can emerge from the cooperative interactions of subsystems, none of which contains a "kicking code" per se. For example, the simultaneous flexions and extensions of the joints of one limb may reflect the simultaneous activation of all the leg muscles, but flexor influences are stronger. Once activated, the regular timing of the movements falls out, given certain time-dependent contractile and elastic properties of the muscle and biodynamic characteristics of the limb segments. Supplied with a continual source of energy, in a "free-running" situation, the system tends to oscillate on its own. The bilateral alternation of movements requires that these intrinsic oscillatory properties of each leg be loosely coupled and capable of mutual entrainment. As I show later in this chapter, there is good evidence for these between-limb dynamic linkages early in ontogeny.

As others have also noted (Prechtl, 1986), it appears that the coordinative patterns of the newborn period—the first month or two—are holdovers from fetal life. Most neurologically based accounts assume that the transition in motor patterns so apparent in the 2nd and 3rd months of life—especially the decline of so-called primitive reflexes, including the newborn stepping pattern—are a result of a maturationally based "remodeling" of the central nervous system. In short, these accounts assume that changes in behavior are driven exclusively by improvements in the central nervous system, improvement that by default must be of genetic origin.

Although there is little doubt as to the major importance at this time of the maturation of the central nervous system, I suggest that the process is much more dynamic and interactive than previously recognized. That is, the demands of the periphery select and mold the preferred neural configurations for action and not the reverse. Consider the dramatic transition from intrauterine to extrauterine life. First, the visual world becomes salient. Second, the infant must contend with gravity, which requires stabilizing the body and overcoming gravity when moving the limbs. In addition, once moving, the limbs and segments have additional inertial and elastic properties that are considerably different from those in a fluid environment. And these properties themselves change as the infant grows in size. Progress toward increasingly adaptive coordinated movement cannot occur in isolation from these aspects of the periphery.

I suggest, then, that the first major catalyst for change in the coordinative patterns of leg movements occurs when the infant is liberated from the confines of the uterine space and faces a gravitational field. This allows for the gradual relaxation of the strong flexor dominance of the newborn period and the exploration of a greatly expanded coordinative state space. These expanded opportunities for movement and the new experiences of perceiving the haptic consequences of moving limbs in gravity are crucial developmental motors driving the infant from the stereotypical and reflexive fetal-neonatal phase into one characterized by more articulated and intentional movements. Indeed, by 3 months, there has been substantial loosening of the tight intralimb synchrony. The infant at this age can produce more complex and articulated movements, although the flexor kick is still a preferred configuration (Thelen, 1985). In addition, infants are considerably more bilaterally asymmetrical at 3 months than in the newborn period (Thelen, Ridley-Johnson, & Fisher, 1983). Finally, by the 3rd month, infants display rudimentary voluntary control of their legs; they can use kicks as operants to control an overhead mobile, for example (Thelen & Fisher, 1983a).

I have made the unconventional claim that this well-documented transition from fetal motor functioning is molded not solely by the maturation of the central nervous system but also by conditions at the periphery. There is compelling evidence that even in the first months, infants are sensitive to the dynamical status of their legs and that the system is capable of responding to that status. First, changing the biomechanical load on the legs by postural manipulations, addition of weight, or submersion directly affected the rate and topography of infant stepping movements (Thelen & Fisher, 1982; Thelen, Fisher, Ridley-Johnson, & Griffin, 1982; Thelen, Fisher, & Ridley-Johnson, 1984). Second, adding weights to one leg of 6-week-old infants changed the bilateral symmetry of both legs in supine kicking (Thelen, Skala, & Kelso, 1987). Finally, when held supported over a moving treadmill, infants as young as 1 month performed well-coordinated, alternating stepping movements, which were distinct from reflexive steps. That infants performed these steps only on the moving treadmill means that information about the moving status of one leg was used to phase the movement of the opposite leg (Thelen & Ulrich, 1988). In fact, infants maintained precisely alternating steps even when one leg was moved at twice the treadmill speed as the opposite leg (Thelen, Ulrich, & Niles, 1987).

Taken together, these experiments demonstrate that the early neuromotor system is not organized as a machine-like assembly of prewired reflexes, but as a responsive, dynamic system in which the sensory consequences of the moving limbs are essential participants in the emergent forms of movement. Constructing categories of movement or coordinative sets through continual exploration of the periphery would seem a more parsimonious developmental strategy than imposing fixed programs from a central command center. During infancy, there are rapid changes in the absolute mass, proportion, and centers of gravity of the limbs and body segments. As infants mature, they also assume vastly different postural sets. Movement of a linked system is entirely dependent on these anatomical and biomechanical parameters, but there is no way that these parameters can be anticipated beforehand. The sensorimotor tracts must be sufficiently plastic to accomodate the consequences of growth. Although maturational changes in muscle strength, bone density, and body composition and proportion favor more independent motor skills, it is the continual exploration of the body-task space through self-generated movement that drives the system into new coordinative forms.

The Challenge of Weight-Bearing and Static Balance

A second major reorganizing context in the coordinative patterns of the lower limbs may well occur as the infant explores the consequences of bearing weight on the feet. This may begin within the first few months if caregivers allow infants to stand while supported. The amount of such supported standing during practice and play appears to be highly culture dependent (Chisholm & Richards, 1978; Super, 1980). Among the Bambara people of Mali, for example, infants are deliberately exercised in a standing posture for many minutes every day practically from birth (Bril & Sabatier, 1986). According to Western norms, independent supported standing is commonly achieved at about 10 months of age.

The task in standing upright is to support the weight on the two limbs and remain vertical. Supporting weight requires sufficient muscle strength, particularly in the dorsal extensor synergies of the legs and trunk, to maintain the legs as relatively rigid columns, as each joint is a site of potential structural collapse. The stiffness in the appropriate muscles appears to develop gradually over the 1st year. When held upright, young infants often collapse with flexions of the hip, knee, and ankle. Later, these collapses appear to become intentional, as infants quite happily bounce up and down in the standing position.

The rhythmical repetition of bouncing may be similar to repetitive kicking: the natural and preferred oscillation of the system under specific maturational and task constraints and energy input (Goldfield, 1988). That is, bouncing, like kicking, may have no prescribed motor program or developmental timetable. Rather, both the temporal and spatial pattern may emerge from the cooperation of the elements in a particular postural and energy context.

This opportunistic and transient behavior may have important consequences, however, in allowing infants to explore the perceptual-motor spaces of standing. Movement in supported standing is important because it brings into play not only the proprioceptive consequences of weight bearing (primarily on the soles of the feet) and the vestibular consequences of changes in head position but also the visual feedback from postural sway, which is simultaneously correlated with changes in the first two modalities. Presumably, similar integration of proprioceptive, vestibular, and visual perceptual spaces occurs in sitting, although adjustments are made in trunk muscle synergies as well as in the legs (Harbourne, Giuliani, & Mac Neela, 1987). The development of appropriate postural synergies may well be, therefore, an emergent result of the perceptual exploration of the space. For example, when Woollacott (in press) subjected infants who were stable independent sitters to perturbations in platform translation, they responded with well-organized muscle synergies as detected by electromyogram. These same infants, however, looked disorganized when similarly perturbed in a supported stand. The new task imposed new demands for the system to explore.

Stoffregen and Riccio (1988) make a compelling argument that humans use their own perceived motion, particularly their acceleration changes as they deviate from the vertical, and information from the support surface as their primary means of maintaining a vertical orientation. These proprioceptive cues naturally correlate with visual flow information during postural sway and adjustments. During development, the continuous movement of the child in each new posture produces repeated, simultaneous entry of multiple sensory modalities. This, in turn, allows the child to construct the correlative maps postulated by Edelman (1987). In fact, the multimodal mapping of postural synergies appears to develop rather gradually, as children do not display fully integrated postural responses until after about 6 years (Woollacott, in press).

Recently, McCollum and Leen (1988) presented a biomechanical model suggesting that infants learn their upright stability limits, that is, the permissible range of movement they can execute and still remain upright, by exploring the consequences of their movements without using biomechanically disadvantageous motor patterns. They showed, for example, that solely for biomechanical considerations, compensatory movements used by adults in certain circumstances are unlikely strategies for infants. For example, flexion at the hip alone to correct for postural perturbations can be effective for adults. Because of their stature, however, infants would not be able to correct for the pendular overshoot once the movement has begun. This observation suggests that successful compensations for postural sway would be either flexion at the ankle or displacements of hip, knee, or ankle in the sagittal plane, and only later would the third strategy be added. In fact, movements of the hip alone are rarely seen in young children (Haas, Diener, Bacher, & Dichgans, 1986). Whether infants try hip strategies and then do not repeat them or simply never try them is unknown.

In either event, the several months of practice in supported or unsupported standing could provide infants with both the strength to provide a firm pillar of support and the multimodality reentry inputs to establish stable categories of coordinative patterns in response to postural perturbations. Under predictable testing circumstances, such as platform translations, these categories appear to be hard-wired synergies. However, despite their rule-driven appearance, they act as softly assembled, but stable, attractors. Different contextual circumstances, therefore, assemble different synergetic patterns. Sufficient strength to prevent joint collapse while standing and stable muscle corrections to postural sway are necessary precursors for the next challenge for leg coordination, upright forward locomotion.

The Challenge of Dynamic Balance

Bipedal walking presents infants with far more complex task demands than does standing alone (Thelen, Ulrich, & Jensen, in press). Consider the problem of moving forward on two legs while remaining upright: Walkers must bear weight on one leg while the opposite leg swings forward and must alternately shift the weight from one stance leg to the other. This requires control of forces and torques acting on a linked system of no small dimension—the linear components of forward propulsion, the inertial properties of the linked joint system, and the forces necessary to counteract gravity. Indeed, walking is often described as a series of intentional falls separated by brief moments of dynamic equilibrium as the center of mass rotates over the base of support (Alexander, 1984).

The demonstration that even very young infants will perform well-coordinated, alternating steps on a moving treadmill suggests that one component of this complex task is available long before walking alone: the ability to generate appropriately patterned movement in response to the sensation of dynamical changes in the legs. But the ability to combine this coordinative performance with the additional coordinative and energy demands of single-leg weight bearing and dynamic balance is much more difficult. Infants who step readily on the treadmill do not step when they must bear their weight alone. Weight bearing disrupts the coordinative pattern, and thus the coordinative pattern becomes less stable before infants shift into the phase of independent walking.

One scenario, then, for the emergence of independent locomotion is that increasing stability in static stance allows infants to explore the coordinative space for shifting and bearing weight on one leg while moving the opposite leg forward, and thus, through repeated practice (likely while supported) to establish more stable categories of compensatory muscle synergies. Once the dynamic condition is established in which the center of mass is over the stiff stance leg and the opposite leg is stretched back, the normal biomechanics of the system allow the pendular swing forward to emerge. The infant must then learn to control the fall by appropriate stiffening of the swing leg before impact.

Recent work has added credence to the "controlled fall" aspect of locomotion. The transition from a static standing posture to a steady gait requires generating muscle forces sufficient to overcome the segmental inertia and damping characteristics of the limb masses and yet maintaining the center of mass within the support surface of the foot, so as to prevent falling over. Breniere, Do, and Bouisset (1987) showed that adults reached a desired steady-state velocity by the end of the first step, a finding suggesting that they were able to plan their muscle forces to account for biomechanical constraints in advance of executing the movement—they precisely controlled their fall. In contrast, infants who had been walking for only 90-100 days reached a steady-state velocity only after two to four steps. Their velocity at the end of the first step depended not on their final forward velocity but only on their individual body mass and inertia (Breniere, Bril, & Fontaine, in press). New walkers thus initiate stepping by disrupting their static balance but are captured, so to speak, by their biomechanics. Only with experience with the sensory consequences of their own body dynamics in the locomotion task do they impose fine-tuned control over their coordinated movement.

Once the system is set in motion and certain dynamical conditions are satisfied, however, the system seems to prefer a cyclical attractor state. Clark and Phillips (1987) found that infants who had been walking 3-10 months showed temporal organization of step cycle phases almost identical to that of mature walkers, that is, by adjustments in the second half of the stance phase to overall speed change. These researchers interpreted these results as the convergence of neural and dynamical constraints. Energy delivered to the system in the stance push-off phase organizes the system in a preferred configuration in both new and mature walkers.

Additional kinematic and electromyogram studies have provided convergent evidence that mature walking is carved out rather than imposed on the system. Intralimb coordination in new walkers retains some elements of a more primitive step, that is, less articulated hip and knee rotations and more contributions from quadraceps flexor muscles to initiation of the swing (Thelen & Cooke, 1987) as well as significant co-contraction of antagonistic muscle groups (Okamoto & Goto, 1985). These elements become more refined and adult-like as infants practice walking, or in dynamic terminology, as the system settles into a more stable attractor. Likewise, new walkers show overall 50% interlimb phasing, but they become much more consistent over the next few months (Clark, Whitall, & Phillips, 1988).

The rapid improvements in gait seen after infants learn to walk presumably arise because the multiple sensory consequences of self-produced motion—visual, vestibular, somatosensory, and proprioceptive—converge in certain thermodynamically stable solutions for overground locomotion. At the same time however, infants continue exploring skilled coordinations for other task-appropriate coordinations: running, climbing, jumping, hopping, scooting, and so on. Each of these requires the correlation of motor output and the task-generated sensory consequences to produce stable, yet flexible, categories of movement.

CONCLUSION

This view of leg coordination reveals a complex story in which elements of movement skills emerge and dissolve as the neuromuscular system changes. The tasks and their sensory consequences also have their own dynamics. Early leg coordination is not a linear progression toward more adult-like configurations of movement. Rather, it consists of topographies like kicking, bouncing, and crawling, which appear as largely transient phenomena, and like upright locomotion, whose onset is a discontinuous phase shift. All these coordinative outcomes are intertwined with dramatic changes of body stature, proportion, and tissue composition.

Only a dynamical, emergent view can account for the nonlinearity of the process and the ability of the motor system to continually recalibrate for growth and the inevitable contextual changes that come with adding new skills. However, such a view does not explain the particulars. This explanation awaits patient descriptive and experimental manipulations to identify the crucial points of transitions to new forms and the parameters that cause the system to seek these new coordinative modes.

ACKNOWLEDGMENT

I am grateful to Beverly D. Ulrich and Jody L. Jensen for their continuing help and support and for their comments on this chapter.

REFERENCES

Alexander, R. M. (1984). Walking and running. *American Scientist, 72*, 348-354.

Bernstein, N. (1967). *Co-ordination and regulation of movements.* New York: Pergamon Press.

Breniere, Y., Bril, B., & Fontaine, R. (in press). Analysis of the transition from upright stance to steady state locomotion for children with under 200 days of autonomous walking. *Journal of Motor Behavior.*

Breniere, Y., Do, M. C., & Bouisset, S. (1987). Are dynamic phenomena prior to stepping essential to walking? *Journal of Motor Behavior*, *19*, 62-76.

Bril, B., & Sabatier, C. (1986). The cultural context of motor development: Postural manipulations in the daily life of Bambara babies (Mali). *International Journal of Behavioral Development*, *9*, 439-453.

Chisholm, J. S., & Richards, M. P. M. (1978). Swaddling, cradleboards and the development of children. *Early Human Development*, *2*, 255-275.

Clark, J. E., & Phillips, S. J. (1987). The step cycle organization of infant walkers. *Journal of Motor Behavior*, *19*, 421-433.

Clark, J. E., Whitall, J., & Phillips, S. J. (1988). Human interlimb coordination: The first 6 months of independent walking. *Developmental Psychobiology*, *21*, 445-456.

Edelman, G. M. (1987). *Neural Darwinism.* New York: Basic Books.

Goldfield, E. C. (1988). *Ontogeny of infant crawling: Biomechanical, neurological, and environmental influences.* Manuscript submitted for publication.

Haas, G., Diener, H. C., Bacher, M., & Dichgans, J. (1986). Development of postural control in children: Short-, medium-, and long latency EMG responses of leg muscles after perturbation of stance. *Experimental Brain Research*, *64*, 127-132.

Harbourne, R. T., Giuliani, C. A., & Mac Neela, J. C. (1987). *A kinematic and electromyographic analysis of the development of sitting posture in infants.* Paper presented at the meeting of the American Academy of Cerebral Palsy and Developmental Medicine, Boston.

Kelso, J. A. S., Holt, K. G., Kugler, P. N., & Turvey, M. T. (1980). On the concept of coordinative structures as dissipative structures: II. Empirical lines of convergence. In G. E. Stelmach & J. Requin (Eds.), *Tutorials in motor behavior* (pp. 49-70). New York: North-Holland.

Kelso, J. A. S., Mandell, A. J., & Shlesinger, M. F. (Eds.). (1988). *Dynamic patterns in complex systems.* Singapore: World Scientific.

Kugler, P. N., Kelso, J. A. S., & Turvey, M. T. (1980). On the concept of coordinative structures as dissipative structures: I. Theoretical lines of convergence. In G. E. Stelmach & J. Requin (Eds.), *Tutorials in motor behavior* (pp. 3-47). New York: North Holland.

Kugler, P. N., & Turvey, M. T. (1987). *Information, natural law, and the self-assembly of rhythmic movement.* Hillsdale, NJ: Erlbaum.

Kuperstein, M. (1988). Neural model of adaptive hand-eye coordination for single postures. *Science, 239,* 1308-1311.

McCollum, G., & Leen, T. K. (1988). *Stability limits.* Manuscript submitted for publication.

Okamoto, T., & Goto, Y. (1985). Human infant pre-independent and independent walking. In S. Kondo (Ed.), *Primate morpho-physiology, locomotor analyses and human bipedalism* (pp. 25-45). Tokyo, Japan: University of Tokyo Press.

Prechtl, H. F. R. (1986). Prenatal motor development. In M. G. Wade & H. T. A. Whiting (Eds.), *Motor development in children: Aspects of coordination and control* (pp. 53-64). Dordecht, The Netherlands: Martinus Nijhoff.

Schöner, G., & Kelso, J. A. S. (1988). Dynamic pattern generation in behavioral and neural systems. *Science, 239,* 1513-1520.

Stoffregen, T. A., & Riccio, G. E. (1988). An ecological theory of orientation and the vestibular system. *Psychological Review, 95,* 3-14.

Super, C. M. (1980). Behavioral development in infancy. In R. H. Monroe, R. L. Monroe, & B. B. Whiting (Eds.), *Handbook of cross-cultural human development* (pp. 181-270). New York: Garland STPM.

Thelen, E. (1984). Learning to walk: Ecological demands and phylogenetic constraints. In L. P. Lipsitt (Ed.), *Advances in infancy research* (Vol. 3, pp. 213-250). Norwood, NJ: Ablex.

Thelen, E. (1985). Developmental origins of motor coordination: Leg movements in human infants. *Developmental Psychobiology, 18*, 1-22.

Thelen, E. (1988). Dynamical approaches to the development of behavior. In J. A. S. Kelso, A. J. Mandell, & M. F. Shlesinger, (Eds.), *Dynamic patterns in complex systems* (pp. 348-369). Singapore: World Scientific.

Thelen, E. (in press-a). On the nature of developing motor systems and the transition to extrauterine life. In W. Smotherman & S. A. Robinson (Eds.), *Fetal behavior*. Bloomfield, NJ: Telford Press.

Thelen, E. (in press-b). Self-organization in developmental processes: Can systems approaches work? In M. Gunnar (Ed.), *Systems in development: The Minnesota Symposium in Child Psychology* (Vol. 22). Hillsdale, NJ: Erlbaum.

Thelen, E., & Cooke, D. W. (1987). The relationship between newborn stepping and later locomotion: A new interpretation. *Developmental Medicine and Child Neurology, 29*, 380-393.

Thelen, E., & Fisher, D. M. (1982). Newborn stepping: An explanation for a "disappearing reflex." *Developmental Psychology, 18*, 760-775.

Thelen, E., & Fisher, D. M. (1983a). From spontaneous to instrumental behavior: Kinematic analysis of movement changes during very early learning. *Child Development, 54*, 129-140.

Thelen, E., & Fisher, D. M. (1983b). The organization of spontaneous leg movements in newborn infants. *Journal of Motor Behavior, 15*, 353-377.

Thelen, E., Fisher, D. M., & Ridley-Johnson, R. (1984). The relationship between physical growth and a newborn reflex. *Infant Behavior and Development, 7*, 479-493.

Thelen, E., Fisher, D. M., Ridley-Johnson, R., & Griffin, N. (1982). The effects of body build and arousal on newborn infant stepping. *Developmental Psychobiology, 15*, 447-453.

Thelen, E., Kelso, J. A. S., & Fogel, A. (1987). Self-organizing systems and infant motor development. *Developmental Review, 7*, 39-65.

Thelen, E., Ridley-Johnson, R., & Fisher, D. M. (1983). Shifting patterns of bilateral coordination and lateral dominance in the leg movements of young infants. *Developmental Psychobiology, 16*, 29-46.

Thelen, E., Skala, K., & Kelso, J. A. S. (1987). The dynamic nature of early coordination: Evidence from bilateral leg movements in young infants. *Developmental Psychology, 23*, 179-186.

Thelen, E., & Ulrich, B. D. (1988, April). *Cryptic development of locomotor coordination: Treadmill stepping in the first year.* Paper presented at the International Conference on Infant Studies, Washington, DC.

Thelen, E., Ulrich, B. D., & Jensen, J. L. (in press). The developmental origins of locomotion. In M. Woollacott & A. Shumway-Cook, (Eds.), *The development of posture and gait across the lifespan*. Columbia, SC: University of South Carolina Press.

Thelen, E., Ulrich, B., & Niles, D. (1987). Bilateral coordination in human infants: Stepping on a split-belt treadmill. *Journal of Experimental Psychology: Human Perception and Performance, 13*, 405-410.

Woollacott, M. H. (in press). Children's development of posture and balance control: Changes in motor coordination and sensory integration. In D. Gould & M. Weiss (Eds.), *Advances in pediatric sport sciences: Behavioral issues*. Champaign, IL: Human Kinetics.

Yates, F. E. (1987). *Self-organizing systems: The emergence of order*. New York: Plenum Press.

SECTION 3

COORDINATION OF ADULT MOTOR BEHAVIOR

Perspectives on the Coordination of Movement
S.A. Wallace (Editor)

KNOWLEDGE-DIRECTED COORDINATION IN REACHING FOR OBJECTS IN THE ENVIRONMENT

Sylvic ATHÈNES*

CNRS Cognitive Neuroscience Unit

and

Alan M. WING

MRC Applied Psychology Unit

ABSTRACT

Reaching for an object in the environment requires coordination between transport and grasp. In this chapter, we review published work on the role of prediction in such coordination. We also present data from a previously unpublished study of reaching in which we introduced positional uncertainty of the target object relative to the hand. Our general conclusion is that the act of reaching is structured in advance to allow for various possible sources of error that are likely to occur during movement. The coordination between transport and grasp depends on prior knowledge gained from previous real-world interactions with objects.

INTRODUCTION: REACHING AS AN EXAMPLE OF COORDINATION

In the study of motor behavior, the term *coordination* refers to the way in which two or more distinct elements are brought together to form a new complex, in which the temporal or spatial characteris-

---

*Address correspondence to: Sylvie Athènes, Unité de Neurosciences Cognitives, CNRS-LNF, 31 Chemin Joseph Aiguier, BP71, 13402 Marseille, Cedex 9, France.

tics of the original elements are mutually constrained. Functionally, the constraint makes behavior efficient; at some level coordination enhances the system whose elements are under study. The elements constrained may involve a single limb, as in squeezing the thumb and finger together, two limbs, as in clapping the hands, or even different people, as in the cooperative lifting of a heavy piece of furniture. If the constraint on coordination is temporal, it may involve simultaneity or successiveness. If the constraint is spatial, the elements may or may not be in proximity. There may be an asymmetry in the constrained elements as, for example, between the hands in assembling a nut and bolt. People often hold one component still—perhaps the bolt—until they have engaged the thread by turning the nut with the other hand. Or there may be reciprocal interactions, as in the coordination exhibited by two people, each with one oar, attempting to row a boat in a straight line by using a synchronized stroke. But in every case of coordination, the behavior of the individual elements changes and usually confers some advantage in attaining a specifiable goal.

Reaching to grasp an object in the environment is an excellent example of coordination. Movements of the fingers, wrist, elbow, and shoulder are readily carried out separately under conscious control. But when the hand takes hold of an object at some distance from the body, changes in the shape and position of the hand are brought together to serve the single goal of first encompassing and then stabilizing the object.

Recent work on reaching has concentrated on the concurrent regulation of coordination and focused on feedback mechanisms used to home in with the hand on the position of an object (e.g., Goodale, Pelisson, & Prablanc, 1988; Paillard, 1982). Such processes are undoubtedly important, for example, in the pursuit and capture of slowly moving objects. Nevertheless, in this work, the aiming component of reaching and grasping is emphasized to such an extent that there is a danger of regarding the action merely as the reduction of distance between the hand and the object.

The problem with looking at reaching behavior in this way is the implication that the psychological boundaries of the action can be defined in terms of the beginning of hand movement and the landing of the hand on the object, that is, in terms of the kinematic trace. However, we argue that in some situations, what happens during the movement can be explained only by what happens outside the limits suggested by the kinematic record. For example, a person using a knife to cut a slice of bread would grasp it by the handle. In contrast, the circus performer might choose to pick the knife up by the blade and throw it. A satisfactory characterization of reaching, of the coordination in approaching and grasping the knife, is likely to elude us if we fail to consider the context of the action. It is important to ask what motivates an action prior to reaching and what role is played by the target object, once it has been grasped, in achieving the goal.

The different uses of an object—in the case of the knife, cutting with edge of the blade or penetration with the tip—lead to consideration of the role of knowledge, derived from prior encounters with knives, in determining coordination in reaching. If nothing else, it seems probable, for example, that a person will take considerably greater care in the approach when picking the knife up by the blade than when picking it up by the handle. What then are the ramifications of this kind of knowledge for kinematic traces? This question raises issues that are addressed later in this chapter. It is immediately clear, however, that it would not make sense to average the kinematic traces for the two ways of picking up the knife. The patterns of coordination appropriate to each situation are different and would give rise to contrasting kinematic traces. Furthermore, the patterns of movement that accomplish reaching in the two cases are unlikely to be mere variants of the same basic pattern but instead are probably qualitatively different actions.

In this chapter, we wish to draw attention to the cognitive determinants of hand-arm coordination and more particularly to the role of prior experience in an individual's preparation of an appropriate pattern of hand and arm use. Such preparation is based on prediction by the individual of likely outcomes of the action. We present empirically based work relevant to these ideas to show how cognitive determinants affect the coordination of hand and arm movements.

TWO VISUAL-MOTOR CHANNELS IN REACHING

Reaching can be logically separated into two components, transport and grasp. On the grounds that there are two distinct visual systems, ambient and focal, Jeannerod (1981) suggested that reaching behavior should be analyzed in terms of two distinct visual-motor channels. This idea contrasts with the "classical" subdivision of reaching behavior into visual and motor functions and may be seen as a development of a slightly earlier formulation by Paillard and Beaubaton (1978). The latter authors described one (subcortical) route via "structures colliculaires" that calibrates transport on the basis of information about the object's location relative to the body and a second (cortical) system including the "système genicolostrie" that, by providing details about the object, permits appropriate preshaping of the hand.

Arbib (1981) adopted the concept of two visual-motor channels embedded within a "schema" framework, a number of semi-autonomous processes acting cooperatively to achieve completion. According to his schematic of the processes involved in reaching and grasping, different aspects of the visual input determine different motor components of reaching (see Figure 9.1). Information about the location of the object to be grasped specifies the ballistic movement of the arm; information about the size and the orientation of the object dictates finger adjustment and hand rotation.

The anatomical separability of the processes controlling transport and grasp is corroborated by the fact that a person can move the arm without moving the fingers or wiggle the fingers without moving the arm. This observation is obvious but not trivial because the arm and fingers could have been "hardwired" together like the pantograph gripper depicted in Figure 9.2. In this pincer device, extension of the arm and the size of the grip are linked in such a way that closing of the grip can be achieved only with lengthening of the arm. The coordination between the two is fixed by the design in such a way that to achieve an opening of the grip appropriate to some object, distance from the object has to be continually adjusted—which makes the device remarkably difficult to control!

Another demonstration of the capacity for independent arm and finger movements, one in which there is a clear goal, is that people can catch a ball without moving the arm. For the more general case in which both transport and grasp are required, Jeannerod (1981) provided some evidence of their independence by showing that they may be functionally dissociated. An apparently spherical object for which the subjects were reaching was transformed, after the start of the movement, into an ellipsoidal object. According to Jeannerod,

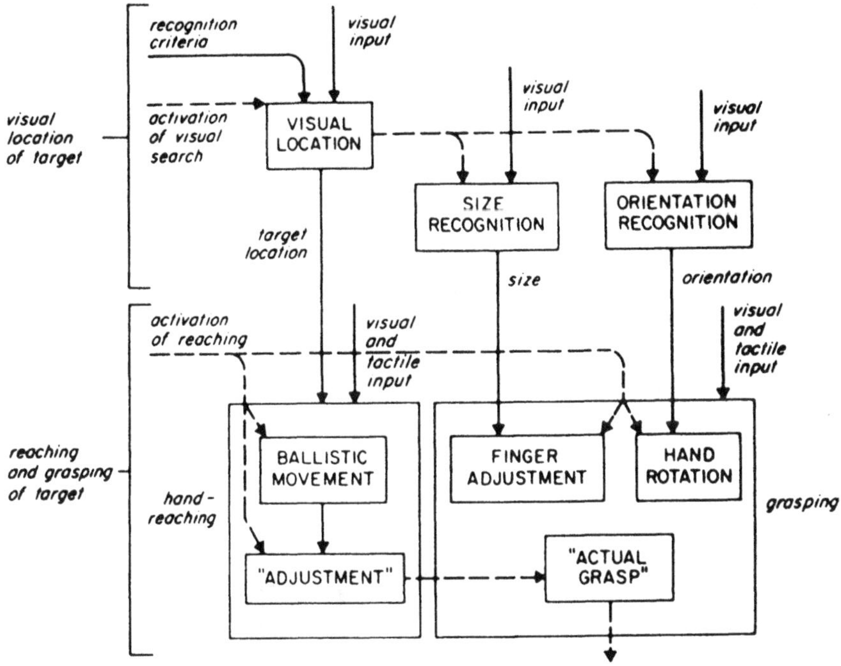

Figure 9.1. Arbib's (1981) diagrammatic representation of the relationship between grasp and transport in reaching. *Note.* From "Perceptual Structures and Distributed Motor Control" by M. Arbib, in V. B. Brooks (Ed.), *Handbook of Physiology: Sec. 1. The Nervous System: Vol. 2. Motor Control,* p. 1468, Bethesda, MD: American Physiological Society. Copyright 1981 by the American Physiological Society. Reprinted by permission.

subjects made appropriate changes to their shaping of grasp without any effect on transport.

INTERDEPENDENCE OF TRANSPORT AND GRASP

Even casual observation of reaching movements indicates some coordination of the motor components. Indeed, the hand has to stay open until it arrives in proximity with the object. One basis for coordination in reaching might be the specification of a simple temporal link between the transport component and the grasp component. Jeannerod (1984) suggested that the beginning of finger

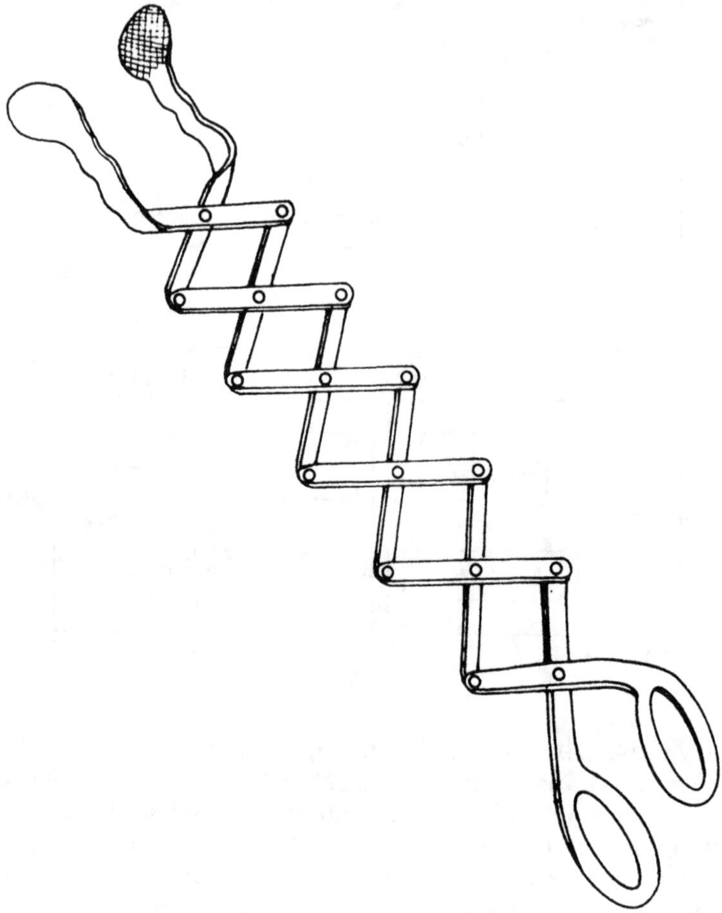

Figure 9.2. Pantograph pincer in which transport and grasp are inextricably linked.

closure corresponds to a breakpoint defined by the onset of low velocity of hand transport.

In the previous section we cited ball catching without arm movement as an example of the potential independence of arm and hand function. However, it should be noted that there is an element of transport in the ball's movement toward the hand. Grasp must be coordinated with this transport component for the catch to succeed. Data from McLeod, McLaughlin, and Nimmo-Smith (1985) suggest that the timing necessary for coordinating the size of the grasp with the approach of the ball is driven by the ball's looming in the visual field as it gets nearer.

In the case of reaching, as opposed to catching, we argue that coordinating the size of the grasp with the position of the hand relative to the object is more complex than a simple temporal link. There is, in addition, what might be termed informational dependence. Wing, Turton, and Fraser (1986) showed that loss of visual information that would have allowed more accurate positioning of the hand during reaching—object size, orientation, and position remaining the same—led to wider opening of the hand. Figure 9.3 illustrates the experimental arrangement that they used. A lightweight cylindrical piece of wooden dowel was placed on end in front of the subject. When ready, the subject had to pick the cylinder up and pass it to the experimenter, whose hand was waiting to receive it on the far side of the table. Relative to its length, the dowel was of small diameter. This meant that it could be easily knocked over if the subject was not careful, and therefore accuracy was more important than speed, which is a general characteristic of normal reaching movements.

A series of trials were videotaped, and for each trial, the maximum grasp aperture was determined. If, after having been allowed to look at the dowel and memorize its position, the subject reached for the dowel with eyes closed, it was found that the hand was opened wider. With closed eyes, hand positioning during reaching was, not surprisingly, less accurate as measured, for example, by variability in midtrajectory. Thus this strategy of opening the hand wider improved the chances of successfully encompassing the object.

A classic finding in the literature on motor control (cf. Sheridan, 1984) is that hand positioning movements carried out rapidly are less accurate in their endpoint than are movements executed more slowly. One interpretation is that with rapid movements, there is less time for visual feedback processing to reduce the hand-target error. When subjects were asked to reach as rapidly as possible, Wing et al. (1986) found that they opened their hands to a wider maximum. As in the case of reaching blind, it seems reasonable to suppose that the increased width of opening served to compensate for the reduced opportunity to home the hand in accurately around

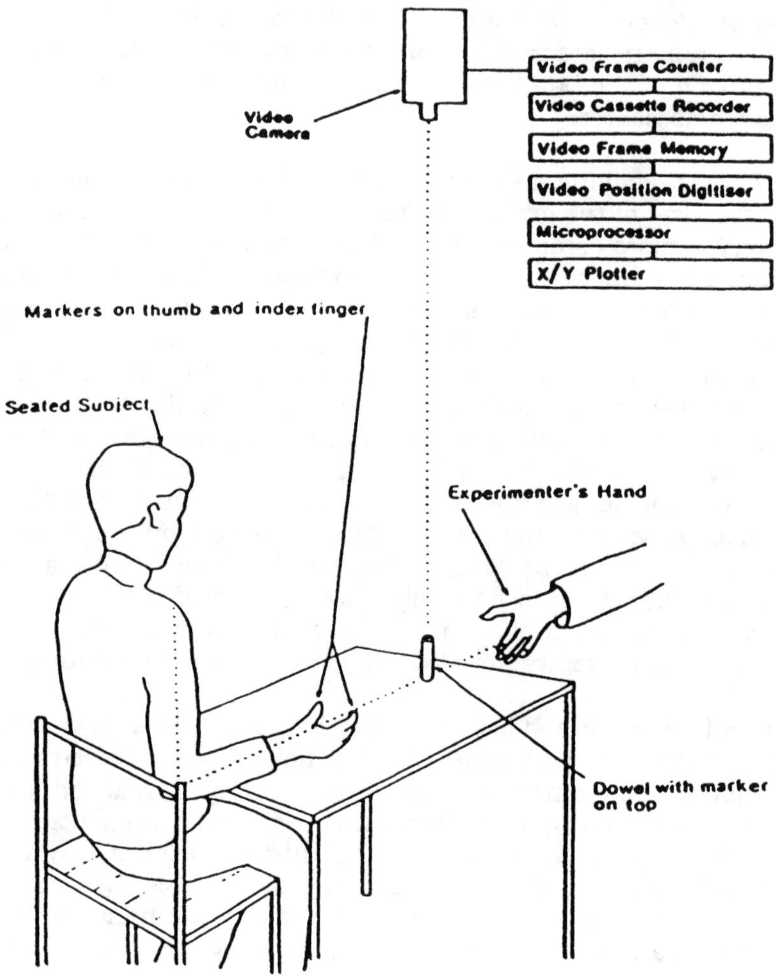

Figure 9.3. The experimental setup for studying coordination in reaching. *Note.* From "Grasp Size and Accuracy of Approach in Reaching" by A. M. Wing, A. Turton, and C. Fraser, 1986, *Journal of Motor Behavior, 18*, p. 249. Published by Heldref Publications. Copyright 1986 by the Helen Dwight Reid Educational Foundation. Reprinted by permission.

the target. This interpretation receives further support from Wallace and Weeks (1988), who subsequently showed that the effect was not due to the change in velocity per se but to the lack of time before the movement was completed.

These observations imply there is informational dependence between the motor components serving transport and grasp in reaching. A strict interpretation of Arbib's (1981) and Jeannerod's (1981) views would be that motor output within each channel is completely specified by the visual input. This explanation implies that coordination depends only on vision and not on a contribution due to reciprocal influence between the two channels at the motor level. In contrast, we are suggesting that a change of the transport phase can induce a change in the patterning of the grasp and that this change takes place (within Arbib's [1981] framework) at the motor level.

PRIOR KNOWLEDGE AS A BASIS FOR COORDINATION IN REACHING

In Wing et al.'s (1986) experiment, the widening of grasp when the experimental conditions constrained the accuracy of hand positioning was viewed as a strategic compensation for an expected decrease in achievable accuracy of the transport component of movement. Wing et al. suggested that subjects used their knowledge about the likely outcome of an action according to the conditions in which it was to be performed. It might be thought that such self-knowledge could influence the transport component of reaching and not just the grasp. Thus, in reaching with the eyes closed, a subject might be expected to move the hand into position with reduced velocity. This strategy would reduce the consequences of an uncontrolled collision of the hand with the target and so improve the chances of successfully grasping it. In Wing et al.'s study, although grasp was wider when subjects reached with eyes closed, the movement time in this condition was no slower than in normal reaching. However, it is possible that if grasp aperture had not been wider, subjects would have reduced the velocity of approach and thereby lengthened movement times.

Another example of the influence of self-knowledge on coordination is provided by a case study of reaching by a girl fitted with a below-elbow artificial hand that provided no tactile feedback (Wing &

Fraser, 1983). The girl was asked to pick up cylindrical pieces of wooden dowel balanced on end and to use either her natural left hand or her artificial right hand. In the later phases of reaching, as the target was neared, the trajectories of the artificial hand revealed a slower approach and a delayed closing relative to the trajectories of the natural hand (see Figure 9.4). Whereas delayed closing would give greater clearance as the hand encompassed the object, the slower approach may be viewed as a strategy to allow more time for processing visual feedback to compensate for the lack of tactile feedback.

The adjustments to grasp or transport that we have discussed so far relate to changes in the state or capabilities of the subject. However, there is no reason that only self-knowledge would be used to strategically alter the component movements in reaching. Prior experience with a range of objects in various environments may lead people to modify the pattern of reaching. People take more care in picking up a full glass, for example by using lower velocities of approach, than they do an empty glass. People develop an ability to predict the behavior of liquid in a glass as a result of learning by trial and error. They may then approach the glass in such a way that impact on contact is avoided.

Evidence of such modifications to reaching based on prior experience with real-world objects was provided by Marteniuk, McKenzie, Jeannerod, Athènes, and Dugas (1987), who showed that the fragility of an object can influence the transport component. They contrasted subjects' performance in picking up a light bulb or a tennis ball. The light bulb was presented with the metal base facing away from the subject so that, from the subject's perspective, it would be comparable in shape and size to the tennis ball. Nevertheless, when subjects reached for the light bulb, the deceleration phase was longer than when they reached for the tennis ball. This adjustment presumably served to reduce the impact on collision. Although not documented, there may also have been widening of grasp to reduce the possibility of uncontrolled collision of the fingers and thumb with the object during the approach.

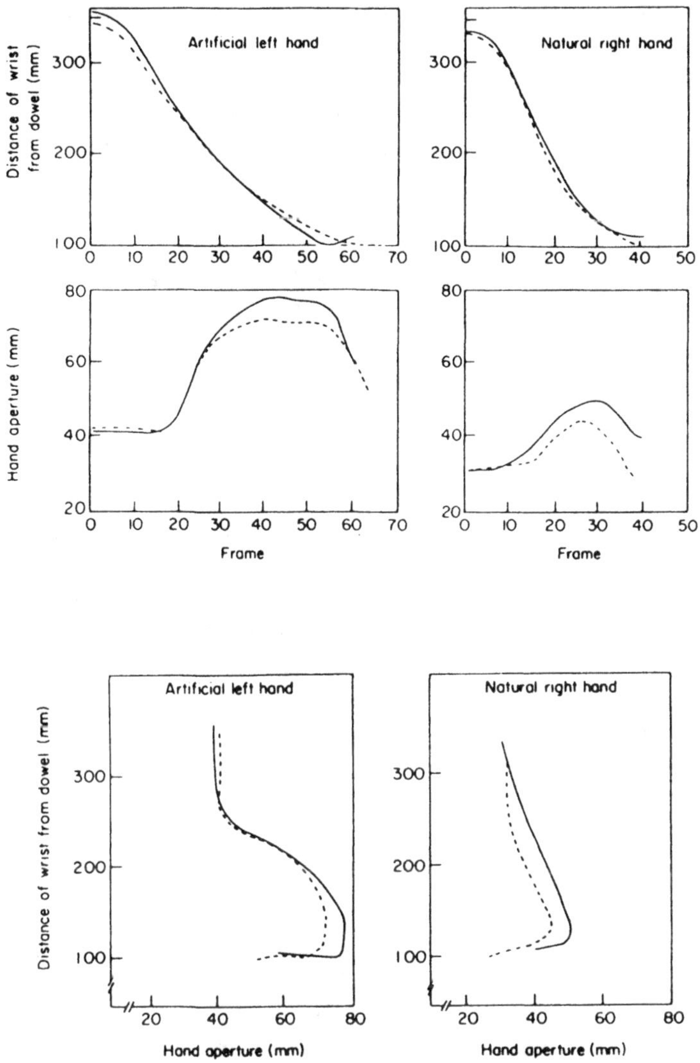

Figure 9.4. Grasp and transport trajectories showing delayed closure on the side with the artificial hand. *Note.* From "The Contribution of the Thumb to Reaching Movements" by A. M. Wing and C. Fraser, 1983, *Quarterly Journal of Experimental Psychology, 35A*, p. 303. Copyright 1983 by the Experimental Psychology Society. Reprinted by permission.

A NEW EXPERIMENT: ADJUSTMENTS OF GRASP FOR WHAT MIGHT HAPPEN

A direct manipulation to provide further evidence of (a) the informational interdependence between the motor components and (b) the influence of the subject's knowledge of the situation was contained in an experiment that we conducted recently. On randomly selected trials, we perturbed the reaching movement after it had been initiated. In some blocks of trials, perturbation was accomplished by the object's being moved to one side. In other blocks, the subject's moving arm was displaced in the opposite direction. In either case, on these trials the sideways displacement of the hand relative to the object was the same and required similar corrective action by the subject. On the remaining trials, the movement was allowed to continue normally to completion.

In this experiment, we were interested in observing whether subjects, knowing that there might be a perturbation, would use a strategy to reduce its likely effects if it should actually occur. In particular, we wanted to know whether the maximum grasp aperture, which the earlier work by Wing et al. (1986) had shown may compensate for uncertainty about an object's position, would be larger in blocks of trials in which perturbations might occur than in blocks that the subject was informed would contain no perturbations.

Because our interest was in anticipatory compensation rather than in the way subjects correct ongoing movement (cf. Abbs, Gracco, & Cole, 1984), measurements were taken only from trials in which there was no perturbation. These were selected from the blocks in which perturbations were randomly intermixed and from two control conditions. In the standard control condition, the subject always reached straight ahead for the object. In the other control condition, subjects reached for an object either centered on the midline or placed either to the left or to the right of midline; straight-ahead reaching trials were selected from this condition. The reason for including this condition was that a context in which some movements (those in the perturbation trials) required a correction with a sideways component might result in enlarged grasp aperture. We wanted to ensure that trial-to-trial directional variation in transport alone would not result in wider maximum hand aperture.

Five right-handed subjects were asked to reach for a light cylindrical object placed 30 cm in front of them along the midline. The cylinder, 8 cm high, had a diameter of 2 cm. In the first two control trial blocks, subjects reached normally. In the first block (five trials), the cylinder was placed straight ahead. In the second block of 15 trials, the cylinder was placed equally often 30° to the left of the center line, 30° to the right, or centrally. In the third and fourth blocks of trials, we introduced two types of perturbation. On randomly selected trials in one block, the subject's arm was briefly pulled (by a string tied around the wrist) to the right during the transport phase. In the other block, on randomly selected trials, the table on which the dowel rested was moved to the left during the transport phase. In both cases, the result of the perturbation (if left uncorrected) was to introduce a 2-cm leftward shift of the target relative to the subject's hand. The instructions for the task stated that subjects should pick up the dowel in spite of the perturbation that would occur on half of the trials. The blocks including perturbation trials consisted of 10 trials—5 perturbed, 5 not perturbed.

The data from those trials in which a perturbation might have occurred but did not after all take place were then compared with the data from the two initial blocks of trials in which there were no perturbations. (The latter trials were run first so that subjects would not have reason to suspect that perturbations would be introduced). A mixed between-within ANOVA with subjects as a between factor and condition as a within factor revealed a significant effect of experimental condition, $F(3, 60) = 10.48$, $p < .01$. The results given in Table 9.1 indicate that the origin of the effect lay in the wider grasp used in the blocks of trials involving perturbation. There were significant differences between subjects in overall grasp size, $F(4, 20) = 10.54$, $p < .01$. In addition, 1 of the 5 subjects showed no differences in maximum aperture, behavior that resulted in a significant Subject x Condition interaction, $F(12, 60) = 2.70$, $p < .01$.

Because the trials analyzed were not actually perturbed trials, we view the increases in grip size as an anticipatory compensation strategy which, as in fast or blind reaching, permits more variability in hand position relative to the object during the approach phase. Thus, grasp is sensitive not only to the characteristics of the object itself but also to the nature of the action directed at the object. In view of this finding, it would be hard to understand coordination in reaching between transport and grasp if it were driven only by vi-

TABLE 9.1

Maximum Aperture Under Different Reaching Conditions (cms)

	No perturbation		Perturbation	
	1-direction reach	3-direction reach	Perturbation of arm	Perturbation of object
Maximum aperture	6.0	5.8	6.4	6.5

sual input. In our view, a model that brings together direct visual input and input from prior knowledge is required. What people know and expect about how movement will proceed makes a significant contribution to coordination in reaching.

THE LIGHT OF EXPERIENCE

The results of our experiment indicate that reciprocal influences at the motor level can modify the reach and grasp pattern without being influenced by the visual input. Drawing on Arbib's (1981) representation, we therefore argue that there is a link between the motor components distinct from links between the visual components. Although the effect of the link is to relate the motor components in reaching, at present we cannot rule out the possibility that the link may be quite indirect, at a cognitive level. One way to resolve this question would be to look at the development of a compensatory strategy after the occurrence of each unexpected perturbation. And, if necessary, the element of surprise could be created by a single incident of perturbation during the experiment. Indeed, repeated perturbations would induce expectancy and thus influence the strategy.

Another way to tackle this issue would be to study coordination in cases in which the degree of subjects' existing knowledge might be expected to be small. Studies of infant reaching provide information about the coordination of reach and grasp. Hofsten (1979), having found that smooth grasping does not appear until after arm transport has been stabilized, questioned the role of experience in adjustment of the hand to the form and size of the target object. As their capacity to stabilize themselves increases simultaneously with knowledge derived from encounters with objects, infants might be expected to show the development of strategies appropriate to their physical development. Moreover, the growth in knowledge available to the infant about object attributes such as mass and stability should help them define the likely consequences of collision. This development, one might expect, should lead to observable changes in the coordination of grasp and transport components of reaching.

Obviously, these suggestions cannot be investigated with the tasks often encountered in the laboratory that deprive subjects of the ability to demonstrate worldly knowledge. In our view, future research on the nature of coordination between transport and grasp in reaching, and perhaps on coordination in general, would improve if it were more firmly grounded on tasks involving realistic goals for objects and chosen for their correspondence to tasks in the real world. Only then will strategies for coordination, built up as they are through the extensive interactions of the subject with the real world prior to coming into the laboratory, be comprehensible when viewed under the behavioral microscope.

REFERENCES

Abbs, J. H., Gracco, V. L., & Cole, K. J. (1984). Control of multimovement coordination: Sensorimotor mechanisms in speech motor programming. *Journal of Motor Behavior, 16*, 195-231.

Arbib, M. (1981). Perceptual structures and distributed motor control. In V. B. Brooks (Ed.), *Handbook of physiology: Sec. 1. The nervous system: Vol. 2. Motor control* (pp. 1449-1480). Bethesda, MD: American Physiological Society.

Goodale, M. A., Pelisson, D., & Prablanc, C. (1988). Large adjustments in visually guided reaching do not depend on vision of the hand or perception of target displacement. *Nature, 320*, 748-750.

Hofsten, C. von. (1979). Development of visually directed reaching: The approach phase. *Journal of Human Movement Studies, 5,* 160-178.

Jeannerod, M. (1981). Intersegmental coordination during reaching at natural visual objects. In J. Long & A. Baddeley (Eds.), *Attention and performance IX* (pp. 153-169). Hillsdale, NJ: Erlbaum.

Jeannerod, M. (1984). The timing of natural prehension movements. *Journal of Motor Behavior, 16,* 235-254.

Marteniuk, R. G., McKenzie, C. L., Jeannerod, M., Athènes, S., & Dugas, C. (1987). Constraints on human arm movement trajectories. *Canadian Journal of Psychology, 41,* 365-378.

McLeod, P., McLaughlin, C., & Nimmo-Smith, I. (1985). Information encapsulation and automaticity: Evidence from the visual control of finely timed actions. In M. I. Posner & O. Marin (Eds.), *Attention and performance XI* (pp. 391-406). Hillsdale, NJ: Erlbaum.

Paillard, J. (1982). The contribution of peripheral and central vision to visually guided reaching. In D. Ingle, M. A. Goodale, & R. M. Mansfield (Eds.), *Analysis of visual behavior* (pp. 367-385). Cambridge, MA: M.I.T. Press.

Paillard, J., & Beaubaton, D. (1978). De la coordination visuomotrice à l'organisation de la saisie manuelle [On visual motor coordination in the organization of manual reaching]. In H. Hécaen & M. Jeannerod (Eds.), *Du contrôle moteur à l'organisation du geste* (pp. 225-260). Paris: Masson.

Sheridan, M. R. (1984). Planning and controlling simple movements. In M. M. Smyth & A. M. Wing (Eds.), *The psychology of human movement* (pp. 47-82). New York: Academic Press.

Wallace, S. A., & Weeks, D. L. (1988). Temporal constraints in the control of prehensile movement. *Journal of Motor Behavior, 20,* 81-105.

Wing, A. M., & Fraser, C. (1983). The contribution of the thumb to reaching movements. *Quarterly Journal of Experimental Psychology*, 35A, 297-309.

Wing, A. M., Turton, A., & Fraser, C. (1986). Grasp size and accuracy of approach in reaching. *Journal of Motor Behavior, 18*, 245-260.

Perspectives on the Coordination of Movement
S.A. Wallace (Editor)

THE COORDINATION OF SIMULTANEOUS ACTIONS

David E. SHERWOOD*

Department of Kinesiology
University of Colorado, Boulder

ABSTRACT

The concept of a coordinative structure is discussed in light of recent experiments involving simultaneous multilimb actions. As in previous experiments, strong temporal relationships between limbs were demonstrated even when task requirements of each limb differed. In such mixed conditions, modulation effects were demonstrated as increased spatial error, increased temporal differences, and reduced correlations compared to same-task conditions. However, practice tended to reduce modulation effects while increasing temporal correlations and reducing temporal differences in groups performing mixed tasks. Although the effect of practice on spatial correlations was inconclusive, the fact that the spatial correlations were consistently lower than temporal correlations suggests that the distinction between essential and nonessential variables is plausible in simultaneous multilimb movements.

Clearly, there are many approaches to the study of human motor coordination, as the variety of techniques described in this book attest. One popular approach has been to study interlimb coordination in cyclical movements such as locomotion (e.g., Hoenkamp, 1978; Shapiro, Zernicke, Gregor, & Diestel, 1981) or two-handed repetitive actions (e.g., Kelso, Holt, Rubin, & Kugler, 1981). In these studies,

---

*Address correspondence to: David E. Sherwood, Campus Box 354, University of Colorado, Boulder, CO 80309-0354, U.S.A.

researchers have manipulated such variables as speed or rhythm in an attempt to identify the "essential" and "nonessential" variables involved in the coordination function of complex actions (Kelso, Putnam, & Goodman, 1983). Essential variables determine the unchanging topological characteristics, whereas nonessential variables can be freely varied without disruption of the internal structure of the action. Studies of these variables might provide clues to the way the nervous system achieves coordination and control of the large number of degrees of freedom present in the human motor system.

By changing distance, load, or velocity, researchers have also attempted to distinguish between essential and nonessential variables in a variety of discrete tasks. The application of this approach to discrete, simultaneous actions is the topic of this chapter.

COORDINATIVE STRUCTURES: A FRAMEWORK FOR STUDYING INTERLIMB COORDINATION

One framework for assessing the contribution of discrete actions within the context of interlimb coordination is the coordinative structure, which Turvey, Shaw, and Mace (1978) define as "a group of muscles, often spanning several joints that is constrained to act as a unit" (p. 563). Theoretically, the coordinative structure simplifies control for the nervous system by controlling groups of muscles as a unit, rather than controlling each muscle or joint individually, and thereby reducing the large number of degrees of freedom inherent in the motor system. In this manner, coordinative structures provide a solution to what Bernstein (1967) referred to as the "degrees of freedom" problem.

When groups of muscles are marshalled together in a coordinative structure, they can be described in terms of essential and nonessential variables (Kelso et al., 1983). The "signature" or expression of a coordinative structure can be identified by manipulation of nonessential variables (e.g., movement speed, overall duration, or overall force) and identification of the aspects of the pattern that remain stable. An example of an essential variable in gait is the invariant relative motion of the thigh and knee across different walking and running velocities, noted by Shapiro et al. (1981).

Although some evidence for coordinative structures exists, (e.g., Kugler, Kelso, & Turvey, 1980), many questions remain unanswered. For example, what are the essential variables in simultaneous limb actions? Are coordinative structures learned with practice, or do they exist on the first attempt of a novel action? Are there structural or task-related limitations that make the formation of coordinative structures difficult when simultaneous actions are required? It is the purpose of this chapter to address some of these questions and bring some recent experimental evidence to bear on the issues raised here.

COORDINATIVE STRUCTURES IN SIMULTANEOUS ACTIONS

Some of the most convincing evidence for coordinative structures in discrete movements comes from the two-handed aiming paradigm used by Kelso, Southard, and Goodman (1979a, 1979b). In a variation of Fitts's (1954) aiming task, the subject made rapid movements of one or two hands in the frontal or sagittal plane to targets of varying sizes. As expected by Fitts' law, movement times increased when subjects moved to a narrow target ("hard" condition) compared to a wide target ("easy" condition). The easy-hard differences in movement times were about 75 ms in single movements and about 80 ms when both hands moved to the same sized target. However, when the hands moved to different-sized targets, the task-related differences in movement times were reduced to about 20 ms. Kelso et al. (1979b) also noted that the limbs were highly synchronized in terms of peak velocity and peak acceleration, a result suggesting that both limbs were temporally constrained to act as a single unit. A later study involving the same task (Kelso et al., 1983) showed strong interlimb correlations for movement time and reaction time across all two-handed conditions, a finding again suggesting that overall duration or movement time are essential variables in this task. Similar results obtained when homologous and nonhomologous muscle groups were involved suggest that the relevant muscles had been functionally combined to operate as a single unit.

Marteniuk and his colleagues (Marteniuk & MacKenzie, 1980; Marteniuk, MacKenzie, & Baba, 1984) have examined the notion of a coordinative structure by varying the intensity requirements in single or two-hand aiming movements. By increasing distance or load in one limb, they determined whether the temporal structure remained constant across changes in intensity. They demonstrated that increasing movement distance from 10 to 30 cm increased the movement time about 60 and 50 ms in single and dual responses, respectively. As in Kelso et al.'s (1979a, 1979b, 1983) studies, the difference between limbs in mixed responses (i.e., 10 cm in one limb and 30 cm in the other) decreased to about 30 ms, a result again suggesting that some common timing mechanism exists across limbs and movement amplitudes. Strong correlations between interlimb movement times (all above .72) were also shown in all two-hand conditions. The same trends were demonstrated in an experiment in which an additional load was added to one limb, but the effects appeared to be stronger with the distance variation. In general, these studies suggest that simultaneous limb movements are not controlled separately but controlled in a way that preserves the temporal structure of the response across variations in the spatial pattern.

MODULATION EFFECTS IN SIMULTANEOUS ACTIONS

In two-hand conditions with the same task requirements, the absolute movement times and response times for each limb have been shown to be very similar (Kelso et al., 1979a, 1979b; Marteniuk & MacKenzie, 1980; Marteniuk et al., 1984). However, the similarity in movement times between the hands, noted in equal-task conditions, was modulated as the task requirements were varied. For example, when the hands moved to wide targets or over less distance, the faster movement time was found in the hand moved to the wider or closer target (Kelso et al., 1979b, Marteniuk & MacKenzie, 1980). But the modulation effect is really more complicated than it appears. When the hands moved to targets of unequal difficulty, the hand moving to the wide target (or over the shorter distance) moved more slowly than it would if both hands had moved to the wide target (or over the same short distance); the hand moving to the narrow target (or over the longer distance) moved faster than if both hands had moved to the narrow target (or over the longer distance). Such modulation effects were also shown in the form of negative constant errors in the hand moving the longer distance and in positive constant errors in the hand moving the shorter distance (Marteniuk et al., 1984).

In a recent experiment, Caldwell and Goodman (cited in Goodman, 1985) examined the modulation effect in a more refined way. Using the two-hand aiming paradigm, they had subjects make lateral movements to same-sized targets 30 cm away. But on 20% of the trials, 70 ms after a tone signaling the subject to initiate movement, another tone signaled the subject to move one hand 20 cm further to a target 50 cm away. Caldwell and Goodman questioned whether or not the temporal structure of the movement would be disrupted by such peripheral input, perhaps leading to independent control and timing of each limb. As in the previous studies, the interlimb differences in movement time and response time were very small in the unswitched trials. Relative to the unswitched trials, the movement times of *both* limbs increased in response to the increased distance in one of the limbs, although the movement time in the unswitched limb was less than that in the switched limb. Again it appears that the temporal pattern is modulated as a whole rather than in each limb independently. It is also interesting to note that the response time in the switched condition was not slower than that in the unswitched condition. This result suggests that the movement was not entirely reprogrammed before it was initiated.

Goodman, Kobayashi, and Kelso (1983), again using the two-hand aiming paradigm, questioned whether or not the same coordinative structure could control movements for which the spatial requirements of the two hands were quite different. In this experiment, one limb was required to clear hurdles between 0 and 40 cm in height, but no hurdle was placed in the path of the other limb. As height of the hurdle increased, the movement time of both limbs increased, although the hurdle-side limb was slightly slower than the nonhurdle-side limb, a finding that suggests common timing across limbs. In addition, the peak vertical displacement in both limbs increased as the height of the hurdle increased, a result suggesting common spatial constraints as well. The effects described here, in which limb actions tend toward similarity, suggest one way in which the motor system achieves economy in the production of movement, perhaps through a coordinative structure.

Another example of modulation effects has been provided by Fagard, Morioka, and Wolff (1985) in a bimanual tracing task requiring either the same or different angular velocities in each hand. Subjects were required to rotate both hands the same direction to move a pen of an X-Y recorder along angles of 45° (same velocity), 67° (right hand velocity greater than left) or 22° (left hand velocity greater than right). Adults and children traced the 45° angle more accurately and in less time than they did the other two angles, with adults showing lower error scores than the children did. However, modulation effects were noted in the directional error data, with performance in both the 67 and 22° conditions deviating toward 45°, this deviation suggesting a tendency for similar actions in both hands. Interestingly, there was no reduction in the directional bias with practice, although the average error about the goal angle did decrease.

The between-hand differences in movement time during mixed task conditions have been interpreted in various ways. Kelso et al. (1983) suggested that the muscles involved in the action act as "soft," non-linear springs in which stiffness decreases with increases in distance, so that longer distances are covered with less average stiffness and result in a slower movement time. Marteniuk and MacKenzie (1980), on the other hand, argued for a model in which the movement time in each limb is modulated by "neural crosstalk" at several levels of the nervous system, with efferent commands destined for the left hand affecting the right hand via ipsilateral connections (Preilowski, 1977). They suggested that the output noted in each limb is an interaction of the descending commands that make the movement times of the two limbs similar, but not identical. Although the issue is far from settled, it does appear that the neural crosstalk model can account for modulation effects arising from variations in both distance and load.

All of the modulation effects noted so far arise when the task requirements of the left- and right-side limbs differ in distance, velocity, or load. Are modulation effects also present when task requirements differ between the upper and lower limbs, for example? The question was addressed in an experiment (Sherwood, 1987) in which the subjects simultaneously moved their upper limbs 16 cm and their lower limbs 9 cm to targets. These subjects were compared with a group moving all limbs 9 cm. In the mixed condition, the lower limbs showed no tendency to overshoot the targets and showed no increase in variable or overall error. In addition, the movement time of the lower limbs was faster than that of the upper limbs, a difference that was maintained throughout the practice session. The results suggest that the two sets of limbs were controlled somewhat independently, with few modulation effects.

PRACTICE AND MODULATION EFFECTS

Although Kelso et al. (1979a, 1979b; 1983) and Marteniuk et al. (1984) did not assess the effect of practice on interlimb coordination in their studies, they did discuss its possible effects. For example, Marteniuk et al. suggested that the learning of bimanual skills involves insulation from neural crosstalk, which implies that the hands may achieve relative independence and show reduced modulation effects. Kelso et al. (1979b) suggested that practice might break down temporal constraints and free the subject from the temporal structure evident in their studies. Another effect of practice might be to strengthen the temporal bonds or invariances between limbs, rather than make them more independent. This viewpoint implies that the between-hand differences in mixed conditions indicate a relatively early stage in the learning process and that continued practice is required to marshall the appropriate muscle groups in the coordinative structure. If practice is required to form the coordinative structure, then one might expect to see smaller, rather than larger, differences in movement time or the kinematics, and the emergence of the essential variables involved in the task.

The problem was recently investigated in my laboratory with a task that required a simultaneous pushing action of all four limbs in the sagittal plane. Twenty right-handed subjects were randomly assigned to different groups. In the *same* group, subjects moved all four limbs to targets the same distance (9 cm) away. In the *different* group, subjects moved the left-side limbs to targets a shorter distance (5 cm) than they moved the right-side limbs (9 cm). Although subjects were prevented from viewing the movement of their limbs, they were given the final position of each limb via a digital display as knowledge of results at the end of the trial. Potentiometers were affixed to the hand levers and the foot pedals to measure limb position and were sampled on-line at 250 Hz. The apparatus described here is shown in Figure 10.1. Both groups were given 125 practice trials, five times as much practice as the subjects were given in the Kelso et al. (1979a, 1979b) and Marteniuk et al. (1984) studies. The main temporal measures taken from the displacement and velocity traces were movement time and the difference in and correlation of the time of peak velocity between left and right limbs. Variable error, overall error, and constant error were also computed from the displacement traces as measures of spatial accuracy. Some repre-

Figure 10.1. Apparatus used in the four-limb studies. (Top left: digital display; top right: foot pedals; bottom: hand levers.)

sentative displacement and velocity traces for one trial for one subject are shown in Figures 10.2 and 10.3, respectively.

The major comparisons of interest were between the right-side limbs in the *same* and *different* groups, in which the left-side limbs were moving either the same or a shorter distance, respectively. Table 10.1 shows the average movement times, peak velocities, amplitudes, and spatial errors for each limb in both groups. The movement times and distances in the *same* group were very similar across limbs, but peak velocity tended to be larger in the right hand and smaller in the right foot. The subjects in the *different* group were successful in varying the distance moved in the left- and right-side limbs. As in previous studies, modulation effects were noted when the limbs moved different distances. For example the right-side limbs in the *different* group showed smaller amplitudes and larger variable and overall error than did the comparable limbs in the *same* group. In addition, the between-hand differences in move-

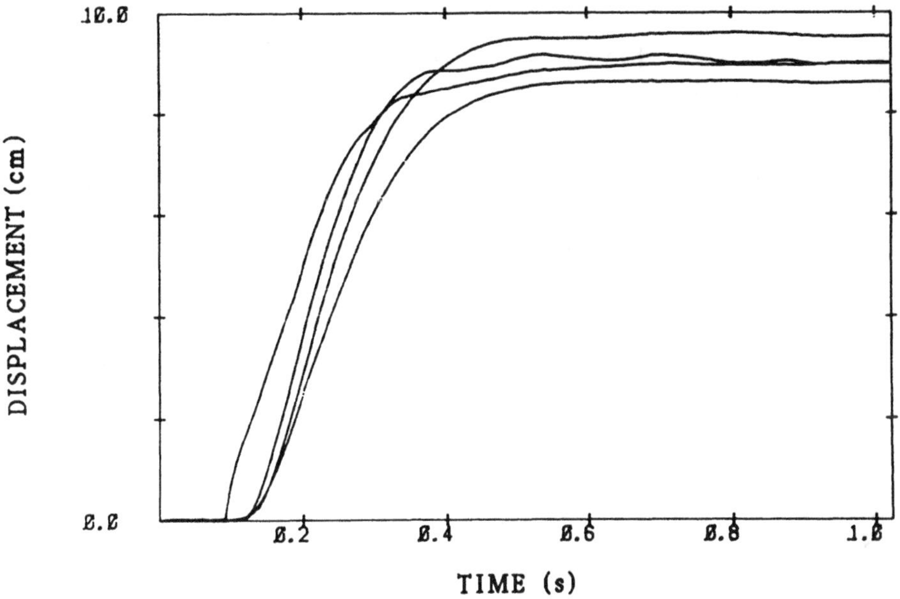

Figure 10.2. *Representative displacement traces of all four limbs of a simultaneous aiming response (Subject 2, Trial 24).*

ment time in the *different* group were much greater than in the *same* group.

Practice, however, did affect the magnitude of the modulation effect. Figure 10.4 shows constant error and Figure 10.5 shows overall error for the right hand and right foot for both groups for each block of trials. Notice the large negative constant errors (indicating undershoots of the target) in the right foot of the *different* group relative to the same limb in the *same* group, particularly on the first few blocks of practice. With additional practice, the constant errors approach zero, a pattern indicating a reduction in the modulation effect and better task performance. The trend of reduced movement distance is also shown in the hands of the *different* group. With practice, both groups reduced overall error, but the errors for the *different* group were always greater than those for the *same* group. It seems that

Table 10.1

Means and Standard Deviations for Each Limb for the Same *and* Different *Groups.*

Group	Amplitude (cm)		Movement time (s)		Peak velocity (cm/s)	
	M	*SD*	*M*	*SD*	*M*	*SD*
Same						
Left hand	9.42	0.67	0.649	0.314	39.7	17.8
Right hand	9.71	0.66	0.652	0.319	46.5	21.6
Left foot	9.18	0.43	0.662	0.306	42.1	16.5
Right foot	8.56	0.40	0.637	0.288	31.3	11.8
Different						
Left hand	5.29	0.56	0.556	0.280	21.6	8.5
Right hand	9.31	1.18	0.764	0.357	38.2	16.5
Left foot	5.13	1.00	0.473	0.174	27.4	8.6
Right foot	8.02	1.05	0.759	0.334	25.5	10.7

practice can reduce modulation effects to some extent but not eliminate them entirely.

Practice also reduced the interlimb difference in movement time. Table 10.2 shows the mean interlimb difference in movement time for both groups. For the upper limbs in the *same* group, there is almost no change in the movement time difference with practice. But the *different* group gradually reduced this difference from 0.238 s on Block 1 to 0.179 s on Block 5, a 25% reduction. The lower limbs showed reduced differences in the last block of practice relative to all blocks but the first. In general, practice appears to make the movement times more similar, particularly in mixed conditions.

Table 10.3 summarizes the mean absolute difference in time of peak velocity for the upper and lower limbs for both groups (the values are based on means computed for each block of 25 trials). In the *same* group, the mean differences were smaller than those in the *different* group, a result suggesting that the temporal structure was stronger

Table 10.1 (continued)

Group	Variable error (cm)		Overall error (cm)	
	M	*SD*	*M*	*SD*
Same				
Left hand	0.556	0.156	0.868	0.448
Right hand	0.527	0.186	0.938	0.537
Left foot	0.739	0.315	0.886	0.336
Right foot	0.589	0.304	0.828	0.374
Different				
Left hand	0.840	0.228	1.026	0.309
Right hand	0.685	0.218	1.321	0.479
Left foot	0.913	0.315	1.228	0.534
Right foot	0.833	0.470	1.518	0.863

Note. *SD* is the between-subject standard deviation in the same units as the respective mean.

when limbs moved the same distance and that modulation effects exist in the temporal data as well. However, the temporal differences between the upper limbs decreased with practice in the *different* group; this finding indicates a strengthening of the temporal structure with practice. The lower limbs in the *different* group showed the same trends as the movement time difference, with the difference in the time of peak velocity at the last block less than all values but the first.

Table 10.4 shows the correlations between the times when peak velocity was reached for the upper and lower limbs for both groups for each block of trials. In general, the correlations are higher in the *same* group and higher in the upper limbs than in the lower limbs. Notice that the temporal correlations for the *same* group tend to show little change with practice and perhaps a slight decrease. The correlations for the *different* group show small but consistent increases with practice. Although the differences or correlations in the *different* group do not reach the level of those in the *same* group,

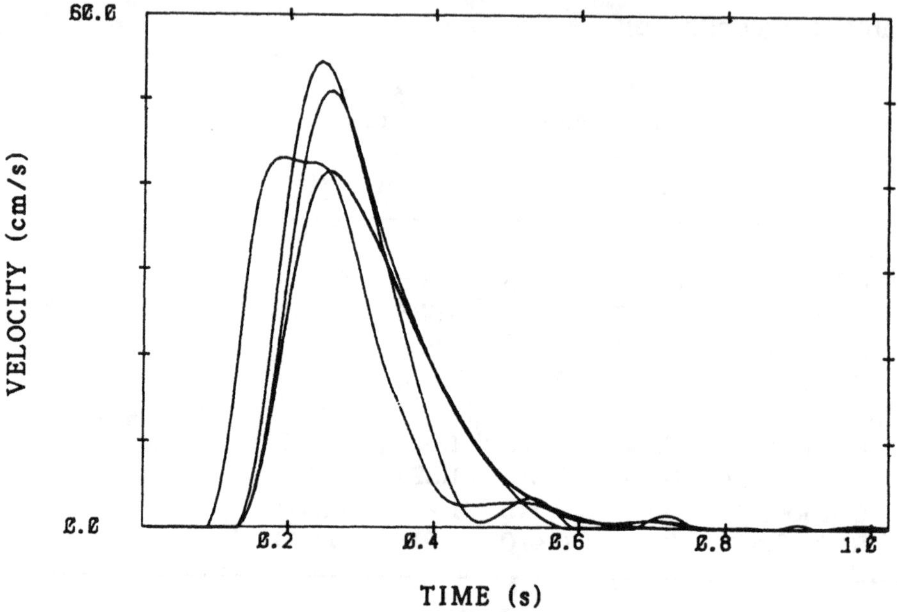

Figure 10.3. Velocity traces for the displacements in Figure 10.2 (Subject 2, Trial 24).

they do suggest a shift toward a unified temporal structure. It appears that practice involves the gradual formation of a coordinative structure, with temporal differences between the limbs decreasing while the temporal correlations between the limbs increase.

Correlations of the distance traveled in each limb (Table 10.5), however, are generally smaller than the temporal correlations (Table 10.4). Relative to the temporal correlations, practice appeared to have a stronger effect on the distance correlations, with the distance correlations decreasing with practice in the *same* group and increasing with practice in the *different* group. The finding that temporal correlations were greater than the spatial correlations is in harmony with the distinction between essential and nonessential variables of a coordinative structure. In fact, Schmidt, Zelaznik, Hawkins, Frank, and Quinn (1979) suggested that the difference between temporal and spatial correlations in two-hand aiming tasks provides one method for discriminating between essential and nonessential variables. In their model, the motor

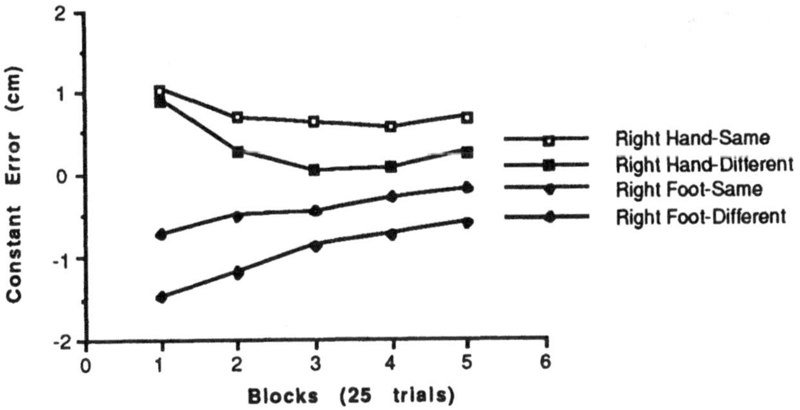

Figure 10.4. Constant error for the right hand and the right foot for each block of trials.

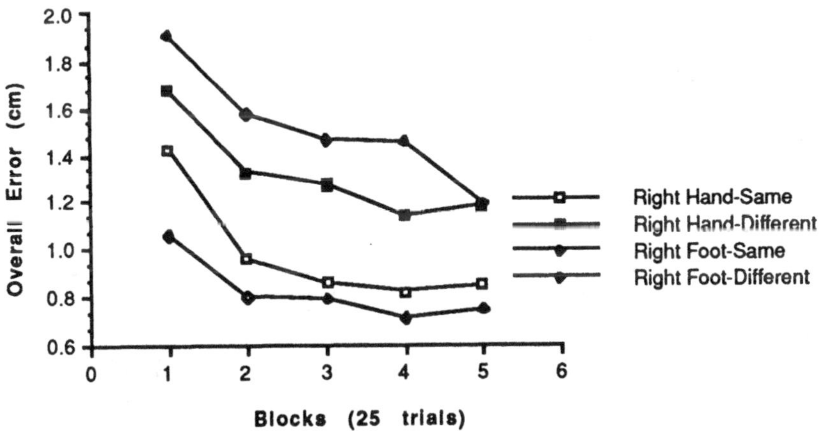

Figure 10.5. Overall error for the right hand and the right foot for each block of trials.

Table 10.2

Mean Differences in Movement Time Between Upper Limbs and Between Lower Limbs for Same *and* Different *Groups for Each Block of Trials*

Block[a]	Same		Different	
	Hands	Feet	Hands	Feet
1	0.003	0.028	0.238	0.228
2	0.001	0.009	0.258	0.296
3	0.011	0.042	0.203	0.324
4	0.003	0.022	0.158	0.301
5	0.004	0.027	0.179	0.280
Mean	0.002	0.025	0.207	0.285

[a]Each block consisted of 25 trials

Table 10.3

Mean Absolute Differences in the Time of Peak Velocity for the Upper and Lower Limbs for Same *and* Different *Groups for Each Block of Trials*

Block[a]	Same		Different	
	Hands	Feet	Hands	Feet
1	0.028	0.067	0.126	0.237
2	0.030	0.111	0.116	0.313
3	0.031	0.115	0.089	0.296
4	0.031	0.103	0.075	0.268
5	0.030	0.109	0.083	0.251
Mean	0.030	0.100	0.098	0.273

[a]Each block consisted of 25 trials

Table 10.4

Mean Correlations of the Time of Peak Velocity Between the Upper Limbs and Between the Lower Limbs for Same *and* Different *Groups for Each Block of Trials*

	Same		Different	
Block[a]	Hands	Feet	Hands	Feet
1	.950	.795	.715	.410
2	.960	.750	.740	.375
3	.960	.780	.738	.458
4	.950	.765	.795	.465
5	.940	.770	.785	.495
Mean	.950	.773	.755	.440

[a]Each block consisted of 25 trials

program first specifies the common properties across each limb (e.g., the overall duration), then specifies the parameters that control the specific movement characteristics of each limb, such as the distance of movement or the direction the limb is to travel. Further, they assume that parameters are independently selected for each limb. Therefore, high within-subject correlations should be noted between variables that are thought to be invariant characteristics applied to both limbs, and low correlations should occur for variables that are specified independently for each limb. Certainly the spatial and temporal correlational results support their model. The results also suggest that motor learning involves the ability to temporally organize the responding limbs to act as a single unit (i.e., to develop the essential variables) and at the same time learn to apply specific parameters independently for each limb.

A recent study of the four-limb task in a reaction time paradigm also suggested that the interlimb difference in reaction time can also be modified by practice (Sherwood, 1987). The subject moved the upper and lower limbs to targets 16 and 9 cm away, respectively, following a warning signal and an auditory initiation stimulus after a variable delay interval (1-3 s). The absolute difference in

Table 10.5

Mean Distance Correlations Between the Upper Limbs and Between the Lower Limbs for Same *and* Different *Groups for Each Block of Trials*

	Same		Different	
Block[a]	Hands	Feet	Hands	Feet
1	.905	.740	.330	.253
2	.815	.600	.380	.262
3	.790	.617	.345	.275
4	.775	.540	.470	.300
5	.812	.630	.562	.300
Mean	.830	.630	.422	.278

[a]Each block consisted of 25 trials

reaction time was computed for each pair of limbs for each block of 25 trials. The mean difference between the hands was 34 ms; the difference between all other pairs was 78 ms. Practice reduced the difference in initiation time an average of 18 ms across each pair over five blocks of trials, with the largest reductions noted between ipsilateral (32 ms) and opposite, unsymmetrical limbs (33 ms). It appears that practice can enhance the synchronicity of simultaneous movements and that temporal constraints can be strengthened with practice.

Perhaps with continued practice, the temporal differences noted between the hands would have decreased further. Some evidence does suggest that in highly practiced performers, coordinative structures are developed. For example, in a recent study of skilled and less-skilled "opposite-field" batters in baseball, an invariant pattern of relative motion was demonstrated between the left wrist and left elbow angle (McIntyre & Pfautsch, 1982). This finding supports the notion of a unified coordinative structure for all skill levels. The less-skilled batters, however, showed a smaller range of motion and greater elbow extension when attempting to hit to the opposite field, a finding suggesting that these may be nonessential variables that distinguish between skill levels. Perhaps with practice or experience, subjects can adjust the nonessential variables to optimize the outcome of the action or vary the task requirements of the two limbs.

INDIVIDUAL DIFFERENCES AND COORDINATIVE STRUCTURES

In some of the studies just discussed, it can be misleading to decide, for example, whether or not movements are made in the same movement time on the basis of the reported mean interlimb difference in reaction or movement time. Further, because of the wide individual differences in performance that have been reported by Marteniuk et al. (1984) and Kelso et al. (1979b, 1983), these means are not a fair test of the notion that a coordinative structure is the mechanism involved in the coordination of aiming actions. An excellent example is Kelso et al.'s (1983) second experiment, in which the subjects were required to move both hands laterally to targets 24 cm away, an 18 cm-high hurdle having been placed in the path of one of the limbs. For three subjects, the movement time for each limb appeared to be independent, with the nonhurdle limb exhibiting a faster movement time and almost no modulation in the vertical displacement of the limb. These results suggest that the movements were not controlled by a coordinative structure. However, for the remaining four subjects, the vertical displacements of the two hands nearly matched, and the movement times were much more similar than for the other subjects. These results suggest that the limbs had been controlled by a coordinative structure. Obviously, in studies of coordination, averaging across subjects can hide valuable information. When possible, individual subjects' data should be presented.

Similar results were obtained in the four-limb experiment described in this chapter. Even when all four limbs moved simultaneously, each limb reached peak velocity at a slightly different time; the timing pattern differed across subjects but was consistent within subjects. That is, each subject developed a unique temporal pattern (as measured by the time when peak velocity was reached in each limb) that was preserved with practice across changes in movement time. Figure 10.6 shows one subject's temporal pattern. The time of peak velocity is plotted for each limb for each block of practice. The consistent temporal order across practice suggests that the sequence

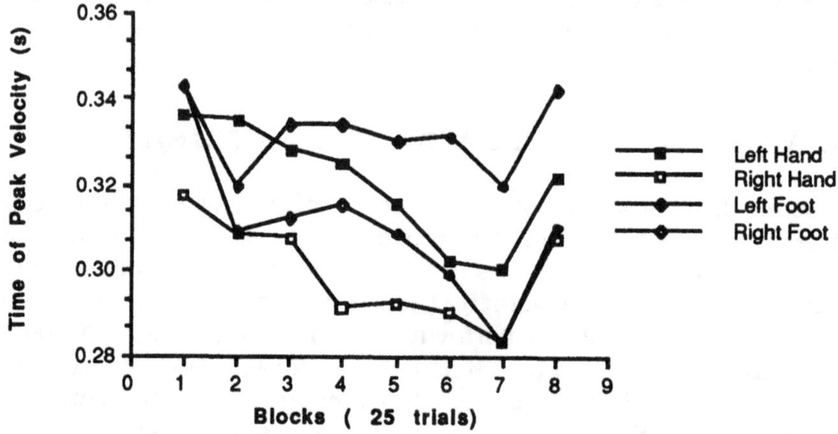

Figure 10.6. Time of peak velocity for each limb plotted for one subject for each block of trials.

of behavior may be an essential variable that is preserved within subjects across changes in overall movement time. The results also suggest that each subject may coordinate the limbs in slightly different ways and that coordinative structures may be very individualistic.

CONSTRAINTS OF MUSCLE GROUPS

The coordinative structure constrains muscles to act as a single unit. Some researchers have questioned whether certain functional muscle groups are relatively easy or difficult to marshall into a coordinative structure (Kelso et al., 1979b, 1983). For example, Kelso et al. (1979b) demonstrated very small interlimb differences in movement time in two-hand aiming movements in three directions, all involving homologous muscle groups (away from midline, toward midline, and forward in the sagittal plane). Later Kelso et al. (1983) replicated the experiment with nonhomologous muscle groups (involving flexion for the left arm and extension for the right arm) and again showed very small differences in movement time between the limbs.

In more complicated simultaneous actions, however, there appear to be advantages in using homologous muscle groups. Fagard (1987) demonstrated in a bimanual tracing task that mirror movements of the hands (homologous condition) were performed faster and more accurately than parallel movements (nonhomologous condition). However, the advantage in making mirror movements was limited to 5- and 7-year-olds, with no advantage shown by age 9. The disappearance of this advantage suggests that children gradually develop a "plasticity" that allows them to overcome the natural constraints and to formulate a new coordinative structure. Using the same-distance condition of the four-limb task described previously, Sherwood and Canabal (in press) demonstrated differential temporal interlimb correlations across pairs of limbs. For example, correlations between opposite, symmetrical limbs were the strongest, followed by those between ipsilateral limbs and opposite, unsymmetrical limbs, a pattern suggesting that coordinative structures are more easily marshalled when symmetrical limbs are involved. Interestingly, the combinations of limbs with the lower initial correlations showed the greater increase in correlation with practice. This improvement suggests that all limbs can be controlled by a coordinative structure if enough practice is provided.

Morasso (1983) studied the temporal synchronicity of bimanual movements in the horizontal plane involving *synkinetic* (spatially similar, nonhomogolous muscles), *homokinetic* (mirror movements, homogolous muscles), and *allokinetic* movements ("free-wheeling, concurrent scribbles" [p. 208]). Temporal coordination of the two limbs was measured by a synchronization coefficient *s*, which was calculated in such a way that if the peaks in the velocity traces occurred at the same time, *s* would be zero. The coefficient would be +1 or -1 if a velocity peak in one limb was paired with a minimum velocity in the other. For synkinetic and homokinetic movement, values for *s* tended to be distributed closely about zero and thus showed strong temporal linkages between the limbs. In allokinetic movements, *s* showed three distinct distributions, grouped around -1, +1, and zero, with more responses in the zero distribution. Apparently, even in seemingly random movements of the two limbs, a high degree of temporal organization was evident and was perhaps a sign that the motor system strives for economy in the coordination of simultaneous actions.

Differences in temporal measures or temporal correlations based on kinematics provide one measure of the ease with which a coordinative structure might be marshalled, but other measures may also be used. For example, measurement of task performance by variable error, overall error, or the number of errors may also contribute to our knowledge about the coordination of actions. If two actions are easily coordinated, then one would expect to find smaller or fewer errors than one would when movements are more difficult to combine. For example, Peterson (1965) devised a task in which each hand could make an aiming movement in either a medial, lateral, distal, or proximal direction from a home key. Various combinations of the movements could be selected to create a number of *same* and *different* movement conditions. A summary of Peterson's results are shown in Table 10.6, in which the number of errors for the hands is shown relative to the action of both hands. Notice that the error rate is lowest along the diagonal, representing conditions in which the hands make identical movements. In mixed conditions, the error rate is lowest when the hands move in either the same or opposite directions in the same plane (e.g., distal-distal, distal-proximal, medial-lateral). The errors are highest when the limbs move in different planes (e.g., lateral-proximal, medial-distal), a result suggesting that pairs of agonistic or antagonistic muscles are more easily marshalled to form a coordinative structure than are other pairs involving different actions.

Error scores may also be higher when it is more difficult to coordinate the actions of the limbs. Figure 10.7 summarizes several experiments using the four-limb aiming task described in this chapter. In the figure, relative error (overall error divided by peak velocity) is plotted by blocks of practice for the right hand or right foot as a function of what the task requirements were for the other limbs. In all cases, data is plotted for limbs attempting to move 9 cm. In general, lower errors are shown when the goal distance is the same rather than different for left and right limbs. The relative errors are higher when the left and right limbs move different distances but not when the feet move a distance different from that moved by the hands. Notice that practice has a large effect on the reduction of error scores in all conditions but that the error scores show the least reduction in the condition in which distance is varied across side.

Table 10.6

Mean Number of Errors in Peterson's (1965) Two-Handed Conditions

Left-hand Movement	Right-hand movement: Distal	Lateral	Proximal	Medial
Distal	19.2	30.0	26.9	28.4
Lateral	26.2	22.7	25.8	23.4
Proximal	20.6	25.6	23.2	25.6
Medial	23.6	20.8	27.7	24.4

Note. All scores in percentages.

SUMMARY

Previous experiments on simultaneous actions of the limbs found a strong temporal relationship between limbs when the task requirements were the same. Studies involving the four-limb task confirmed this relationship. But when the task requirements differed across limbs, larger temporal differences between limbs emerged, particularly when distance was varied across left and right sides. In mixed task conditions, modulation effects were demonstrated as increased spatial error, increased temporal differences, and reduced correlations compared to same task conditions. However, practice tended to reduce modulation effects while increasing temporal correlations and reducing temporal differences in groups performing mixed tasks. Although the effect of practice on spatial correlations was inconclusive, the fact that the spatial correlations were consistently lower than temporal correlations suggests that the distinction between essential and nonessential variables is a plausible one in simultaneous multilimb movements.

Further, there are wide individual differences in task performance in simultaneous actions. Some subjects show strong modulation effects and near-synchronous movements in mixed conditions;

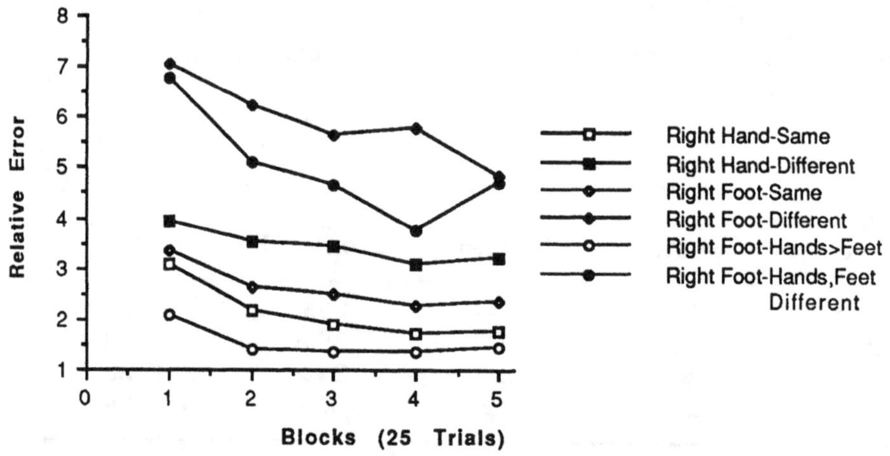

Figure 10.7. Relative spatial error (overall error divided by peak velocity) for the right hand and right foot as a function of task requirements of the other limbs.

others show relative independence in the actions of the limbs. It could be that coordinative structures are very personalized, with each subject exhibiting a distinctive pattern that does not change appreciably with practice. The evidence also suggests that coordinative structures are more easily formed between opposing pairs of muscle groups (e.g., agonists and antagonists for a given action) than between unrelated muscle groups. Whether this situation changes with practice has not been determined.

REFERENCES

Bernstein, N. (1967). *The coordination and regulation of movement.* New York: Pergamon.

Fagard, J. (1987). Bimanual stereotypes: Bimanual coordination in children as a function of movements and relative velocity. *Journal of Motor Behavior, 19*, 355-366.

Fagard, J., Morioka, M., & Wolff, P. H. (1985). Early stages in the acquisition of a bimanual motor skill. *Neuropsychologia, 23*, 535-543.

Fitts, P. M. (1954). The information capacity of the human motor system in controlling the amplitude of movement. *Journal of Experimental Psychology, 47*, 381-391.

Goodman, D. (1985). Synergies and functional constraints in a theory of action. In D. Goodman, R. B. Wilberg, & I. M. Franks (Eds.), *Differing perspectives in motor learning, memory, and control* (pp. 319-340). New York: North-Holland.

Goodman, D., Kobayashi, R. B., & Kelso, J. A. S. (1983). Maintenance of symmetry as a constraint in motor control. *Canadian Journal of Applied Sport Sciences, 8*, 238.

Hoenkamp, E. (1978). Perceptual cues that determine the labeling of human gait. *Journal of Human Movement Studies, 4*, 59-69.

Kelso, J. A. S., Holt, K. G., Rubin, P., & Kugler, P. N. (1981). Patterns of human interlimb coordination emerge from the properties of non-linear, limit cycle oscillatory processes: Theory and data. *Journal of Motor Behavior, 13*, 226-261.

Kelso, J. A. S., Putnam, C. A., & Goodman, D. (1983). On the space-time structure of human interlimb coordination. *Quarterly Journal of Experimental Psychology: Human Experimental Psychology, 35*, 347-375.

Kelso, J. A. S., Southard, D., & Goodman, D. (1979a). On the nature of human interlimb coordination. *Science, 203*, 1029-1031.

Kelso, J. A. S., Southard, D., & Goodman, D. (1979b). On the coordination of two-handed movements. *Journal of Experimental Psychology: Human Perception and Performance, 5*, 229-238.

Kugler, P. N., Kelso, J. A. S., & Turvey, M. T. (1980). On the concept of coordinative structures as dissipative structures: I. Theoretical lines of convergence. In G. E. Stelmach & J. Requin (Eds.), *Tutorials in motor behavior* (pp. 3-47). New York: North-Holland.

Marteniuk, R. G., & MacKenzie, C. L. (1980). A preliminary theory of two-handed coordinated control. In G. E. Stelmach & J. Requin (Eds.), *Tutorials in motor behavior* (pp. 185-197). Amsterdam: North-Holland.

Marteniuk, R. G., MacKenzie, C. L., & Baba, D. M. (1984). Bimanual movement control: Information processing and interaction effects. *Quarterly Journal of Experimental Psychology: Human Experimental Psychology, 36*, 335-365.

McIntyre, D. R. and Pfautsch, E. W. (1982). A kinematic analysis of the baseball batting swings involved in opposite-field and same-field hitting. *Research Quarterly, 53*, 206-213.

Morasso, P. (1983). Coordination aspects of arm trajectory formation. *Human Movement Science, 2*, 197-210.

Peterson, J. R. (1965). Response-response compatibility effects in a two-hand pointing task. *Human Factors, 7*, 231-236.

Preilowski, B. (1977). Phases of motor skills acquisition: A neuropsychological approach. *Journal of Human Movement Studies, 3*, 169-181.

Schmidt, R. A., Zelaznik, H. N., Hawkins, B., Frank, J. S., & Quinn, J. T., Jr. (1979). Motor-output variability: A theory for the accuracy of rapid motor acts. *Psychological Review, 86*, 415-541.

Shapiro, D. C., Zernicke, R. F., Gregor, R. J., & Diestel, J. D. (1981). Evidence for generalized motor programs using gait pattern analysis. *Journal of Motor Behavior, 13*, 33-47.

Sherwood, D. E. (1987, June). *Distance variations and interlimb coordination in a complex motor task.* Paper presented at the meeting of the North American Society for the Psychology of Sport and Physical Activity, Vancouver, B. C.

Sherwood, D. E., & Canabal, M. Y. (in press). The effect of practice on the control of sequential and simultaneous action. *Human Performance.*

Turvey, M. T., Shaw, R. E., & Mace, W. (1978). Issues in the theory of action: Degrees of freedom, coordinative structures, and coalitions. In J. Requin (Ed.), *Attention and Performance VII* (pp. 557-595). Hillsdale, NJ: Erlbaum.

Perspectives on the Coordination of Movement
S.A. Wallace (Editor)

COORDINATION OF MOTOR TASKS IN HUMAN GAIT

David A. WINTER

Department of Kinesiology
University of Waterloo

ABSTRACT

The detailed kinematic and kinetic patterns in human gait and their variability within and between subjects provide considerable insight into the control structure of the central nervous system. Kinematic variables such as joint angles and trajectories of the center of mass of the trunk and of the heel and toe are consistent during both stance and swing. Kinetic patterns, such as joint moments of force, are a measure of the final integrating motor pattern of the central nervous system. However, these moment patterns, especially at the hip and knee, are extremely variable during stance and very consistent during swing. The high hip and knee variability is not random, but rather a stride-to-stride trade-off between the hip and knee. Two major and independent gain synergies are documented as a result of these findings. A support synergy, originally identified in 1980, appears during stance, when a total extensor moment pattern (called support moment) is evident from all three joints of the lower limb, independent of gross pattern changes at the hip or knee. A balance synergy is evident from the anterior-posterior trade-offs between the hip and knee that allow control of the forward and backward angular acceleration of the head, arms, and trunk (HAT). Thus the dynamic balance of HAT is the major stride-to-stride responsibility of the hip extensor-flexors during stance; but in

*Address correspondence to: David A. Winter, Department of Kinesiology, University of Waterloo, Waterloo, Ontario N2L 3G1, Canada.

the presence of considerable stride-to-stride variability, the knee moment pattern compensates to keep the total support moment nearly constant. Another regulating subtask, the safe and efficient trajectory of the foot during swing, has been documented as a small toe clearance (less than 1 cm) followed by a gentle heel contact. The complexity of this multisegment endpoint control demands anticipatory (feedforward) control by the central nervous system and is overlaid on the two stance-phase regulating subtasks.

THE CENTRAL NERVOUS SYSTEM AND ITS PLANT

Walking and running are the most common of all coordinated human movements. They are two of the more difficult movement tasks that we learn, but once learned they become almost subconscious. Only when the neuromuscular and skeletal systems are disturbed by injury, gradual degeneration, or fatigue do we realize how limited our understanding is of the complex underlying biomechanics and motor control mechanisms.

The skeletal and articulating systems are the framework on which the motor control system must function, and these systems have degrees of freedom and constraints of which the central nervous system must be aware. The mechanical coupling between adjacent anatomical segments and the interactions across many segments must also be coordinated for a smooth and safe gait to be achieved. The central nervous system must know the characteristics of the muscle actuators for the timing and magnitude of the neural patterns to be correct. All this must occur in the presence of a constant gravitational field. In other words, the central nervous system must know the plant that it is controlling: the characteristics of that plant and its instantaneous state. Peripheral sensors keep the central nervous system updated (with a small delay) as to the state of the musculoskeletal system, but the central nervous system itself must store information about the characteristics of the muscles and skeletal system to be able to generate the appropriate neural patterns and thereby accomplish the desired kinematic patterns.

Some researchers, especially those studying the gait of quadrupeds, claim that a central pattern generator exists in the spinal column. This hard-wired pattern generator, they claim, is merely modulated by higher centers to achieve the different motor patterns associated with different walking speeds. Other researchers refer to higher center control where prescribed motor patterns are stored for each type of gait (walking, running, walking backwards, walking down stairs). Still others focus on the peripheral sensors as a dominant feedback controller of the total system. The purpose of this chapter is not to focus on any one of these theories but to examine in detail the many kinematic and kinetic patterns of walking, thereby gaining insight into the way the central nervous system coordinates the task of gait, and to make some inferences about the importance of these theories.

DESCRIPTIVE MEASURES OF HUMAN GAIT

An understanding of human gait requires a detailed description of the movement itself: timing, trajectories, and kinematics. Because of the number of segments involved and the number of degrees of freedom, scores of time-related variables would be needed to describe just one stride completely (Winter, 1983a). It is important to document some of the more important of these variables as a framework for understanding how this coordination takes place.

Cadence, Stride and Velocity Measures

Natural cadences reported in the literature had averages varying from 101 to 122 steps/min. For 936 pedestrians, Drillis (1958) reported a mean cadence of 112 with a wide range from 78 to 144. In analysis of over a hundred young adults, our laboratory has found an average cadence of 107 (min = 101, max = 122). Differences between male and female cadence have been reported by many laboratories, the cadence ranging from 6 to 11 steps/min higher for females than for males. Du Chatinier, Molen, and Rozendal (1970) reported the average cadence for females ($N = 57$) to be 122 and for males ($N = 72$) 116. Finley and Cody (1970) reported that 472 females had an average cadence of 116.5 ($SD = 11.7$) and that 434 males had an average cadence of 110.5 ($SD = 10.0$). Molen and Rozendal (1972) reported for about 500 young adults that male cadence averaged 113 compared to 124 for females. Murray, Kory, and Clarkson (1969) in a study of older men noted that both cadence and stride length were less than they were for younger adults.

Stance and swing times as reported in scores of papers have been quite consistent for natural cadences: Stance was 58-61% of stride period, swing was 42-39%. If perfect symmetry is assumed, two double support times of 8-11% of stride period would occur. It will be seen later that this short period is critical to transferring weight from one limb to the other and to controlling balance and the posture of the upper body. In fast walking, these double support times decrease to about 6% of stride; and, of course, in running, double support disappears and is replaced with a free flight phase.

Stride length and velocity are the two remaining, but not independent, basic variables. Lamoreux (1971) summarized the results of his and others' work and noted that between cadences of 80 and 120, stride length and cadence each varied as the square root of velocity and were therefore linearly related. Above cadences of 120, step length leveled off and only cadence increased. In a study of elderly women, Finley, Cody, and Finizie (1969) noted significant differences in velocity and step length from a group of younger women. The older group's mean velocity was 1.57 mph (2.53 km/h) versus 1.83 mph (2.95 km/h) for the young women, and this was almost entirely accounted for by reduced stride lengths (30 in vs. 37 in; i.e., 76.2 cm vs. 94 cm).

KINEMATICS

The term *kinematics* is used to describe the movement itself independent of the forces, internal and external, that cause the movement. Kinematics are output variables resulting from those input forces. Kinematic variables have been reported more often than any other variables and have been quantified with a wide range of measurement devices including goniometers, accelerometers, cinematography, television, and optoelectric devices. Regardless of the way data are collected, the number of kinematic variables (positions, velocities, accelerations) required to describe one gait stride is very high. For example, to describe the movement of the foot in one plane requires six linear measures and three angular measures. Thus, if we consider the human body to be a 15-segment system (2 feet, legs, thighs, upper arms, forearms, and hands, plus head, trunk, and pelvis), we need 135 curves to describe its movement in the plane of progression. These variables are absolute, and from them we can calculate a range of relative variables such as joint angles and angular velocities. Direct measurement systems, such as goniometers, yield joint angles, which cannot yield absolute variables in any of the reference planes. Thus, only imaging systems yield variables that can be used for a kinetic analysis (which requires the direction of the horizontal and gravitational vectors).

Only a few of the large number of variables, both absolute and relative, are reported here. Those selected are both readily available for statistical reporting and indicative of some function at the kinetic level, and thus have relevance to the motor control of this complex, multisegment system.

It is important to note how stride-related curves were processed for reporting intra- and intersubject averages. Because gait is a repetitive event over the stride period and because all variables vary considerably over that period, the time base for reporting is the stride period, reported as 100%, with heel contact set at 0% and the second heel contact at 100%. Because stance was close to 60% in all cases, toe-off was set to 60%. Thus stance was separately normalized to 0-60%, and swing to 60-100%. Each variable within those periods was broken into 2% intervals, and ensemble averages were computed such that a mean and standard deviation were calculated at each 2% interval. A coefficient of variation (*CV*), somewhat similar to single measure scores, was calculated as follows:

$$CV = \frac{\sqrt{\left(\frac{1}{N}\sum_{i=1}^{N}\sigma_i^2\right)}}{\frac{1}{N}\sum_{i=1}^{N}|X_i|} \tag{1}$$

where: N is the number of intervals over the stride,

X_i is the mean value of the variable at the ith interval, and

σ_i is the standard deviation of variable X about X_i.

In effect, this *CV* yields over the stride period an average variability-to-signal ratio, which is very useful in comparing the variability of different variables and relating that to the many motor functions in the task of walking. Also, for certain variables, the square of the numerator of this equation is the mean variance over the stride period, which can be used to calculate covariances between pairs of key variables.

Relative Kinematics

It is important to quantify joint angles as descriptors of the gait pattern because these angles form a framework for all other kinetic events that are taking place. Intrasubject averages are necessary to ascertain how consistent the output pattern is relative to the causative motor patterns. The variability of these patterns is critical to the reliability of these data for representing what is going on from stride to stride and from day to day for any given subject.

Figure 11.1 presents the joint angle data for nine repeat trials for the same subject walking her natural cadence over a period of days (Winter, 1983a). The conventions *+ve* and *-ve* indicate flexion and extension respectively. The average variability (shown by dotted lines) over the stride period is less than 2° at all joints. The *CV*s at the ankle and knee are quite low, whereas the *CV* at the hip is slightly more variable. Such findings may not appear to be critical by themselves, but when related to the variability seen in the motor patterns, they give some insight into two motor synergies essential to the task of walking—support and balance—which will be reported and discussed later.

The actual mean patterns reported in Figure 11.1 provide some insight into this subject's walking patterns, now described very briefly. At heel contact, the ankle is slightly dorsiflexed, and it plantarflexes for the first 8% of stride as the foot is lowered to the floor. Then from 8 to 44%, the leg rotates over the foot and reaches 16° of dorsiflexion prior to a rapid plantarflexion during push-off (44-60% for this subject). Immediately after toe-off, the foot rapidly dorsiflexes so that the toe will clear the ground in mid swing. The knee is fully extended at heel contact and flexes to about 20° during the weight acceptance period (0-15% stride); then it extends slowly during mid-stance (15-40%) and flexes rapidly during push-off, reaching about 50° at toe-off and a maximum of 63° early in swing. Then during the remainder of swing, the leg is extended rapidly prior to the next heel contact.

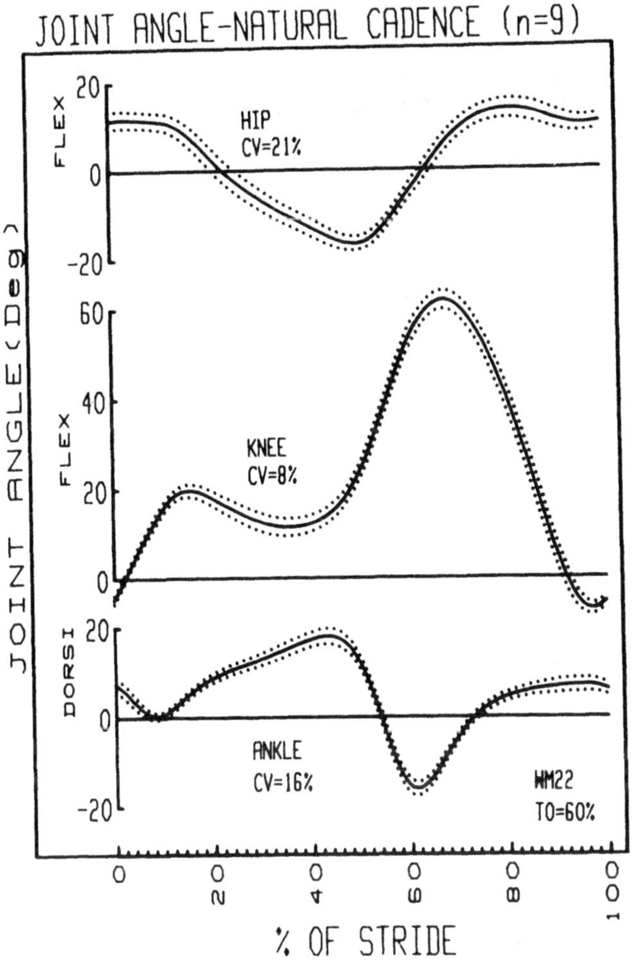

Figure 11.1. Joint angle curves over the walking cycle (heel contact = 0%, toe-off = 60%) for nine trials repeated days apart with the same subject. Average curve is plotted as a solid line with dotted line shown at ±1 standard deviation. *Note.* From *Biomechanics and Motor Control of Human Gait* (p. 22) by D. A. Winter, 1987, Waterloo, Canada: University of Waterloo Press. Copyright 1987 by David A. Winter. Reprinted by permission.

The periods of knee flexion and extension during stance functionally define important phases. Initial knee flexion defines the weight acceptance phase, during which the downward movement of the body mass is decelerated and arrested. Mid stance is the period of time between weight acceptance and push-off, and coincides with the time that the knee extends slowly. The start of push-off, as defined by the start of ankle plantarflexion, coincides almost completely with the start of knee flexion, at about 40% of stride. The hip angle history over the stride period relates to the third major joint in the system and the motor patterns around that joint that are related to controlling the large mass of the head, arms, and trunk (HAT). The HAT segment represents 2/3 of body mass, which is located about 2/3 of body height above ground level. It is the large inertial load of HAT that must be controlled, and therefore the kinematics of the hip joint during stance are related to the task of keeping HAT erect within a small number of degrees. No statistics for this task are yet available in the literature; however, for the single subject presented in Figure 11.1, the HAT segment varied ±2° over the gait cycle.

In a similar fashion, the ensemble average of a group of subjects can be calculated; such an average for 19 adults walking their natural cadence is plotted in Figure 11.2. The general shape and magnitude of this intersubject average is similar to that already described in Figure 11.1. However, the variability at each joint ranged from about 2 1/2 to 4 1/2 times that for an individual. The average standard deviations for these subjects in degrees were about 4° at the ankle, 5° at the knee, and 6° at the hip. As will be shown, the increase in this variability from distal to proximal is matched by a similar trend and more pronounced variability in the moment-of-force profiles. The electromyogram patterns also exhibited exactly the same trend in variability (Winter & Yack, 1987). The lower variability at the ankle indicates that the motor tasks of the ankle muscles are somewhat fixed over the stride. The flexors and extensors of the hip, in contrast, must accomplish several tasks, and these can change on a stride-to-stride basis. The most variable of these tasks is the balance of the trunk, which is being continuously corrected on each stride, and the primary muscle groups responsible are the hip muscles.

The average joint angle curves change with cadence. Figure 11.3 is a plot from three cadence groups: natural cadence; slow cadence, de-

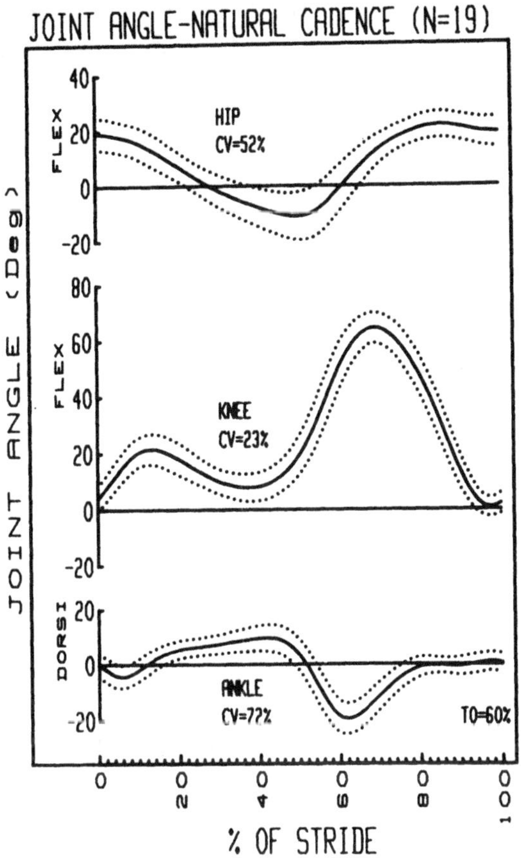

Figure 11.2. Joint angle curves over the stride period for 19 subjects walking their natural cadence. *Note.* From *Biomechanics and Motor Control of Human Gait* (p. 23) by D. A. Winter, 1987, Waterloo, Canada: University of Waterloo Press. Copyright 1987 by David A. Winter. Reprinted by permission.

fined as natural cadence minus 20 steps/min; and fast cadence, defined as natural cadence plus 20 steps/min. On this normalized time scale, it is apparent that, except for the knee angle, the curves are almost identical. The knee angle showed minor differences during weight acceptance: The fast walkers flexed to 25°, the natural cadence group to 20°, and the slow walkers to about 15°. The correlations of these curves were high and varied from .95 to .995

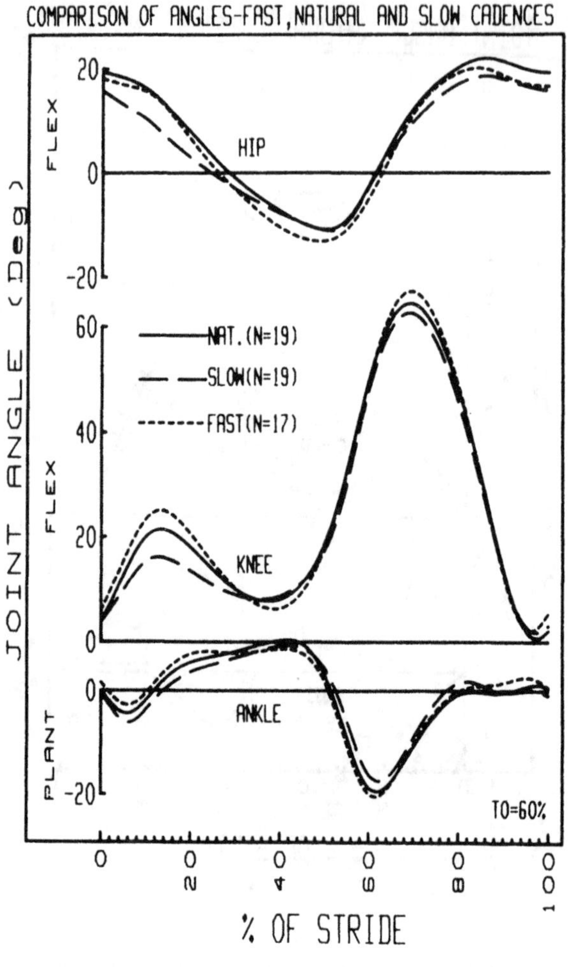

Figure 11.3. Average joint angle curves over the stride period for three cadence groups. Slow cadence was 20 steps/min less than each subject's natural cadence; fast cadence was 20 steps/min higher. Time base for the stance period has been normalized to 60% and swing to 40%. *Note.* From *Biomechanics and Motor Control of Human Gait* (p. 24) by D. A. Winter, 1987, Waterloo, Canada: University of Waterloo Press. Copyright 1987 by David A. Winter. Reprinted by permission.

(Winter, 1983a), showing that essentially the same joint trajectories prevail over a wide range of cadences and that the same pattern is accomplished in less or more time.

From these plots, we would predict that the joint angular velocities must also be quite similar in shape and that their amplitude must be closely correlated to cadence. Figure 11.4 overlays the three joint angular velocities for the three cadence groups (Winter, 1987b). The shapes of these curves were quite similar, the correlations ranging from .98 to .995. The slope of the linear regressions is very closely related to the ratio of cadences. For example, the fast cadence group walked 17% faster than the natural cadence group; and the ankle velocity increased 18%, the knee velocity 20%, and hip velocity 21%. Thus the joint angular velocities are almost perfectly related to cadence, which means that the shortening and lengthening velocities of the muscles is also similarly related. Thus the spindle receptors, which detect muscle velocity, have the potential for providing precise cadence-related feedback to the central nervous system to control the motor profiles.

SELECTED TRAJECTORIES

Although we could examine scores of trajectories from each segment in the spatial reference system, a few important areas may be selected for study of the mechanics of gait and the tasks involved. It is critical to examine how the dominant mass of the HAT, with its somewhat unstable location well above ground level, is moved across the ground. From an energetics standpoint, it is important that this segment not make excessive energy demands on the lower limbs. It should be highly conservative of its energy over the stride period, and this has been shown to be quite true (Ralston & Lukin, 1969; Winter, Quanbury, & Reimer, 1976). Kinematically, this conservation is quite evident in two trajectory plots: Potential energy is related to the vertical trajectory of the center of mass of HAT, and kinetic energy is related to the horizontal velocity of this center of mass. Figures 11.5 and 11.6 for the natural cadence group show this relationship (both figures have been normalized to the mean height above ground and the mean forward velocity). The figures show that these two curves are virtually out of phase. The vertical displacement reaches a minimum at 4% and 54% of stride (during the middle of double support), while the forward velocity reaches a maximum at 6% and 56%. The percentage conservation

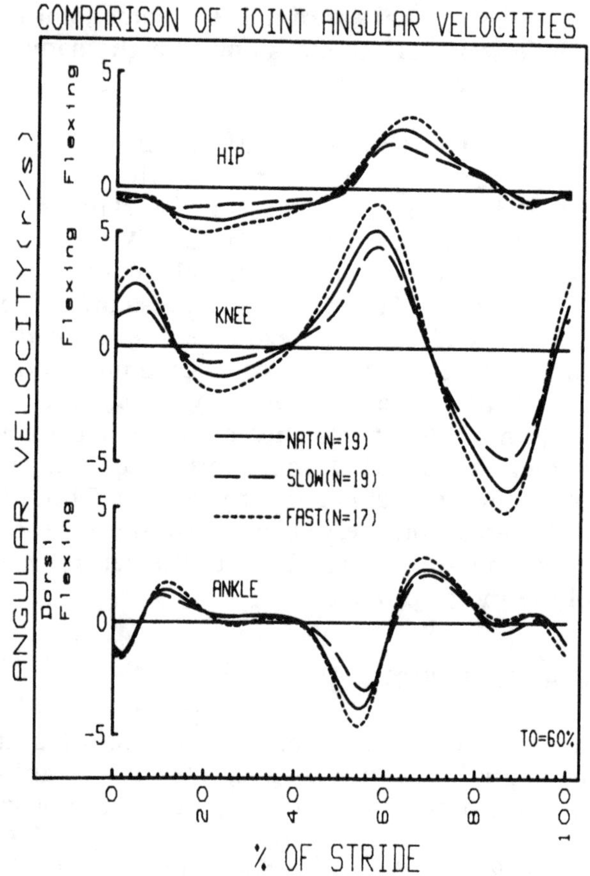

Figure 11.4. Average joint angular velocities for the three cadence groups reported in Figure 11.3. *Note.* From *Biomechanics and Motor Control of Human Gait* (p. 26) by D. A. Winter, 1987, Waterloo, Canada: University of Waterloo Press. Copyright 1987 by David A. Winter. Reprinted by permission.

for this segment has been calculated to be about 80% (Winter et al., 1976).

Thus the large mass of HAT acts as an inverted pendulum during stance, a function that is critical to the efficiency of walking. However, many researchers have carried this finding to a ridiculous extreme and have used it as an excuse to model the lower limb as a

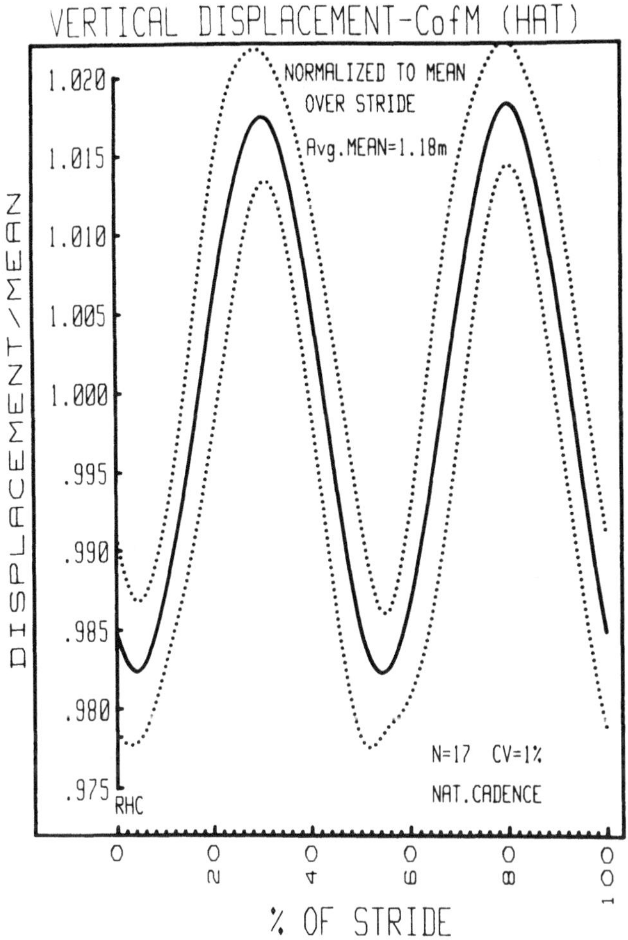

Figure 11.5. *Vertical displacement of the center of mass of the head, arms, and trunk (HAT) for 17 subjects walking their natural cadence. The displacement was normalized for each subject prior to averaging so that the mean height above ground was set at 1.000. The mean height above ground was 1.18 m.* Note. From Biomechanics and Motor Control of Human Gait *(p. 19) by D. A. Winter, 1987, Waterloo, Canada: University of Waterloo Press. Copyright 1987 by David A. Winter. Reprinted by permission.*

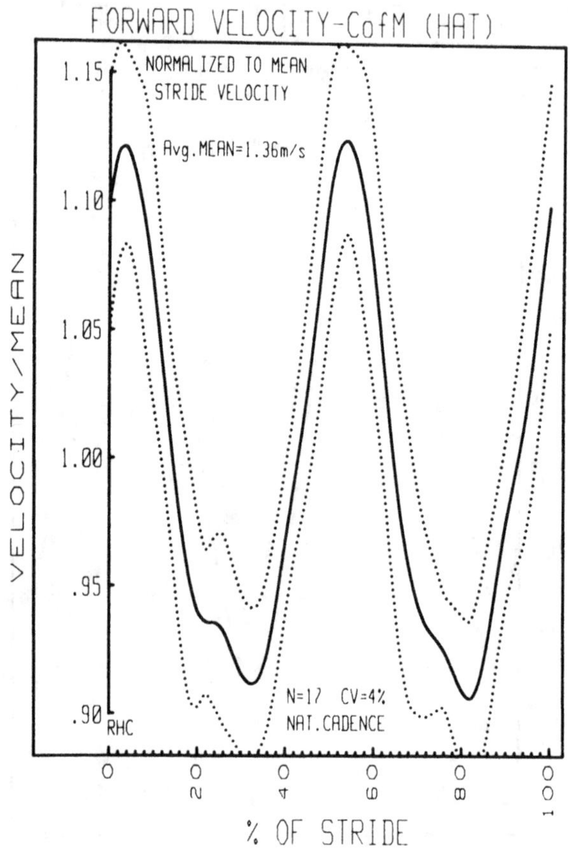

Figure 11.6. Horizontal velocity of the center of mass of the head, arms, and trunk (HAT) for the same 17 subjects as in Figure 11.5. The average velocity for each subject was normalized to 1.00 prior to averaging. The average forward velocity was 1.36 m/s. That this velocity curve is virtually 180° out of phase with the vertical displacement curve indicates considerable conservation of energy within HAT over the gait cycle. *Note.* From *Biomechanics and Motor Control of Human Gait* (p. 20) by D. A. Winter, 1987, Waterloo, Canada: University of Waterloo Press. Copyright 1987 by David A. Winter. Reprinted by permission.

rigid segment with no feet. Such constraints are invalid because they negate the dominant roles of the ankle and knee muscles in generating and absorbing energy over the stride period.

A second set of trajectories provides considerable insight into the role of the motor system during swing. Figures 11.7 and 11.8 show the trajectory and velocities of the heel and toe marker over the stride period for 14 subjects walking their natural cadence. From these curves we can learn much about the task of moving the lower limb forward and achieving a safe trajectory of the end points (heel and toe). The number of degrees of freedom and the length of the segments involved in that end-point control make the task quite formidable. The more important degrees of freedom that influence the trajectories in the plane of progression are flexion and extension at the ankles and knees and all three degrees of freedom at both hips. To this must be added the inversion and eversion of the swinging foot. Thus a seven-segment chain (two feet, legs, and thighs, plus pelvis) with 11 degrees of freedom must be controlled in order to achieve a safe end-point trajectory; that is, a safe toe clearance in mid stance and a relatively gentle heel contact. The average toe trajectory for these 14 subjects walking their natural cadence showed that the toe has a clearance of 0.87 cm during mid swing. The horizontal velocity at this time is greater than 4 m/s; so we apparently achieve this difficult task with no conscious effort. At the end of swing, the heel is prepared for contact with the ground (Figure 11.8) by having its velocity reduced to virtually zero in both horizontal and vertical directions. The trajectory of the heel is such that it reaches a height of about 25 cm during early swing and drops rapidly during mid swing to about 1 cm about 10% before heel contact. During this last 10%, the heel moves almost horizontally and decreases its velocity from 4 m/s to almost zero before heel contact.

The trajectory just prior to heel contact is like that of an airplane just about to touch down, with the difference that horizontal velocity is reduced to near-zero just before contact. Thus, to refer to this event as heel strike is erroneous. It is also erroneous to refer to lower limb motor control (as compared with upper extremity and hand control) as a gross motor task. The length of the segments and the number of degrees of freedom involved mean that control of the foot trajectory is indeed a fine motor task.

KINETICS

Kinetics by definition deals with those variables that cause the specific walking or running pattern we observe or measure with our

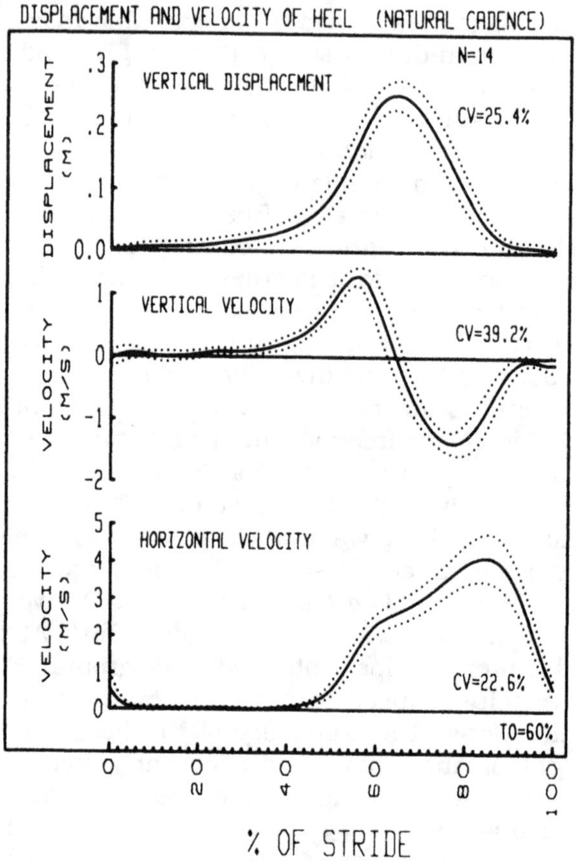

Figure 11.7. Displacement and velocities of the heel for 14 subjects walking their natural cadence. The horizontal velocity exceeds 4 m/s late in swing but reduces to virtually zero in both the vertical and horizontal directions prior to heel contact. Thus a gentle contact is achieved rather than the heel "strike" described by many researchers. *Note.* From *Biomechanics and Motor Control of Human Gait* (p. 17) by D. A. Winter, 1987, Waterloo, Canada: University of Waterloo Press. Copyright 1987 by David A. Winter. Reprinted by permission.

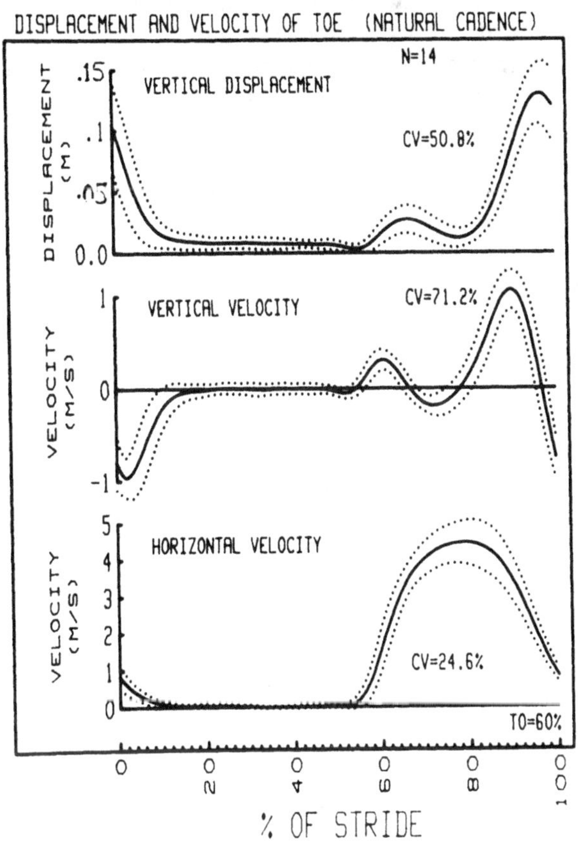

Figure 11.8. Displacement and velocities of the toe for the same 14 subjects as in Figure 11.7. The horizontal velocity reaches about 4.5 m/s just as the toe reaches its lowest clearance of less than 1 cm in mid swing. The combination of this safe but low toe clearance plus the near-zero heel velocity at heel contact is evidence of a very fine end-point control task. *Note.* From *Biomechanics and Motor Control of Human Gait* (p. 18) by D. A. Winter, 1987, Waterloo, Canada: University of Waterloo Press. Copyright 1987 by David A. Winter. Reprinted by permission.

cameras. As such, we are concerned with individual muscle forces, the net moments generated by those muscles at each joint, the mechanical power patterns (rate of energy generation or absorption by muscles, or rate of transfer between segments). We must invoke

Newton's laws and the law of conservation of energy to interpret what is happening at each phase of the gait cycle, and we must bear in mind the principle of indeterminacy. Many combinations of muscle forces can result in the same movement pattern; many combinations of agonist and antagonist activity can result in the same moment of force at a given joint. Similarly, it has been shown that an infinite number of combinations of moments of force at the hip, knee, and ankle can result in the same kinematics of the stance limb (Winter, 1984). This principle demonstrates the tremendous flexibility and adaptability of the neuromuscular system.

Ground Reaction Forces

The easiest kinetic variables to record are ground reaction forces. They are measured by a force platform and reflect the net vertical and horizontal (shear) forces acting on the surface of the platform. As such, they are the algebraic summation of the mass-acceleration products of all body segments while the foot is in contact with the platform. The vertical force reflects the acceleration due to gravity as well as accelerations seen by the camera.

The ground reaction forces for 19 subjects walking their natural cadence are presented in Figure 11.9. The forces are normalized to body mass in order to reduce the intersubject variability, and stance for all subjects was set to 60% of stride. Body weight is shown in Figure 11.9, and because of the normalization technique, it is 9.81 N/kg for all subjects. The horizontal reaction force reaches -2 N/kg during weight acceptance, and this negative pattern during the first half of stance indicates a net deceleration of the total body mass. This deceleration is balanced by an equal and opposite forward acceleration during the latter half of stance, reaching a peak at 50% of the stride period, which is the middle of the push-off phase. Theoretically, if all subjects were walking at a constant velocity, the area under the positive and negative impulses would be equal. The magnitude of the positive and negative peaks is about the same for a given cadence but increases with cadence: 1.5 N/kg for slow walking, 2 N/kg for natural cadence, and 2.5 N/kg for fast walkers. Similarly, for the vertical reaction forces, there are fluctuations about body weight (note that greater than 9.81 means the body is being accelerated upwards and that for less than 9.81, there is a net downward acceleration). For the natural cadence group, the first peak is 10.8 N/kg, indicating that at the end of weight acceptance

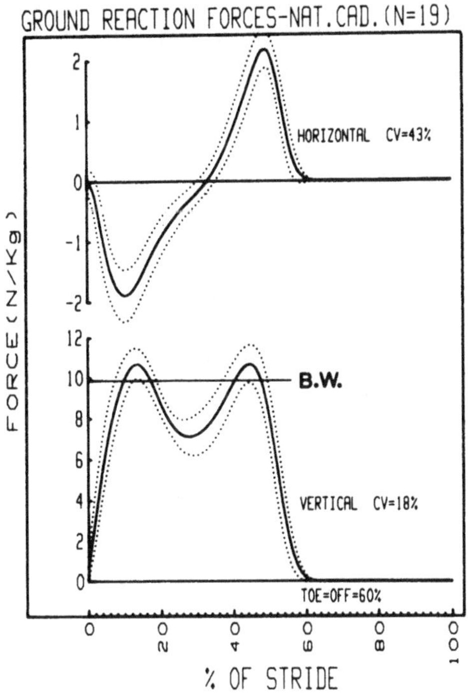

Figure 11.9. Averaged horizontal and vertical ground reaction forces for the same 19 subjects as in Figure 11.2. These ground reaction forces reflect the horizontal and vertical acceleration and deceleration of the body's center of mass during weight bearing. Prior to averaging, each reaction force was normalized by division by body mass. Thus body weight appears at 9.81 N/kg on the vertical reaction force curve. *Note.* From *Biomechanics and Motor Control of Human Gait* (p. 30) by D. A. Winter, 1987, Waterloo, Canada: University of Waterloo Press. Copyright 1987 by David A. Winter. Reprinted by permission.

(15% of stride), the body is being accelerated upwards. In mid-stance (about 30% of stride) the minimum is 7.1 N/kg, indicating a net downward acceleration; and then during push-off (about 47% of stride), there is a net upward acceleration. The magnitudes of these peaks and valleys vary considerably with cadence (Winter, 1987b). For slow walkers, the peak was 10 N/kg and the valley was 8.5 N/kg; but for fast walkers, the peak increased to over 12.5 N/kg and the valley decreased to 5.2 N/kg. Thus, as cadence increases, there is a drastic increase in both vertical and horizontal body decelerations and accelerations. The *CV*s for these ensemble-averaged profiles were about the same for all three cadence groups and were not very high. For the vertical reaction force, *CV*s ranged from 15 to 19%, and for the horizontal forces, from 43 to 64%.

Moment-of-Force Patterns

Moments of force are the net result of all muscular, ligament, and friction forces acting to alter the angular rotation of the joint. In normal gait, the joint angles do not reach their extreme limits, and friction forces are minimal. Thus, the net moment, as calculated, can be interpreted as due to muscle forces only, and therefore under complete control of the central nervous system. The moments of force are calculated with an inverse dynamics of the link segment model (Bresler & Frankel, 1950; Winter, 1979). These moments are internal, meaning that the interpretation of, for example, a net flexor moment at the knee is that the knee flexor muscles were creating a net moment in excess of the knee extensor muscles. Only via electromyogram analyses can we ascertain that only agonists were active in creating the net moment, or that a cocontraction existed.

The convention for reporting polarities of moments of force varies considerably with individual researchers. Biomechanically, a moment is counterclockwise if positive, and clockwise if negative. However, such conventions result in different polarities depending on whether we are looking at the right or left limb or whether the subject is walking from left to right or vice versa. For purposes of interpretation, an extensor moment is positive and a flexor moment is negative; the argument for this choice will become evident shortly.

Intrasubject Averages

Figure 11.10 presents the moment of force patterns for a subject walking her natural cadence; each trial was recorded on separate days (this subject's data appear in Figure 11.1, where her average joint angle curves were plotted). In addition to the hip (M_h), knee (M_k) and ankle (M_a) moments, there is a support moment plotted, where $M_s = M_a + M_k + M_h$, with the individual joint polarities as shown. This support moment summation was first defined by

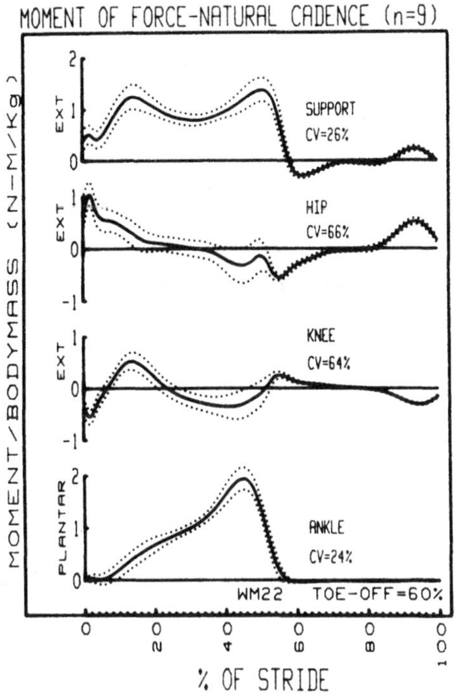

Figure 11.10. Averaged joint moment-of-force curves for the nine repeat trials on the same subject as in Figure 11.1. The *CV* for each curve was calculated for the stance period only and shows high variability at the hip and knee and low variability for the ankle and support moments. *Note.* From *Biomechanics and Motor Control of Human Gait* (p. 32) by D. A. Winter, 1987, Waterloo, Canada: University of Waterloo Press. Copyright 1987 by David A. Winter. Reprinted by permission.

Winter (1980) and has been shown to be consistently positive during stance for all subjects (Winter, 1984) and all assessments of pathological gait. This support moment documents a basic total limb synergy evident in stance and relates to the total extensor thrust of all three joints and to the strength with which the lower limb pushes away from the ground. The similarity in the shape of M_S to the vertical ground reaction force is obvious.

It is worth examining these individual curves, all of which have been normalized by division by body mass. For this subject, the ankle had a very small dorsiflexor moment for the first few percentage points of stance; here the dorsiflexors controlled the lowering of the foot to the ground after heel contact. Then for the balance of stance, the plantarflexors came on and increased their activity, which peaked during push-off. From 6% to 44% (Figure 11.1), the ankle dorsiflexed (i.e., the leg rotated forward over the foot); thus the plantarflexors eccentrically acted to control the forward rotation of the leg, and thus in turn controlled the angle of the knee. Then at 44% to 60%, there was a rapid ankle plantarflexion caused by the very high plantarflexor moment. This high-intensity concentric contraction is called push-off, and it has been shown to be responsible for almost all energy generation during the gait cycle (Winter, 1983a, 1983b). Then, immediately after toe-off, there was a small dorsiflexor moment that rapidly dorsiflexed the foot to achieve toe clearance during mid swing.

The knee moment pattern for this subject showed a short flexor pattern for the first 6%, whose peak coincided with the extensor peak at the hip. This means that there was strong and early hamstring activity. Then the knee extensors turned on to aid the control of knee flexion during weight acceptance and early mid-stance. Then from about 25% of stride until late stance (52%) there was a net flexor moment at the knee, a moment due to the gastrocnemius muscles (already quite active as plantarflexors at the ankle). Then, just before toe-off, the gastrocnemius muscles turned off, and the net motor activity reversed to extensor until mid swing. This activity controlled the amount of knee flexion during late push-off and decelerated the backward rotating leg early in swing. Then during late swing, the knee flexors acted to decelerate the forward rotating leg; and because of the simultaneous extensor pattern seen at the hip, this deceleration can be attributed entirely to the hamstrings.

The hip profile was extensor during the first half of stance in order to control the amount of hip flexion (and thereby assist in controlling the amount of knee flexion). During the latter half of stance and first half of swing, the hip flexors activated to reverse the direction of the backward rotating thigh and then to assist in the forward rotation of the thigh during late stance and early swing.

Then during late swing, the hamstring pattern already noted decelerated the forward rotating thigh and leg.

Support Moment Synergy

If we look at the support moment, we see that for the first half of stance (0-30%), all three joints contribute to the net extensor pattern. This means that all three motor patterns contribute to controlling the amount of knee flexion: the ankle plantarflexors by controlling the forward rotation of the leg over the foot, the knee extensors by direct control, the hip extensors by controlling collapse of the thigh from the proximal end. This basic motor synergy is seen in all subjects and all assessments of pathological gait.

The variability of these motor patterns is also extremely important in ascertaining the consistency of this synergy. High variability is quite evident during stance, but it drops drastically during swing. As shown in Figure 11.1, the kinematics of this lower limb is extremely consistent over the stride, and one would expect consistent kinetics over the total stride. But this is not so during the stance period. An examination of the biomechanics of stance has shown that the kinematics of the thigh, leg and foot can remain virtually the same with different combinations of moments at the three joints (Winter, 1984). Examination of these curves separately shows that the greatest variability is at the knee and hip. If the motor patterns at these joints varied randomly with respect to each other, then the sum of these curves would have even higher variability. But the variability of M_S was greatly reduced. Thus the stride-to-stride moments did not vary in a random manner; rather, there is some considerable cancellation of variability taking place in this M_S summation. Therefore, some significant covariance exists between the motor patterns at these three joints.

Covariance of Joint Motor Patterns

Figure 11.11 documents the degree of covariance between the joint moment patterns for the subject presented in Figure 11.10, along with the degree of covariance from a second subject who was assessed over 10 repeat trials but whose data were collected minutes apart rather than days apart. The calculation of these mean vari-

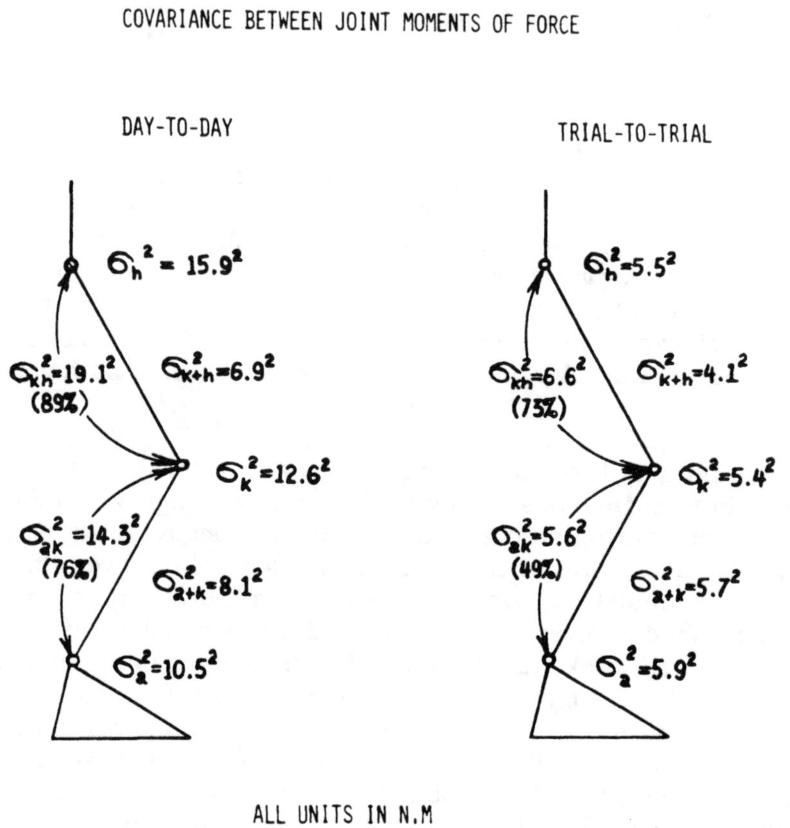

Figure 11.11. Variance and covariance of the joint moments of force for repeat trials for one subject days apart (left) and minutes apart (right). The day-to-day variances at the hip and knee are high, but there is considerable covariance between the moment profiles at the hip and knee (89%) and between the moment profiles at the knee and ankle (76%). The variances and covariances are less for the trials analyzed minutes apart because the subject did not exhibit nearly as much variability over such a short span of time.

ances and covariances is based on the following equations, with all units in $(N.m)^2$:

$$\sigma^2_{h+k} = \sigma^2_h + \sigma^2_k - \sigma^2_{hk} \quad (2)$$

where: σ^2_h and σ^2_k are the mean variances over stance at the hip and knee,

σ^2_{h+k} is the mean variance of the sum of hip + knee moment profile,

σ^2_{hk} is the covariance between hip and knee moment patterns.

The term σ^2_{hk} can be expressed as a percentage of the maximum possible covariance, which would be 100% if $\sigma^2_{h+k} = 0$, and would mean that the variability in the knee moment was completely out of phase with that at the hip. Thus the percent covariance, *COV*, is given by

$$COV = \frac{\sigma^2_{hk}}{\sigma^2_h + \sigma^2_k} \times 100\% \quad (3)$$

In the first set of repeat trials the *COV* between the hip and knee was 89%. A similar calculation can be made at the knee and ankle, and this yields a *COV* of 76%. These high covariances are extremely strong evidence of tight coupling between the motor patterns at these adjacent joints and are not surprising when we consider the opposite and cancelling functions of the hamstrings and rectus femoris muscles at the hip and knee, and of the gastrocnemius muscle at the knee and ankle. Therefore, this tight coupling is partially due to the anatomy of these biarticulate muscles. Anatomically, the potential for force generation can be considered to be approximately proportional to the physiological cross-sectional area of each muscle. However, the single joint muscles constitute 2/3 of the physiological cross-sectional area of all the muscles crossing the hip and knee joints (Wickiewicz, Roy, Powell, & Edgerton, 1983). Thus, the magnitude of *COV* is well in excess of that possible anatomically, and therefore must be part of a neural control pattern.

The lower *COVs* from the second subject's trials (73% between the hip and knee and 49% between the knee and ankle) appear to be due to the lower variability (adaptability) seen over these repeat trials that took place minutes apart. Thus it appears that we adopt patterns that have larger differences on a day-to-day basis than they do over shorter periods of time. These day-to-day and minute-to-minute alterations are largely very deterministic and reflect the plasticity of the motor control system. It is now important to determine the nature of these stride-to-stride trade-offs, and it will be seen that they are related to a second motor synergy, that of balance.

Stride-to-Stride and Subject-to-Subject Differences

Figure 11.12 is presented to delineate the nature of these stride-to-stride differences (Winter, 1984). Two of the nine trials for this subject were selected along with the mean of the nine strides (solid line). WM22D (i.e., subject WM22, Trial D) is shown by the long dashed line; WM22J, by the short dashed line. During Trial J, the subject had a dominant hip extensor pattern and a knee flexor pattern, whereas Trial D was biased in the reverse direction: hip flexor and knee extensor. Thus the dominant muscle pattern for Trial J was hamstrings (posterior muscles) in contrast to that of Trial D, which was primarily rectus femoris (anterior muscles). The net extensor pattern acting on the thigh for both of the strides was about the same, but in one case it was being accomplished by anterior muscles and in the other by posterior muscles. Thus this trade-off between anterior and posterior did not influence the net extensor (support pattern).

Far more dramatic differences are present in the comparison of two subjects. Figure 11.13 is a similar set of moment patterns for two subjects walking their natural cadence. Both subjects had similar lower limb kinematics; and, as can be seen, the support moments were almost identical. WM20A had a dominant hip extensor pattern, knee flexor pattern, and a higher plantarflexor moment, whereas WM24B was biased to a hip flexor pattern, knee extensor,

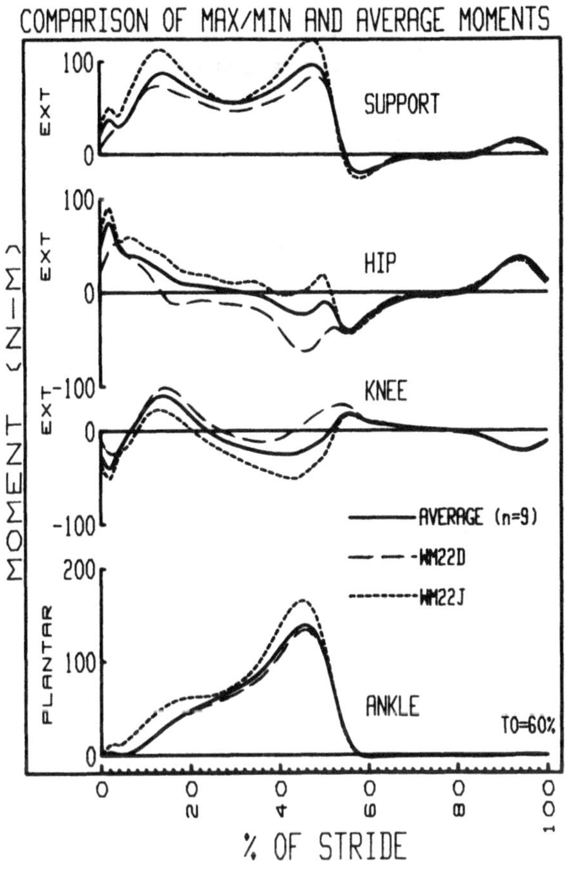

Figure 11.12. Two trials selected from the nine repeat trials reported in Figure 11.10 to show how the moments of force varied. The major changes showed a well-defined bias change at the hip and knee such that on one day (WM22D) the subject's motor patterns were biased towards the anterior muscles (hip flexors and knee extensors), and on another day (WM22J), towards the posterior muscles (hip extensor and knee flexors). *Note.* From "Kinematic and Kinetic Patterns in Human Gait: Variability and Compensating Effects" by D. A. Winter, 1984, *Human Movement Science, 3,* p. 62. Copyright 1984 by Elsevier Science Publishers B. V. (North-Holland). Reprinted by permission.

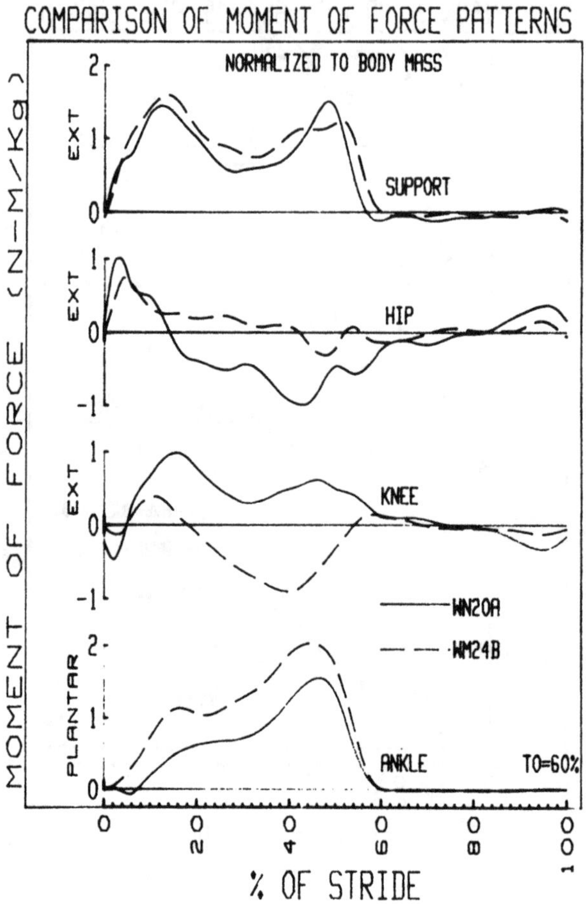

Figure 11.13. Two trials selected from a number of subjects walking their natural cadence. These trials showed drastic differences in moments at the ankle, knee, and hip and yet showed a very similar support moment profile. As with the intrasubject trials (Figure 11.12), the difference between the two subjects is a bias towards the anterior muscles (WM20A) or the posterior muscles (WM24B).

and a lower plantarflexor profile. The major difference between these two subjects was a bias of WM20A towards greater activity in the posterior muscle group and of WM24B towards greater activity in the anterior muscle groups (or, in the case of plantarflexors, a much lower level of activity). In spite of these considerable differences at all three joints, the total extensor synergy, as indicated by the support moment, was essentially the same. This comparison is further evidence contradicting Pierrot-Deseilligny, Bergego, and Mazieres's (1983) generalization that the quadriceps are of paramount importance during stance phase of human gait. In fact, I have noted patients with knee pathologies who walk fairly normally but compensate without their use of quadriceps during all of stance (Winter, 1981).

The question remains, Are these changes seen across all subjects or across many strides of the same subject? Figure 11.14 is a plot of the mean knee moment versus the mean hip moment during stance for the nine repeat day-to-day trials of Subject WM22. As can be seen, the slope of the linear regression is effectively -1, meaning that there is a one-for-one trade-off for each individual walking trial. Trials plotted within the upper left quadrant documented a dominant hip extensor and knee flexor pattern; within the lower right quadrant, a dominant knee extensor and hip flexor pattern. Whether this trend extended over many subjects walking a range of cadences can be determined from a plot of the same variables in Figure 11.15. The regression for 54 separate subject trials was also effectively -1. This plot documents two major synergies, one of which has already been discussed. The one-for-one trade-off of moments between the knee and hip shows that there is a constant net extensor moment acting on the thigh during stance, and this finding further explains why the support moment (which is the sum of this thigh moment plus ankle moment) is so consistent, in spite of major individual differences at the knee and hip. The second major synergy is related to the trade-off between anterior and posterior muscles and has been shown to be highly correlated with the balance of HAT (Winter, 1987a).

Balance Synergy

On a stride-to-stride basis, there is consistency in the kinematics of the ankle and knee joints (*CVs* of 9% and 10% from Figure 11.1) for an individual subject. On these same repeat analyses, the head horizontal acceleration and trunk angular acceleration had *CVs* of 106% and 124% respectively. Because HAT represents 2/3 of the body mass, these upper body accelerations represent a large ballistic load, which must be controlled by the lower limb muscles. Presumably, this control must be exerted during stance. To deceler-

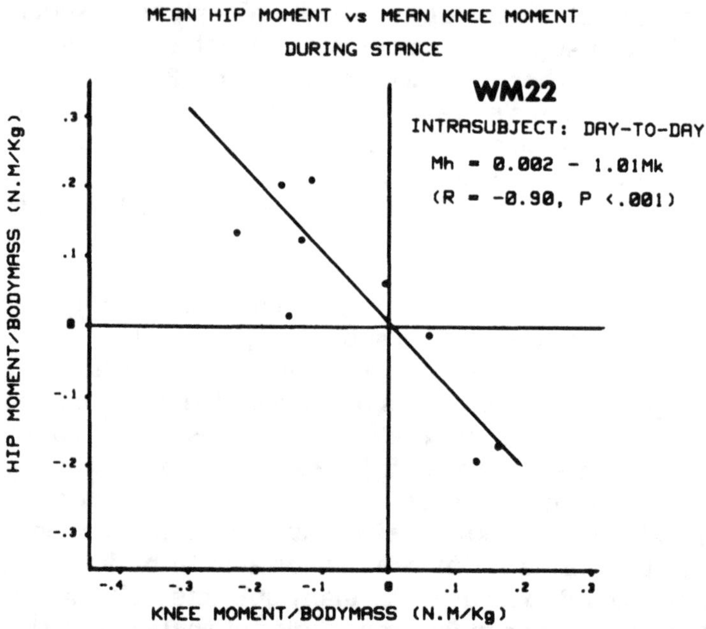

Figure 11.14. Mean knee versus mean hip moment during stance for the nine repeat trials with the same subject (WM22). The slope of the regression is effectively -1, which means that from trial to trial, there is a one-for-one trade-off of moments of force between the knee and hip.

ate the trunk or pull it more upright, the gluteus maximus and hamstrings would have to be more active. Conversely, to accelerate the trunk forward or to correct its posture in a forward direction, the iliopsoas, rectus femoris, and vasti muscles would have to be more active during stance. Thus balance and posture control is an anterior-posterior regulating task that changes from stride to stride, and these changes are evident in the random selection of strides that were measured and plotted in Figures 11.13 and 11.14. However, in the process of correcting balance, there was no significant change in the net extensor (support) pattern. Thus the control of collapse of the lower limb appears to be quite independent of the control of posture and balance.

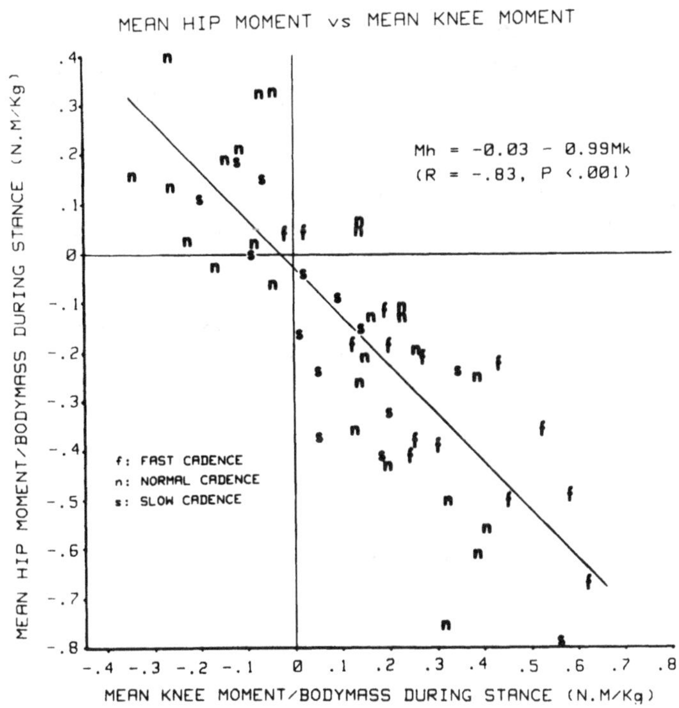

Figure 11.15. Mean knee versus mean hip moment during stance for 54 subjects walking slow, natural, and fast cadences. The slope of this linear regression is -1, which means that normal subjects operate along this same line showing different biases towards the posterior or anterior muscles crossing the hip and knee.

From a biomechanical perspective, the easiest way to control HAT would be by the hip flexor and extensor muscles, and this control would take place during stance, primarily during single support. This has, in fact, been shown to be true (Winter, 1987a). For Subject WM22's nine repeat trials, the correlation between the hip moment and the angular acceleration of the trunk was .93 during single support, and the slope of this regression was within 10% of a textbook calculation of the moment of inertia of HAT. This correlation dropped as the correlation period was expanded to include double support, a finding that is not surprising when one considers that both ipsilateral and contralateral hip muscles can now control HAT. Only when a simultaneous bilateral analysis is done can the balance of HAT be documented. However, it would be easy to hypothesize from initial analyses that the control of HAT is transferred from one limb to the other during each double support period.

Thus the primary role of the hip muscles is to control posture and balance of HAT. The balance synergy here is closely linked to the support synergy such that the knee motor pattern makes up the difference to supply a sufficient extensor pattern to control, along with the ankle plantarflexors, knee flexion. Because the balance control system must continuously alter the anterior-posterior motor patterns on a stride-to-stride basis, the hip and knee motor patterns will vary considerably. But because of the one-to-one trade-off between the hip and knee, the total extensor contribution to the support process is almost constant.

SUMMARY: COORDINATION OF MOTOR TASKS

From the support and balance synergies described in this chapter and from the detailed description of the end-point control of the foot during swing, we can now theorize about the essential building blocks of the human locomotor system. Support and balance have been shown to be essentially independent and are primarily stance phase tasks. It could be theorized that swing control is independent of both balance and support because most of the control of the swinging limb occurs at a different time in the gait cycle. The control of walking could be considered the integration of three independent control systems, and successful gait would require all three systems to operate correctly. At the individual muscle level, we might see an algebraic summation of the requirements of all three tasks. At any given time, a given muscle might be responding and producing a net tension that contributes to two or three of the tasks. During early stance, for example, the hip extensors could be contributing simultaneously to control of the balance of HAT and to the control of knee collapse. Or during push-off, the ankle plantarflexors could be contributing to the initiation of swing but also assisting in the balance of HAT in late stance (double support). This theory implies that if researchers look at individual joint motor patterns by themselves, it would be impossible to identify control strategies; a look at the total limb is essential. Certainly, the structure of the neural networks can accommodate such an integration. The motor neuron pool is an excellent summation point for all the excitatory and inhibitory contributions from the balance, support, and swing control systems.

At this time, we cannot say whether this task-related theory is more compatible with theories of central versus peripheral control than are theories proposed heretofore, or how this theory may support or deny the hypothesized central pattern generator theory. However, this task-related theory is functionally related and strongly supported by the experimental evidence presented in this chapter. What is now needed is further testing of the theory.

REFERENCES

Bresler, B., & Frankel, J. P. (1950). The forces and moments in the leg during level walking. *Transactions of the ASME, 72*, 27-36.

Drillis, R. (1958). Objective recording and biomechanics of pathological gait. *Annals of the New York Academy of Sciences, 74*, 86-109.

Du Chatinier, K., Molen, N. H., & Rozendal, R. H. (1970). Step length, step frequency and temporal factors of the stride in normal walking. *Proceedings Koninklijke Nederlandse Academie van Wetenschappen, Series C, 73*, 214-226.

Finley, F. R., & Cody, K. A. (1970). Locomotive characteristics of urban pedestrians. *Archives of Physical Medicine and Rehabilitation, 51*, 423-426.

Finley, F. R., Cody, K. A., & Finizie, R. V. (1969). Locomotion patterns in elderly women. *Archives of Physical Medicine and Rehabilitation, 50*, 140-146.

Lamoreux, L. W. (1971). Kinematic measurements in the study of human walking. *Bulletin of Prosthetics Research, 10*, 3-84.

Molen, N. H., & Rozendal, R. H. (1972). Fundamental characteristics of human gait in relation to sex and location. *Proceedings Koninklijke Nederlandse Academie van Wetenschappen, Series C, 75*, 215-223.

Murray, M. P., Kory, R. C., & Clarkson, B. H. (1969). Walking patterns in healthy old men. *Journal of Gerontology, 24*, 169-178.

Pierrot-Deseilligny, E., Bergego, C., & Mazieres, L. (1983). Reflex control of biped human gait in man. In J. E. Desmedt (Ed.), *Motor control mechanisms in health and disease*. New York: Raven Press.

Ralston, H. J., & Lukin, L. (1969). Energy levels of human body segments during level walking. *Ergonomics, 12*, 39-46.

Wickiewicz, T. L., Roy, R. R., Powell, P. L., & Edgerton, V. G. (1983). Muscle architecture of the human lower limb. *Clinical Orthopedics and Related Research, 179*, 275-283.

Winter, D. A. (1979). *Biomechanics of human movement*. New York: Wiley.

Winter, D. A. (1980). Overall principle of lower limb support during stance phase of gait. *Journal of Biomechanics, 13*, 923-927.

Winter, D. A. (1981). Use of kinetic analyses in the diagnostics of pathological gait. *Physiotherapy Canada, 33*, 209-214.

Winter, D. A. (1983a). Biomechanical motor patterns in normal walking. *Journal of Motor Behavior, 15*, 302-330.

Winter, D. A. (1983b). Energy generation and absorption at the ankle and knee during fast, natural and slow cadences. *Clinical Orthopedics and Related Research, 175*, 147-154.

Winter, D. A. (1984). Kinematic and kinetic patterns in human gait: Variability and compensating effects. *Human Movement Science, 3*, 51-76.

Winter, D. A. (1987a). Balance and posture in human walking. *IEEE Engineering in Medicine and Biology, 6*, 8-11.

Winter, D. A. (1987b). *Biomechanics and motor control of human gait*. Waterloo, Canada: University of Waterloo Press.

Winter, D. A., Quanbury, A. O., & Reimer, G. D. (1976). Analysis of instantaneous energy of normal gait. *Journal of Biomechanics, 9*, 253-257.

Winter, D. A., & Yack, H. J. (1987). EMG profiles during normal human walking: Stride-to-stride and inter-subject variability. *Electroencephalography and Clinical Neurophysiology, 67*, 402-411.

SECTION 4

COORDINATION AND MOVEMENT DISORDER

Perspectives on the Coordination of Movement
S.A. Wallace (Editor)

MOVEMENT DISORDERS AND THE NEURAL BASIS OF MOTOR CONTROL§

James G. PHILLIPS, Friedemann MÜLLER, and George E. STELMACH*

Motor Behavior Laboratory
University of Wisconsin-Madison

ABSTRACT

An examination of movement disorders can contribute to the understanding of processes involved in the coordination of movement because the impairment of a particular process, as seen in a movement disorder, provides convergent evidence that that process occurs in the coordination of normal movements. The distributed nature of brain structures can make inferences of function difficult; however, a number of movement disorders can be related to specific brain structures and to specific aspects of movement organization. There are sufficient data to discuss the roles of the basal ganglia and cerebellum in the coordination of movement, and there is a fast-growing neuropsychological and neurophysiological literature pertaining to the function of the basal ganglia and Parkinson's disease. Nevertheless, our understanding of the cerebellum's contribution to movement still relies heavily upon clinical observations and animal experiments. Move-

---

*Address correspondence to: George E. Stelmach, Motor Behavior Laboratory, 2000 Observatory Drive, University of Wisconsin-Madison, Madison, WI 53706, U.S.A.

§The preparation of this chapter was supported by a grant from the National Institute of Neurological Diseases and Stroke NS 17421. Friedemann Müller was supported by Grant Mu 750/1-1 from the Deutsche Forschungsgemeinschaft.

ment researchers need to integrate their work with the neurophysiological data base by providing detailed descriptions of how and when a structural deficit causes an impairment of function. It is postulated that the impaired function observed in certain disorders of movement suggests that computational processes are involved in the coordination of movement.

SIGNIFICANCE OF MOVEMENT DISORDERS RESEARCH

Problems in the coordination of movement can arise as a result of damage to specific brain structures. The resulting movement disorders stimulate and challenge research because these disorders dramatize the brain's computational role in the initiation and control of movement. The study of movement disorders can increase our understanding of the role of the brain in the coordination of movement, and it can also contribute to the understanding of cognitive processes involved in movement organization. This chapter, a discussion of current research on the functions of the basal ganglia and cerebellum, shows the potential of research in this area.

Trauma or disease can cause a variety of disorders of movement, which may involve problems in the strength, timing, or sequencing of muscle activity or problems with reflex responses and their functional coordination with voluntary movement (Grimm & Nashner, 1978; Lee & Tatton, 1978). Although disturbances of movement can arise as a result of damage to peripheral structures such as muscles and afferent nerve fibres or as a result of damage to the central nervous system, it is the damage to central control mechanisms that may potentially provide information about processes involved in the coordination of movement. We therefore focus on movement disorders resulting from damage to central structures.

To establish the significance of damage to a brain structure, any consideration of structural impairment requires an assessment of motor functioning. It is necessary to understand which underlying functions are affected by structural damage (Grimm & Nashner, 1978) and to document the functional correlates of movement production (Rolls, 1983). Researchers face the task of supplementing the neurophysiological data base with detailed descriptions of the functional consequences of structural damage.

It is possible to make inferences about motor coordination from the disturbances of function seen in movement disorders. Damage to a structure can cause a loss or disturbance of function, and such changes point to the existence of that function in the normal coordination of movement. There are a variety of movement disorders affecting several facets of motor coordination. For instance, the pattern of activation of muscles during ballistic movements is affected differently by specific movement disorders (Hallett, 1983). Whereas some diseases, such as Parkinson's disease, affect the scaling of movement parameters centrally, others, such as athetosis, interfere with the orderly sequencing of muscle activity. Still others, such as Tourette's syndrome, disturb the voluntary initiation of movement while leaving an apparently normal pattern of muscle activity (Hallett, 1983; Hallett & Marsden, 1981). These patterns of deficits suggest that there are mechanisms controlling the energization, scaling, and sequencing of movements. An examination of movement disorders can therefore reify, and provide converging evidence for, hypothetical mechanisms discussed by theories of motor coordination (e.g., Margolin, 1984; Wing, 1984; Wing, Keele, & Margolin, 1984).

A variety of brain structures—for example, the motor cortex, basal ganglia, and cerebellum—are associated with the generation and coordination of movement. Of these structures, the cerebellum and basal ganglia have been thought to have important roles in coordination (Brooks, 1986; Holmes, 1922). For example, the basal ganglia and cerebellum have been closely associated with the programmed control of movements (Eccles, 1973; 1979), the basal ganglia having a role in the initiation of movements and the cerebellum having a role in the storage of motor programs. More recently, Brooks (1986) has suggested that the cerebellum is involved in the orderly activation of muscles and control of posture and has implied that the basal ganglia contribute to the scaling of movements for environmental demands. Not only are these structures potentially important to the coordination of movement, but current advances in pharmacology and in brain imaging have also generated considerable interest in localizing and examining the function of brain structures in vivo (Posner, Pea, & Volpe, 1982). In particular, there are localizable problems with specific neurotransmitter systems associated with the basal ganglia, and there are anatomically well-localized problems with the cerebellum.

However, attempts to determine the function of particular brain structures must address two issues: the degree to which deficits can be *localized* within a distributed control system (Stelmach & Diggles, 1982) and the degree to which such distributed control systems can *compensate* for damage (Grimm & Nashner, 1978). These issues pertain to both structural damage and functional deficits.

Localizing Damage Causing a Movement Disorder

Edelman and Mountcastle (1979) argued from an examination of neural anatomy and observations of the effects of brain damage that the brain is a distributed control system in which functions are distributed across a network of modular elements. The brain can be thought of as a large number of serial and parallel connections. Edelman and Mountcastle suggested that information flow through this system may follow a number of different pathways, the dominance of one path or another being a dynamic and changing property of the system. The redundancy of such a distributed system means that local lesions degrade rather than destroy system function, and this accounts for some of the difficulty in localizing brain function.

This redundancy may be seen in the effects of damage to a particular brain structure. It is striking to observe how much damage must occur before noticeable signs and symptoms of disease appear (Grimm, 1983). For example, it is estimated that at least 80% of dopaminergic cells in the substantia nigra must be lost before signs of Parkinson's disease are observed. Even with such damage, considerable compensation seems to occur among remaining control systems (Grimm & Nashner, 1978).

Care must therefore be taken when a function is attributed to a brain structure. If damage results in some disturbance of function, it is naive to attribute this disturbance specifically to the damaged structure because damage has implications for the functioning of other structures as well. Consideration must be given to the inputs that structures such as the basal ganglia and cerebellum relay, the way they process these inputs, and the structures that are modulated by their outputs. It is thus necessary to consider the neural flow of activity during the production of movement (e.g., DeLong et al., 1984).

Given these considerations, Marsden (1982) suggested that damage to a brain structure produces both effects upon the structure itself and effects upon other structures that it innervates. There is thus a loss of the functions associated with the damaged structure (negative symptoms), and there is an excess of activity in other functions of the central nervous system (positive symptoms) due to a loss of inhibitory control. Marsden suggested that to understand the function of a particular brain structure that has been damaged, researchers should focus on the functions that have been impaired because of that damage.

Identifying Functional Impairments of a Movement Disorder

Movement disorders can cause abnormalities in a wide range of movement, for example, walking, reaching, writing, and speaking. The form of the disturbance can indicate the functions that are lost (e.g., Margolin, 1984). However, researchers attempting to *localize function* must take care in assessing functional impairment. They must distinguish motor deficits proper from changes in performance due to other cognitive processes; and they must also decompose performance to distinguish which *specific* motor processes are impaired and which aspects of motor performance change as a consequence (biomechanical or compensatory) of this impairment. Considerable methodological expertise is therefore required before unequivocal inferences can be made concerning functional impairment (Wing & Miller, 1984). If specific motor impairments are not identified, or if confounding variables are not acknowledged, inferences concerning functional impairment can be faulty. A case in point is cognitive functioning in Parkinson's disease.

Although motor deficits are the most readily apparent aspects of Parkinson's disease, this disease may affect cognitive processes as well. Until the advent of L-dopa therapy, motor impairments had made it difficult to assess any cognitive impairment in Parkinson's disease. However, treatment with L-dopa has improved the mobility of patients with Parkinson's disease, and motor impairments now tend to be underestimated because patients are usually tested while medicated. There is now an increased interest in investigating cognitive deficits in Parkinson's disease (Knight, Godfrey, & Shelton, 1988; Mortimer, Pirozzolo, Hansch, & Webster, 1982; Rajput, Offord, Beard, & Kurland, 1984; Rogers, Lees, Trimble, & Stern, 1986). Although the increased mobility of patients with the disease means that cognitive deficits can now be assessed (Korczyn et al., 1986), this research should be viewed with caution because the patients' cognitive performance is undoubtedly influenced by residual motor impairments, depression associated with the disease itself, or the depression caused by the diagnosis of a chronic debilitating disease (Bieliauskas, Klawans, & Glantz, 1986; Dakof & Mendelsohn, 1986; Gotham, Brown, & Marsden, 1986).

Because most methods of assessing cognitive capacities have motor components and depend upon the patient's speed of performance, results are difficult to interpret unless the contributions due to motor slowing are controlled (e.g., Girotti et al., 1986). For instance, experiments using tasks with substantiated motor components have reported visuo-spatial deficits in Parkinson's disease (Bowen, Burns, Brady, & Yahr, 1976; Bowen, Hoehn, & Yahr, 1972) whereas experiments that have controlled motor components have not found visuo-spatial deficits (Brown & Marsden, 1986; Della Sala, Lorenzo, Giordano, & Spinnler, 1986; Stelmach, Phillips, & Chau, in press).

Wing and Miller (1984) have discussed how methods of decomposing performance can be used to examine movement disorders. They assumed that psychomotor performance can be described as a series of processes or stages. Inferences can be made about a particular stage of processing when experimental tasks are devised that vary in complexity. For example, if two tasks systematically vary in their amounts of response preparation, the difference in performance between these tasks may indicate the amount of preparation required. Movement disorders can be examined with this method of decomposing performance. For example, the performance of normal controls and patients with Parkinson's disease can be compared over the two tasks. If the difference in response time between the two tasks (which is used as an index of preparation) varies between the two groups, then it can be inferred that Parkinson's disease affects preparatory processes.

Deficits in a specific motor process may have biomechanical or computational consequences for other motor processes. It is thus necessary to distinguish the major impairment from any reactions to that impairment. For example, a slowness or inaccuracy in the performance of simple movements tends to break down the smooth execution of more complex movement sequences. There are also good reasons to suppose that *compensatory* strategies occur in movement disorders. Grimm and Nashner (1978) suggested that patients with such disorders rely more upon other control systems. Cerebellar patients may move more slowly to compensate for their inability to control the extent of their movement, and patients with Parkinson's disease may rely more upon perceptual feedback and less upon programmed control (De Ajuriaguerra, 1975; Flowers, 1976; Marsden, 1982). Such strategic changes lead us to question the use of the subtractive procedures outlined by Wing and Miller (1984). The assumption underlying these procedures is that increasing task complexity simply adds a stage of processing to the task rather than altering the way the task is performed. If there are grounds to believe that strategic changes in performance are occurring, then subtractive procedures are not appropriate (Chase, 1978). Such phenomena are a challenge to traditional techniques and demonstrate the need for a broader approach that assesses human capabilities in conjunction with subtractive procedures (Gopher & Sanders, 1984).

The remainder of this chapter illustrates the issues raised in this section with a discussion of the roles of the basal ganglia and cerebellum in the coordination of movement.

PARKINSON'S DISEASE AS A MODEL OF BASAL GANGLIA FUNCTION

Marsden (1982) has suggested that the movement disorder caused by Parkinson's disease provides the best model for inferring the motor functions of the basal ganglia. The basal ganglia mediate between higher and lower brain structures, receiving inputs from cortical areas and the substantia nigra and innervating thalamic and midbrain nuclei (DeLong, Georgopoulos, & Crutcher, 1983; Gunilla, Oberg, & Divac, 1985; Tatton, Eastman, Bedingham, Verrier, & Bruce, 1984). Although Parkinson's disease seems to affect more than one neurotransmitter system, the most specific disturbance is associated with changes in dopaminergic functioning, in particular a loss of the dopaminergic cells in the substantia nigra that project to the striatum (Forno, 1982; Hornykiewicz, 1982). This loss causes disturbances of basal ganglia function.

Symptoms of Parkinson's Disease

In general, although patients with Parkinson's disease initially may demonstrate their symptoms unilaterally, as the disease progresses, symptoms become more pervasive (Hoehn & Yahr, 1967) and affect all movements bilaterally. The primary symptoms of Parkinson's disease are

> *akinesia*: a lack of activity or volition, seen in a slowness in the initiation of movement (De Ajuriaguerra, 1975);
>
> *bradykinesia*: a slowness in the execution of movement and rapid fatigue of movement (Hallett & Khoshbin, 1980);
>
> *rigidity*: resistance to the passive stretch of muscles (McLellan, 1981);
>
> *tremor*: a tremor at rest around 4 to 5.5 Hz (Gresty & Findley, 1981).

Although these symptoms are considered to be cardinal symptoms, in the course of the disease other problems, such as postural instability, tertiary gait, axial apraxia, and monotonous speech, gain prominence. The primary symptoms are relatively independent of one another (Teräväinen & Calne, 1980; Zetusky, Jankovic, & Pirozzolo, 1985). Marsden (1982) suggested that positive symptoms such as tremor and rigidity are due to the loss of the inhibitory influences within the basal ganglia. He reasoned that the real function of the basal ganglia might be best understood by examination of the loss of functioning, that is, the negative symptoms of akinesia and bradykinesia (Marsden, 1982; 1984; 1985).

Clinical Consideration of Basal Ganglia Function

Clinically, Parkinson's disease appears to be a disorder of the scaling of movement. Movements are slow, but patients do not exhibit the obvious abnormalities in the selection and sequencing of muscle activity that are seen in chorea, athetosis, or spasticity. Similarly, although handwriting tends to be slower or smaller than normal, patients with Parkinson's disease do not show the serious disorders of shapes of handwritten letters that may be seen, for example, in patients with some forms of neuropsychological dysfunction such as dysgraphia (Margolin, 1984). This difference suggests that the motor program is intact but that patients with

Parkinson's disease have problems controlling specific parameters of the movement (Marsden, 1982).

Examination of electromyogram patterns suggests that the cause of this slowness of movement is not an obvious cocontraction of agonist and antagonist, as is seen in spasticity (Hallett, 1983). Indeed, peripheral and spinal mechanisms such as muscle spindles (Burke, 1985) and patterns of reciprocal inhibition (Obesa, Quesada, Artieda, & Martinez-Lage, 1985) function normally in patients with Parkinson's disease, albeit probably with a higher level of background activity. These findings suggest that symptoms of Parkinson's disease may be associated with hyperactivity of supraspinal loops (Cody, Macdermott, Matthews, & Richardson, 1986; Delwaide, 1985).

Slowness in Parkinson's disease is also not due to an inability to produce greater amounts of muscle activity. Although the muscles may be rigid and have a higher tonic level of activity, Berardelli, Dick, Rothwell, Day, and Marsden (1986) have shown that patients with Parkinson's disease can produce increased muscle activity in response to central commands and that central commands do not saturate at the muscle level. Instead, Hallett, Shahani, and Young (1977) have observed that these patients have normal triphasic patterns of muscle activation (agonist, antagonist, agonist) but that they require more cycles of activity to produce their movements. It seems that there are central deficits in the scaling of the size and/or duration of bursts of muscle activity.

These observations suggest that Parkinson's disease produces dysfunctions in the preparation of movement parameters. This hypothesis is supported by the prolonged response times of patients with the disease and observations that the period before onset of muscle activity (premotor time) is prolonged in these patients (Evarts, Teräväinen, & Calne, 1981; Sheridan, Flowers, & Hurrell, 1987) whereas the period between start of muscle activity and onset of movement (motor time) is normal. These deficits in preparatory processes have implications for the execution of movement.

It has been found that in patients with Parkinson's disease, muscle activity is not scaled appropriately for task demands. Although the burst of initial agonist activity in normal subjects is related to movement amplitude (Brown & Cooke, 1981), this is not the case for patients with Parkinson's disease (Baroni, Benvenuti, Fantini, Pantaleo, & Urbani, 1984; Berardelli, Dick, Rothwell, Day, & Marsden, 1986; Hallett et al., 1977). Neither is the initial acceleration of movements scaled appropriately for task demands in Parkinson's disease. Whereas in normal subjects, movements of greater amplitude involve greater initial acceleration (Megaw, 1974), patients with Parkinson's disease do not scale their initial acceleration in this way (Baroni et al., 1984; Draper & Johns, 1964; Flowers, 1976; 1978). In addition, researchers (Berardelli, Accornero, Argenta, Meco, & Manfredi, 1986; Draper & Johns, 1964; Ingvarsson, Johnels, Lund, & Steg, 1986) have suggested that patients with Parkinson's disease have problems intentionally modulating the speed of their movements and are less able to speed up their performance.

The foregoing evidence suggests that although patients with Parkinson's disease do not have problems with their overall motor program, they have problems specifying some of the parameters for these programs (Marsden, 1984). The task of researchers has been to document how Parkinson's disease affects the selection and preparation of response parameters.

Basal Ganglia and Preparatory Processes

Preparatory processes serve to reduce the amount of on-line movement control that is required for movement. If a movement is prepared in advance, as in a motor program, less of the movement needs to be monitored during its execution. Preparatory processes therefore play an important role in the smooth execution of movements (Lashley, 1951). The computational complexity of movement preparation is usually inferred from movement latencies (reaction times) under the assumption that the brain requires more time to prepare more complex movements. Stelmach, Worringham, and Strand (1986) have reasoned that if patients with Parkinson's disease have disproportionate problems preparing a movement dimension, this difficulty should be evident in longer response times.

A number of experimenters have considered the effects of Parkinson's disease upon response preparation. Early experiments compared response times in simple and choice tasks or examined patients' performance on tracking tasks. The focus of these experiments was whether patients with Parkinson's disease can use predictive information to prepare their movements and, in particular, whether patients have problems preparing a specific parameter of movement such as movement extent.

Ability to prepare for movement in advance may be evaluated to some extent by comparing response times for choice tasks, in which there is some response uncertainty, with those for simple tasks, in which there is no response uncertainty and for which the response can be prepared in advance. Whereas normal subjects are much faster in simple tasks than in choice tasks, differences in response times for these two tasks are much smaller in patients with Parkinson's disease (Bloxham, Mindel, & Frith, 1984; Evarts et al., 1981; Girotti et al., 1986; Pullman, Watts, Juncos, Chase, & Sanes, 1988). However, it is difficult to determine from a simple-choice comparison whether patients' problems are due to perceptual or motor processes.

Della Sala et al. (1986) investigated whether Parkinsonian patients have difficulties perceiving predictive information and demonstrated that they do not. Della Sala et al. used a nonspeeded task with minimal movement to examine patients' ability to extrapolate and predict where one line would intersect with another. Patients with Parkinson's disease were as accurate as normal subjects in this task, which implies that the problems these patients have are in the utilization of predictive information to prepare their movements.

Parkinsonian patients' ability to use predictive information was demonstrated by Day, Dick, and Marsden (1984) and Stern (1986). Day et al. compared performance in tracking tasks when patients were unaware of the repetitive nature of the task and when patients were aware of repetition. That patients made fewer tracking errors when they were aware of the repetition in the task showed that they could use predictive information. Patients' ability to use this information was reduced when their medication was discontinued.

Unfortunately, it is difficult to draw conclusions from either tracking performance or comparisons of response times in simple and choice tasks. The major problem with these experiments is that factors were not systematically manipulated, and detailed inferences about preparatory processes therefore cannot be made. It is difficult to discern from these experiments which specific parameters patients have problems preparing. We therefore focus upon more comprehensive experiments (i.e., Rafal, Inhoff, Friedman, & Bernstein, 1987; Sheridan et al., 1987; Stelmach et al., 1986).

Using Rosenbaum's (1980) precueing paradigm, Stelmach et al. (1986) examined Parkinson's patients' ability to prepare the parameters of direction, arm, and extent. Although patients took longer to prepare movements and were slower in executing them, no specific parameter was disproportionately difficult for the patients to prepare. In addition, the patients seemed to prepare their parameters at about the same rate as normal subjects did.

Similar results were found by Sheridan et al. (1987) using simple and choice tasks that tested patients' ability to prepare movements of different lengths and movements to targets of different amplitudes. Although Sheridan et al. found that patients with Parkinson's disease had longer premotor times, a result suggesting that problems were central in origin, no specific parameter was disproportionately difficult for patients to prepare.

Rafal et al. (1987) examined the ability of patients with Parkinson's disease to prepare repetitive movement sequences of different lengths. Rafal et al. found that both response times and movement times were longer and that as with normal subjects, patients required more time to prepare longer movement sequences. On the other hand, patients did not display disproportionate deficits in the preparation of any specific movement parameter.

Difficulties encountered in identifying preparation deficits for specific parameters in Parkinson's disease patients have led some researchers to suggest that patients have nonspecific deficits in their ability to prime or alert themselves (Brown & Marsden, 1988; Cools, Van Den Bercken, Horstink, Van Spaendonck, & Berger, 1984; Stern & Mayeux, 1986). Indeed, Bloxham et al. (1984) and Bloxham, Dick, & Moore (1987) have found that patients with Parkinson's disease benefit less from warning signals than normal subjects do. Whereas normal subjects can use warning signals to improve their reaction times, patients with Parkinson's disease cannot. This finding suggests that patients' deficits are not in the selection of response parameters (Stelmach et al., 1986) but in later stages of the response process such as response priming or the initiation of muscle activity (Sanders, 1980).

Open- and Closed-Loop Control of Movement and Parkinson's Disease

Given that clinical and experimental studies suggest that patients with Parkinson's disease should have difficulty preparing parameters of movement extent, it is surprising that experiments assessing preparatory processes have not linked a specific parameter to Parkinson's disease. Indeed, Evarts et al. (1981) reported that the findings of prolonged response times in patients with Parkinson's disease were less reliable than were findings of prolonged movement times. This may be because Parkinson's disease affects a lower level of the preparatory process, such as response energization. On the other hand, it is possible that the disease has a qualitative effect upon the control of movement. Assessment of the preparatory processes of patients with Parkinson's disease is thus complicated because there is evidence that patients do not prepare their movements in the same manner as normal subjects do (Marsden, 1984).

The function of preparatory processes is to reduce the requirement for ongoing control of movement. However, both Flowers (1976) and Marsden (1982) have suggested that the effect of Parkinson's disease on the preparation of movement is that patients tend to rely less upon the advance preparation (open-loop control) of movement and more upon feedback guidance (closed-loop control).

This shift in mode of control may be seen in an experiment by Stelmach, Worringham, and Strand (1987). Stelmach et al. (1987) considered how normal subjects and patients with Parkinson's disease prepared sequences of repetitive tapping movements. In such tasks, the response time of normal subjects increases with the number of taps in a response sequence, a result showing that normal subjects program their responses in advance. However, in Stelmach et al.'s (1987) experiment, patients' response time did not increase with the number of taps in a response sequence. In addition, patients showed longer movement times, and this finding suggests that patients controlled their movements during execution rather than preparing them completely in advance.

These inconsistencies in the literature can be reconciled to some extent because it appears that patients with Parkinson's disease can prepare responses in advance for simpler tasks but have more difficulty with more complicated tasks. Whereas patients have problems tracking unpredictable targets (Flowers, 1976; 1978), they have greater success with more regular targets (Bloxham et al., 1984). Similarly, Stelmach, Phillips, and Chau (in press) found in a repetitive tapping task that patients with Parkinson's disease prepared two taps in advance when there was a straightforward link between stimulus and response but did not prepare two taps in advance for a more difficult task in which the link between stimulus and response was more complicated. It is in more complicated movement tasks that patients probably shift from open-loop control of movement to closed-loop control.

Potential changes in the mode of motor control can create difficulties in assessing the adequacy of preparatory processes in patients with Parkinson's disease, especially because the amounts of preparation are usually inferred from the resulting motor responses. In some situations, however, it is possible to examine the preparation independently of the execution of movement (Sanes & Evarts, 1985). Before the execution of voluntary movements, there are preparatory postural adjustments, and it is possible to examine them to see whether patients with Parkinson's disease do prepare their movements. Dick et al. (1986) found that these patients do indeed make preparatory postural responses; however, these postural adjustments may not be adequate because they are reduced in size.

In summary, experiments that assess preparation for movement in patients with Parkinson's disease have shown that these patients may prepare their movements but that preparation may not be adequate. This inadequacy is probably not a problem of response selection but of response energization and means that patients may not prepare more complex movements in advance.

Assessment of Abilities of Patients with Parkinson's Disease.

If strategic changes in performance accompany Parkinson's disease (Flowers, 1976; Marsden, 1982), then some of the subtractive, stage-related approaches to interpreting data have serious problems (Wing & Miller, 1984). An assumption in subtractive approaches is that increasing the difficulty of the task merely adds processing steps rather than causing qualitative changes in the way the task is performed. Because Parkinson's disease causes problems in preparatory processes such that patients tend to rely less upon open-loop control and more upon closed-loop control, increasing the difficulty of the task may alter how tasks are performed rather than simply adding more processing steps. Hence subtractive methods of studying the performance of Parkinsonian patients are to some extent questionable. Our laboratory has therefore taken a different approach in the last few years to the study of deficits due to Parkinson's disease.

Instead of manipulating task difficulty and observing performance, we have focused upon patients' ability to control their performance. Because compensatory strategic changes in task performance may be occurring, it is necessary to identify which functional deficits occur as a direct result of Parkinson's disease and to distinguish these from any subsequent, compensatory strategic changes in performance. We made this distinction by examining patients' ability to intentionally control aspects of their performance such as force and speed. We reasoned that patients would be able to intentionally control those parameters used in their compensatory strategies but not those parameters associated with the primary deficits caused by Parkinson's disease.

Several experiments tested patients' ability to modulate the force, speed, and spatial or temporal aspects of their movements. These experiments tended to show that patients' higher (intentional) control mechanisms are intact but that control of lower elements of performance is impaired.

Experiments by Stelmach and Worringham (1988a) and Stelmach, Teasdale, Phillips, and Worringham (in press) tested the ability of patients with Parkinson's disease to intentionally control their production of force. Previous experiments had implied that these patients were deficient in the control of forces in relation to task requirements (Baroni et al., 1984; Berardelli, Dick, Rothwell, Day, & Marsden, 1986). Our experiments showed that patients with Parkinson's disease were relatively accurate in their control of forces compared with normal subjects. However, patients required more time to produce these forces and were more variable in their production of forces.

As mentioned previously, researchers had reported that patients with Parkinson's disease could not intentionally speed up their movements (Berardelli, Accornero, Argenta, Meco, & Manfredi, 1986; Ingvarsson et al., 1986) and could not modulate their movements in accordance with task demands. Teasdale, Phillips, and Stelmach (1989) examined patients' ability to control the speed of their movements. Teasdale et al. determined a baseline movement speed by asking subjects to move as quickly as possible. Subjects were subsequently asked to move at various percentages of this maximum speed (10% faster, the same speed, 30% slower, and 60% slower). Although patients' movements were slower and showed irregularities in their muscle activity, patients showed control of their movement times and could speed up their movements when required. This finding contradicts the observations of Ingvarsson et al. (1986) and Berardelli, Accornero, Argenta, Meco, and Manfredi (1986). Sanes (1985) and Teasdale and Stelmach (1988) suggested that patients' velocity control problems depend upon task difficulty and that these problems are less severe under less demanding task constraints.

The foregoing experiments produced evidence that the ability to intentionally control the speed and force of movements is intact in patients with Parkinson's disease. However, these findings contradict suggestions that Parkinson's disease is the result of an inability to intentionally control an aspect of movement. Moore (1987) presented a means of reconciling these observations in a discussion of the differences between weakness and bradykinesia. In weakness, reduced strength can be detected by a comparison of peripheral feedback and efference; and once reduced strength has been detected, subjects can compensate (within limits) for reduced power. Moore suggested that these comparisons occur at a conscious, cortical level for patients with weakness but at a subconscious, subcortical level for bradykinesia. Our experiments (Stelmach, Phillips, & Chau, in press) showed that the motor impairments caused by Parkinson's disease are not the result of impairments in higher-order cortical processes. Although their range of forces may be restricted, patients with the disease can clearly intentionally control their force output with accuracy comparable to that of normal subjects. We suggest that these patients are aware of their output problems but have problems at a subcortical level in activating their muscles. Although patients can control their force output, they showed a high degree of variability and jerkiness in their control of movements. This finding suggests that Parkinson's disease is associated with problems in initiating and maintaining muscle activity, that is, the energization of movements.

Complex Movements and the Basal Ganglia

The effects of Parkinson's disease are not limited to simple movements. As the disease progresses, it tends to affect all forms of movement: handwriting (McLennan, Nakano, Tyler, & Schwab, 1972), speech (Canter & van Lancker, 1985), and walking (Knutsson & Martensson, 1986). However, there is a need to decompose performance to show which specific processes are impaired because deficits in the control of lower levels of motor organization could cause some of the problems observed with complex movements.

There are two possible explanations for Parkinsonian patients' problems with complex movements (Marsden, 1985). Parkinson's disease could affect either *higher* or *lower* levels of movement organization, and either effect could influence the performance of complex movements. Marsden suggested that inaccuracies in each simple movement may accumulate during a movement sequence so that the error in the first movement creates problems for the second movement and causes a progressive degradation of the movement sequence. That is, problems in each part of the sequence cause the entire sequence to break down. Alternatively, Marsden suggested that patients with Parkinson's disease may have problems sequencing their movements. Data from a finger-tapping experiment (Stelmach et al., 1987) strongly argue against the first hypothesis. A complex movement can be considered to be composed of a sequence of simpler movements, and patients with Parkinson's disease may have problems switching from one simple movement to the next. From this point of view, patients would have problems organizing these simpler parts into a complex movement.

There is evidence that patients with Parkinson's disease have problems coordinating more than one movement at a time. For example, Benecke, Rothwell, Dick, Day, and Marsden (1986; 1987) required that subjects simultaneously flex their arm and squeeze a manipulandum with the same limb. Patients with Parkinson's disease had problems superimposing and sequencing movements, as was evident in the dramatic increases in movement times when patients had to perform both movements at once. However, coordination of two movements has been examined more systematically by Stelmach and Worringham (1988b) in a bimanual aiming task. Although patients with Parkinson s disease were slower in executing bimanual movements, they did not exhibit disproportionately more difficulty controlling symmetrical and asymmetrical movements than normal subjects did. This finding implies that patients' problems lie largely in the combination or sequencing of separate movements of the same limb into a whole.

Nakamura, Nagasaki, and Narabayashi (1978) and Nagasaki and Nakamura (1982) have examined the control of sequences of movements in patients with Parkinson's disease. Wing et al. (1984) had suggested that Parkinson's disease causes problems in the timing of movement sequences. Canter and van Lancker (1985) and Nakamura et al. (1978) have shown in speech and in a variety of other movements that patients with Parkinson's disease are unable to accurately time responses of repetitive movements. For example, Nakamura et al. (1978) required that patients perform finger taps in synchrony with a periodic stimulus. Patients had difficulty tapping at certain frequencies. At some critical frequency between 2.5 and 5 Hz, control of tapping deteriorated, and subjects showed a hastened tapping at about 5 to 6 Hz that was independent of the stimulus with which they were to tap in synchrony. Irrespective of the type of movement (grasping, arm lifts, arm extensions), patients with Parkinson's disease had difficulty timing their movements and tended to perform hastened responses at about 5 Hz (Nagasaki & Nakamura, 1982). Although this phenomenon may be related to Parkinsonian tremor and could be the result of a tremor entraining rhythmic movements, it nevertheless demonstrates how a simple impairment at a lower level of movement organization could affect complex movements.

Gait, Posture, and the Basal Ganglia

Experiments on gait and posture have examined repetitive movement sequences in more detail, and these experiments suggest that it is partly a reduction in the size of Parkinsonian movements that causes the hastening or "festination" of movements in patients with Parkinson's disease. Knutsson and Martensson (1986) discussed gait in these patients and suggested that some of the abnormalities of posture and gait may be the result of poor preparatory postural adjustments, Parkinsonian rigidity, and adaptation to cope with impaired postural reflexes.

Preparatory postural adjustments occur before the execution of voluntary movements. Bazalgette, Zattara, Bathien, Bouisset, and

Rondot (1986) examined the preparatory postural adjustments preceding voluntary arm movements. Bazalgette et al. found that patients with Parkinson's disease did not make preparatory postural adjustments before initiating voluntary movement and that any adjustments that did occur were not specifically related to the form of the voluntary movement. However, Dick et al. (1986) have examined preparatory postural adjustments associated with voluntary arm movements in more detail. These researchers examined the muscle activity in patients' postural muscles and showed that patients with Parkinson's disease did have appropriately timed preparatory postural responses; however, these responses occurred less often and were of smaller functional amplitude than those of normal subjects. Although patients with Parkinson's disease do indeed make preparatory postural adjustments, these are reduced in size.

Several temporally distinct *reflex components* may be elicited when preactivated muscles or voluntary movements are perturbed (Lee & Tatton, 1978). Researchers have suggested that the different latencies of these components reflect different motor control systems that, although somewhat difficult to localize, are associated with different anatomical structures (Lee & Tatton, 1978). These reflex components have different functional roles. Early components of these reflexes have been related to the spinal stretch reflex, but later so-called long-latency components of these reflexes have been related to higher structures in the central nervous system (Diener & Dichgans, 1986; Lee & Tatton, 1978; Marsden, Rothwell, & Day, 1983); the long-latency components of these reflexes are more important in the control of posture (Evarts, 1985).

Experiments that have assessed long-latency reflexes in patients with Parkinson's disease have found that, although the latency of stretch reflexes are normal in these patients, there are abnormalities in the gain of the longer-latency components (Berardelli, Sabra, & Hallett, 1983; Chan, Kearney, & Melvill-Jones, 1979; Cody et al., 1986; Lee, Murphy, & Tatton, 1983; Mortimer & Webster, 1978). These abnormalities have also been related to the poorer balance and rigidity in these patients (Mortimer & Webster, 1978, Traub, Rothwell, & Marsden, 1980).

Knutsson and Martensson (1986) noted that Parkinsonian rigidity causes patients to lean forward and that impaired postural reflexes

(Traub et al., 1980) cause poorer balance in these patients. Knutsson and Martensson (1986) suggested that some of the abnormalities of posture and gait may be the result of Parkinsonian rigidity and adaptation to cope with impaired postural reflexes. They observed that the patterns of muscle activity are maintained during walking but that it is abnormalities in the amounts of muscle activity that are causing problems with walking. Patients with Parkinson's disease show a diminished stride length and an increased duration for each stepping cycle (Knutsson, 1972), and Murray, Sepic, Gardner, and Downs (1978) have associated these deficits with reductions in the motion at each joint.

The result of these tendencies upon Parkinsonian gait is best described by James Parkinson himself:

> The propensity to lean forward becomes invincible, and the patient is thereby forced to step on the toes and forepart of the feet, whilst the upper part of the body is thrown so far forward as to render it difficult to avoid falling on the face. In some cases, when this state of the malady is attained, the patient can no longer exercise himself by walking in his usual manner, but is thrown on the toes and forepart of the feet; being at the same time, irresistibly impelled to take much quicker and shorter steps, and thereby to adopt unwillingly a running pace. In some cases it is found necessary entirely to substitute running for walking; since otherwise the patient, on proceeding only a few paces, would inevitably fall. (Parkinson, 1817, pp. 6-7)

This observation suggests that maintaining the size of individual movements can cause problems with the overall movement sequence. However, these problems are clearly task-specific (Teasdale & Stelmach, 1988). Parkinson's disease is characterized by bradykinesia, and festination is seen only in certain forms of movement. For example, Ingvarsson et al. (1986) used a device to measure movement in three-dimensional space and examined Parkinsonian patients' ability to perform a sequence of movements in which they were required to bend down, pick up an object, and lift and place the object on top of a shelf. Whereas postural, erecting, stepping, and manual phases of the movement tended to overlap for normal subjects, the patients' movements were slower, there were pauses between each phase of the movement (phases no longer overlapped), and the patients had trouble speeding up their movements. Difficulty in maintaining movement amplitude means that when possible, patients with Parkinson's disease break a movement down into component parts, reducing the effects of inaccuracies in each part of a movement. When patients are constrained to maintain a movement sequence, this difficulty results in a festination of the sequence, as it is composed of smaller "incomplete" movements.

Functions of the Basal Ganglia

What does Parkinson's disease tell us about basal ganglia function? The loss of function observed in patients with the disease suggests that the basal ganglia have a role in the initiation and maintenance of muscle activity. This role does not seem to be related to intentional cortical processes because these processes do not appear to be impaired in Parkinson's disease. Problems in maintaining muscle activity cause reductions in the size and speed of movements, making preparatory responses less effective. The underlying deficits manifest themselves in different ways, depending upon task constraints.

CEREBELLUM AND COORDINATION

Although some researchers (e.g., Fahn, Calne, & Shoulson, 1983) limit a discussion of movement disorders to diseases of the basal ganglia such as Parkinson's disease, chorea, and dystonia, disorders resulting from cerebellar damage are of particular interest. Damage to the cerebellum produces abnormalities in the scaling of movements and in the coordination of voluntary movements with postural reflexes. This indicates that the cerebellum plays a very important role in the coordination of movement (Gilman, Bloedel, & Lechtenberg, 1981). The focus of this section is upon problems in the posture and coordination of movements in patients with cerebellar disorders. We do not address abnormalities of eye movements such as saccadic dysmetria, impairment of smooth pursuit, and gaze nystagmus; or speech disorders such as scanning dysarthria or slow slurring speech.

Our knowledge of the cerebellum's role in the coordination of movement in man is still limited: Few studies report careful examinations of the functioning of the human cerebellum. This is astonishing considering that the cerebellum is a major brain structure, clearly anatomically distinct from the rest of the brain, yet comprised of more neurons than the rest of the brain (Rothwell, 1987). Even though the histology and physiology of the animal cerebellum has been extensively explored (for a review see Ito, 1984), obvious phylogenetic differences in the size, connections, and motor function of the cerebellum necessitate that care be taken in extrapolation from animal studies to human behavior (Massion & Sasaki, 1979).

Localizing Cerebellar Damage

Some diseases have poorly localized effects upon the cerebellum. For example, vascular accidents, multiple sclerosis, and generalized degenerative diseases such as olivo ponto-cerebellar atrophy affect different parts of the cerebellum and extracerebellar structures in a nonspecific fashion. In addition, nonspecific cerebellar damage may be caused by toxic substances such as alcohol and heavy metals and by tumor infiltration, inflammation, or vascular accidents. However, other disorders affect clearly defined areas and make it possible to infer specific functional properties of these areas.

Whereas disorders of the basal ganglia affect a particular neurotransmitter system, lesions of the cerebellum very often do not show similar pathophysiologic properties. For example, although Parkinson's disease is the result of a progressive degeneration of specific types of neurons, degenerative diseases of the cerebellum may affect sections of the cerebellum without affecting a specific neurotransmitter system, the errors in metabolism usually being unknown.

In basal ganglia disorders such as Parkinson's disease, damage usually is inferred from clinical manifestations of disorder long before imaging techniques can show that there is any pathology. In contrast, the size of the cerebellum and its obvious macroanatomical separation from the rest of the brain make it easier to localize cerebellar damage with new intra vitam methods such as computerized tomography scans and magnetic resonance imaging (but see Diener et al., 1986). Such techniques make it possible to localize damage even within specific sections of the cerebellum.

Three major portions of the cerebellum can be distinguished, in a simplified scheme, both from histoanatomical observations of specific afferent and efferent connections of different parts of the cerebellum and from obvious phylogenetic differences. The *vestibulocerebellum* incorporates the flocculonodular lobes and receives its main input from the vestibular organs and vestibular nuclei in the medulla oblongata. The *spinocerebellum* occupies the central longitudinal regions of the anterior and posterior lobes and receives the major input through spinocerebellar tracts. The *pontocerebellum*, often called neocerebellum, which gains size late during phylogeny, comprises the lateral hemispheres. It receives input through the pontine nuclei from the cerebral cortex.

The output of the cerebellum is transmitted through nuclei that are specific for the different subdivisions of the cerebellum. Although many studies do not distinguish between locations of cerebellar lesions, there is converging evidence that postural problems are associated with lesions in the vestibulocerebellum and the anterior lobe of the spinocerebellum (Dichgans & Diener, in press). Neocerebellar lesions do not predominantly produce problems of balance, but they can be associated with problems of movement planning and execution.

Unfortunately, few studies quantify experimentally the deficits associated with cerebellar diseases. Some of the information about cerebellar functions thus comes instead from artificially induced lesions in animal cerebellums. And, although much basic neurophysiological knowledge is derived from lower animals, extreme care is necessary when inferences are made about similar structures in humans. This is especially true for the cerebellum because this structure has changed markedly during evolution. For that reason, the study of cerebellar diseases in humans plays an important role in the understanding of cerebellar functions during movement.

Clinical Observations of Cerebellar Function

Gordon Holmes's (1939) classic work on cerebellar function implicated the cerebellum in the regulation of muscle tone, the coordination of movement, and the control of posture and gait. Holmes supported these claims by careful observation of functional impairments resulting from cerebellar lesions caused by gunshot wounds. Minor lesions, especially of the cerebellar cortex, can be compensated for fairly well, but broader lesions or lesions of interrelating nuclei or the peduncles result in disturbances that are unique for cerebellar damage. Holmes described those pathologic features as abnormalities of rate, range, and force of movement (Holmes, 1922). His is still considered the best short characterization of a cerebellar movement disorder.

The most important clinical symptoms of cerebellar damage are

Dysmetria: Ballistic movements are irregular in acceleration and deceleration; the movement of a limb may be stopped prematurely, so that the patient undershoots the target, or it may be stopped too late, so that the patient overshoots the target. The patient finally reaches the target by a series of correcting movements.

Dysdiadochokinesia: Repetitive alternating movements are severely impaired by irregularities of speed and force.

Intention" or terminal tremor: This symptom is frequently observed. In contrast to Parkinsonian tremor, it is absent during rest but is provoked by a voluntary movement, primarily during the final parts of a movement, when fine adjustments of the movement are demanded.

Posture: Postural instabilities are among the most common features of cerebellar disorder and range from a wide base stance and gait to an inability to stand or sit without support.

Ataxia: Although this term is very often used to describe the abnormalities of movement and gait in cerebellar patients, there is no definition of ataxia more precise than that of a lack of coordination in absence of a muscular weakness.

The Cerebellum and the Initiation of Movement

Prolonged delays before the onset of movement (Holmes, 1939; Rondot, Bathien, & Toma, 1979) and before the onset of muscle activity

(Beppu, Suda, & Tanaka, 1984) imply that the cerebellum has a role in the initiation of movement. However, there is insufficient evidence upon which to base firm inferences of cerebellar function from studies of response times. After a brief discussion of response time studies involving invasive techniques in monkeys, we try to infer the functions of the cerebellum from the effect of cerebellar lesions on motor control.

The importance of the cerebellum in initiating movements is emphasized by single cell recordings in monkeys. During the initiation of a movement, neurons in the dentate nuclei (the nuclei through which the output from the neocerebellum is transmitted) change their firing rate at about the same time or before neurons in the motor cortex start firing (Lamarre & Chapman, 1986; Lamarre, Spidalieri, & Chapman, 1983; Thach, 1978). Studies in which the function of the dentate and interpositus nuclei was blocked by permanent lesions or reversible cooling showed an increase in response time (Meyer-Lohmann, Hore, & Brooks, 1977) and even a delay in the change of firing rate in the motor cortex (Meyer-Lohmann et al., 1977). Although the experimental evidence does not allow us to attribute to the cerebellum sole responsibility for the planning of a movement, it suggests that the cerebellum plays a major part in that process.

The Cerebellum and Ballistic Movements

Fast ballistic movements are initiated by well-timed triphasic activation patterns of agonist and antagonist muscles in normal subjects (Hallett, Shahani, & Young, 1975a). Studies of patients with cerebellar diseases (Hallett, Shahani, & Young, 1975b) and studies of cerebellar malfunction experimentally induced by voluntary ingestion of considerable amounts of alcohol (Marsden et al., 1977) have shown that the triphasic pattern, though still observable, is altered: The duration of bursts seems to be prolonged for the first agonist muscle burst and often for agonist and antagonist together, and thus results in the clinical picture of overshooting. A smaller group of the subjects showed a prolongation of the antagonist burst, which was paralleled by an undershooting of the movement. These abnormalities are not due to deficits in motor unit control because motor unit recruitment is not deficient in cerebellar disorders (Hallett, 1983).

Rapid alternating movements require a precise termination of muscle activation. Whereas normal subjects usually terminate the activity of the antagonist well before the agonist is activated, 12 out of 16 subjects in Hallett's experiment behaved differently, their performance indicating that the problem in dysdiadochokinesia is due to an inability to precisely control the succession of antagonistic muscle activity (Hallett et al., 1975b). This interpretation is supported by a series of single cell recordings in monkeys, which indicate that Purkinje cell firing has a role in the control of antagonistic muscle cocontraction, whether via inhibition or activation (Thach, 1978).

The Cerebellum and Visual Guidance of Movement

Cerebellar lesions cause problems in both the initiation and execution of movements. Although it may be difficult to make specific functional inferences, cerebellar lesions cause obvious discontinuities and jerkiness in slow movements, as shown by clinical evidence (finger-nose test, for example) and recent experiments using tracking tasks (Beppu, Nagaoka, & Tanaka, 1987; Beppu et al., 1984; Morrice, Lee, Becker, White, & Hoffer, 1987). Besides the increase in reaction time, cerebellar patients in the experiments by Beppu et al. (1984) showed an inability to scale the initial acceleration phase of tracking movements, which led them to over- or undershoot the target, as well as an inability to switch to a normal smooth tracking pattern. Patients were unable to maintain a constant target velocity and instead produced jerky or saccadic movement patterns, as if they continuously had to repeat corrective responses. When allowed to go at their own pace, they could produce fairly smooth movements.

To investigate these deficits, Beppu et al. (1987) removed the error information during the tracking task by suppressing either target display or movement display. Cerebellar patients then exhibited far fewer undulatory movement patterns, a result that reinforces the interpretation that these undulations are corrective responses. Morrice et al. (1987) showed that the addition of a viscous load to the tracking movement dampened the jerkiness of the movement, whereas adding an inert load did not alter the pathologies. This finding supports the view that the major deficit lies in precisely controlling the braking mechanism, which is normally supplied by the activation of the antagonist.

Complex Movements and the Cerebellum

Cerebellar disorders cause obvious problems in the coordination of complex movements. Although such problems could be due to difficulties in the higher-level coordination of movements, this seems unlikely. The limited available evidence indicates that cerebellar patients do not have problems performing two movements at once. In contrast to Parkinson's disease, in which the simultaneous execution of two motor acts is severely disturbed (Benecke et al., 1986; Shimizu, Yoshida, & Nagatsuka, 1986), cerebellar ataxia (caused by spinocerebellar degeneration) does not cause any additional disturbance of simultaneous performance of tapping and diadochokinesia tasks when compared to the already irregular performance of one task alone (Shimizu et al., 1986).

On the other hand, problems of coordination may be the result of difficulty in scaling muscle activity or coordinating voluntary movement with postural reflexes. Complex movements involving more than one joint, such as touching certain objects with the tips of the index finger at some distance, require close control over the timing and amounts of activity in each of the muscles involved. In cerebellar patients, finger paths of those complex movements are nonlinear and highly variable, and they do not display the symmetrical, bell-shaped velocity profile that characterizes normal movements. Three-dimensional recording demonstrates an uncoupling of the normal, precise relationship of movements in the different joints involved (Lee, Becker, Morrice, & White, 1987). It is apparent how such problems of undershooting or overshooting could contribute to the ataxia seen in cerebellar disorders.

Posture and the Cerebellum

Coordination is especially important in maintaining upright posture and calls for the well-timed activation of postural muscles during or before the execution of those voluntary movements that would otherwise cause a destabilization of stance or gait. Postural instability can be assessed by means of a force-measuring platform. Postural problems are among the most characteristic features of cerebellar disorders. Although lesions of the neocerebellum do not produce serious postural problems (Dichgans & Diener, in press), there are distinct syndromes of cerebellar ataxia of stance and gait associated with lesions in the vestibulocerebellum and parts of the spinocerebellum. Selective atrophy of the midline structures of the anterior lobe is seen in patients with severe alcohol abuse over many years. It results in a wide-based stance and gait with instability of the trunk, while the upper limbs are only slightly handicapped. Posturographic studies (Dichgans & Mauritz, 1983; Diener, Dichgans, Backer, & Gompt, 1984) have shown a specific sway pattern, predominantly in the anterior-posterior direction. The main frequency is around 3 Hz; other more diffuse cerebellar lesions produce a less pronounced instability.

Pure lesions of the vestibulocerebellum are rare but can be observed in children and are caused by medulloblastoma in the flocculonodular lobe. Such lesions result in the disturbance of eye movements and severe difficulty in maintaining equilibrium of the trunk, for which vision can only partially compensate. Because these patients often fall in any direction, even when sitting with their eyes open, Dichgans and Diener (in press) concluded that neither the visual contribution to postural stabilization nor the vestibular graviceptive set value for spatial orientation is accessible to patients with lesions in this, the phylogenetically oldest part of the cerebellum.

Investigators can assess control of posture by moving the surface supporting a subject's body, thereby creating postural disturbances. This technique results in three reflex responses of the leg muscles (Lee & Tatton, 1975), which can be distinguished by their latency. Although the origins of these responses are still being debated, the long-latency component, sometimes called *M3*, is probably of supraspinal origin, under the modulatory control of the cerebellum. Although the latency of long-latency reflexes in cerebellar patients seems to be within normal range, the duration and the muscular activity (expressed as the integral of the electromyogram) are increased (Dichgans, Diener, & Müller, 1985; Friedemann, Noth, Diener, & Bacher, 1987). Nashner's (1976) and Nashner and Grimm's (1978) findings have demonstrated additional pathologic features, either of the muscular response itself or because of a lack of adaptation over several trials. However, this difference is probably due to differences in the size and speed of perturbations used in the different paradigms as well as in patient selection. Similarly, Traub et al. (1980) looked at postural reflex activity in lower limbs during perturbations of voluntary upper limb movements. They did not find differences in postural reflex activity in mildly affected cerebellar patients but did find an absence of reaction in 2 patients with very severe postural instability. The cerebellum thus seems to be involved in the regulation of the size and duration of long-latency postural reflexes.

Functions of the Cerebellum

The cerebellum is a complex structure with a variety of roles in the coordination of movements. Although researchers still rarely distinguish between different locations of cerebellar damage, an increasing amount of the emerging data indicate that damage to localized, specific parts of the cerebellum causes different disturbances (Dichgans & Diener, 1984). Disorders in the initiation and parameterization of movements seem to be caused primarily by lesions of the neocerebellum. The paramedial regions of the cerebellum are important for postural control and coordination of voluntary movements with postural mechanisms.

CONCLUSIONS

Research on Movement Disorders

In this chapter, we have discussed the potential contribution that research on movement disorders can make to understanding the role of brain structures in the coordination of movement. Movement disorders can thus potentially dramatize and reify issues in the coordination of movement and provide a window through which the role of brain structures involved in the coordination of movements can be examined. For instance, an examination of some movement disorders caused by damage to basal ganglia or the cerebellum suggests that the basal ganglia have a role in the initiation and maintenance of muscle activity and that the cerebellum has roles in the parameterization and coordination of voluntary movement associated with postural mechanisms.

Such attribution of the role of a brain structure in the coordination of movements must be made carefully, however. Inferences must be predicated on evidence that a *specific* brain structure is injured; that is, investigators must locate the damage causing the movement disorder. Investigators may then establish the role(s) and significance of a structure by relating damage to a specific functional impairment(s).

The distributed nature of brain structure, and the strategic changes that influence motor control, make research in this area difficult:

1. It is necessary to decide which symptoms of a movement disorder will provide the pertinent information about a given brain structure. The role of a structure can best be understood in terms of those impaired functions that are observed. For example, akinesia and bradykinesia have been considered the symptoms of Parkinson's disease that are most instructive of basal ganglia function.

2. Because problems of motor coordination can be confounded with deficits in other cognitive processes, identification of functional impairments must proceed cautiously.

3. Functional impairments caused by structural damage must be distinguished from any biomechanical or compensatory reactions to the damage. Inferences about specific motor deficits must be made punctiliously because complex problems of motor coordination can be the result of simpler problems in the energization or parameterization of movements. In addition, because normal functioning is disrupted, patients may adopt compensatory strategies (Grimm & Nashner, 1978).

Although these issues are formidable, it is a stimulating time for research on movement disorders. We believe that closer cooperation between neurology and cognitive psychology will overcome methodological difficulties. Advances in the localization of brain structure and function can further our understanding of the brain's role in the coordination of movement. However, researchers must be prepared to learn new research techniques and consider new phenomena. The discussion of basal ganglia and cerebellar function showed that researchers must consider issues in the localization of structural impairments and adopt sophisticated experimental designs to allow unequivocal attribution of functions.

The study of movement disorders requires not only that researchers consider traditional phenomena such as reaction time and movement time but also that they adopt a wider range of techniques, such as electromyography and movement kinematics. With these techniques, researchers will be able to probe phenomena such as long-latency reflexes, preparatory postural adjustments, balance, and tremor and will be able to better describe and understand any functional deficits. One of the most valuable results of exposure to new problems, phenomena, and methods is that this exposure will enable researchers to assess the relevance and limits of their theories (Moray, 1984). For example, research on movement disorders has exposed the limitations of subtractive methodologies and has shown the necessity for greater understanding of the neurophysiological and biomechanical basis of movement.

Implications for the Neural Basis of Motor Control

A study of movement disorders focuses upon the neurophysiological substrates of movement coordination and can demonstrate the brain's computational role in motor coordination. Neurophysiologists have traditionally thought that movement coordination is the result of computations performed by a hierarchical system of brain structures, in which the basal ganglia have a role in the initiation of motor programs and the cerebellum has a role in their storage (Eccles, 1973; 1979). Indeed, damage to these structures causes disorders in the initiation and execution of movements and results in the patient's greater reliance upon feedback. Some of the difficulty in identifying specific functional deficits suggests, however, that there are considerable redundancy and flexibility in the coordination of movement (distribution of control; see Pew, 1984). Earlier conceptualizations emphasized the higher-level roles of these structures in movement coordination, but current research suggests that the roles of these structures in coordination is at a somewhat lower level. For example, although clinical evidence implies that the basal ganglia have a role in movement parameterization, they actually seem to have a lower-level role in the initiation and maintenance of muscle activity. Similarly, although the cerebellum had at one time been considered the "storage house" for motor programs (Eccles, 1973), the lack of dramatic errors in motor programs observed in cases of cerebellar damage suggests that the cerebellum has a lower-level role in specifying parameters such as movement extent, and another role in the control of the mechanisms that guide posture.

Similarly, current theories of movement coordination seek to avoid explanatory problems caused by invocation of a homunculus and seek to reduce the computational loads placed upon a central processor by the degrees of freedom problem (Kelso, 1981; Schmidt, 1975). The study of movement disorders demonstrates computational roles for the brain in the coordination of movement; however, the brain performs its calculations in a distributed fashion (Stelmach & Diggles, 1982). Movement coordination is clearly not the responsibility of only one brain structure because control processes may be anatomically and functionally distinct. Indeed, "homuncular" functions such as intentional control may be left intact, although there may be clear functional deficits. Postulation of a distributed processing system largely avoids these problems.

REFERENCES

Baroni, A., Benvenuti, F., Fantini, L., Pantaleo, T., & Urbani, F. (1984). Human ballistic arm abduction movements: Effects of L-dopa treatment in Parkinson's disease. *Neurology, 34*, 868-876.

Bazalgette, D., Zattara, M., Bathien, N., Bouisset, S., & Rondot, P. (1986). Postural adjustments associated with rapid voluntary arm movements in patients with Parkinson's disease. In M. D. Yahr & K. J. Bergmann (Eds.), *Advances in neurology: Vol. 45. Parkinson's disease* (pp. 371-374). New York: Raven Press.

Benecke, R., Rothwell, J. C., Dick, J. P. R., Day, B. L., & Marsden, C. D. (1986). Performance of simultaneous movements in patients with Parkinson's disease. *Brain, 109*, 739-757.

Benecke, R., Rothwell, J. C., Dick, J. P. R., Day, B. L., & Marsden, C. D. (1987). Simple and complex movements off and on treatment in patients with Parkinson's disease. *Journal of Neurology, Neurosurgery, and Psychiatry, 50*, 296-303.

Beppu, H., Nagaoka, M., & Tanaka, R. (1987). Analysis of cerebellar motor disorders by visually-guided elbow tracking movement: Part 2. Contribution of the visual cues on slow ramp pursuit. *Brain, 110*, 1-18.

Beppu, H., Suda, M., & Tanaka, R. (1984). Analysis of cerebellar motor disorders by visually guided elbow tracking movement. *Brain, 107*, 787-809.

Berardelli, A., Accornero, N., Argenta, M., Meco, G., & Manfredi, M. (1986). Fast complex arm movements in Parkinson's disease.

Journal of Neurology, Neurosurgery, and Psychiatry, 49, 1146-1149.

Berardelli, A., Dick, J. P. R., Rothwell, J. C., Day, B. L., & Marsden, C. D. (1986). Scaling of the size of the first agonist electromyogram burst during rapid wrist movements in patients with Parkinson's disease. *Journal of Neurology, Neurosurgery, and Psychiatry, 49*, 1273-1279.

Berardelli, A., Sabra, A. F., & Hallett, M. (1983). Physiological mechanisms of rigidity in Parkinson's disease. *Journal of Neurology, Neurosurgery, and Psychiatry, 46*, 45-53.

Bieliauskas, L. A., Klawans, H. L., & Glantz, R. H. (1986). Depression and cognitive changes in Parkinson's disease: A review. In M. D. Yahr & K. J. Bergmann (Eds.), *Advances in neurology: Vol. 45. Parkinson's disease* (pp. 437-438). New York: Raven Press.

Bloxham, C. A., Dick, D. J., & Moore, M. (1987). Reaction times and attention in Parkinson's disease. *Journal of Neurology, Neurosurgery, and Psychiatry, 50*, 1178-1183.

Bloxham, C. A., Mindel, T. A., & Frith, C. D. (1984). Initiation and execution of predictable and unpredictable movements in Parkinson's disease. *Brain, 107*, 371-384.

Bowen, F. P., Burns, M. M., Brady, E. M., & Yahr, M. D. (1976). A note on alterations of personal orientation in Parkinsonism. *Neuropsychologia, 14*, 425-429.

Bowen, F. P., Hoehn, M. M., & Yahr, M. D. (1972). Parkinsonism: Alterations in spatial orientation as determined by a route-walking test. *Neuropsychologia, 10*, 355-361.

Brooks, V. B. (1986). *The neural basis of motor control.* New York: Oxford University Press.

Brown, R. G., & Marsden, C. D. (1986). Visuospatial function in Parkinson's disease. *Brain, 109*, 987-1002.

Brown, R. G., & Marsden, C. D. (1988). Internal versus external cues and the control of attention in Parkinson's disease. *Brain, 111*, 323-345.

Brown, S. N. C., & Cooke, J. D. (1981). Amplitude and instruction-dependent modulation of movement related electromyogram activity in humans. *Journal of Physiology, 316*, 97-107.

Burke, D. (1985). Muscle spindle feedback in Parkinson's disease. In P. J. Delwaide & A. Agnoli (Eds.), *Clinical neurophysiology in Parkinsonism: Contribution to assessment and pathophysiology* (pp. 1-8). Amsterdam: Elsevier.

Canter, G. J., & van Lancker, D. R. (1985). Disturbances of the temporal organization of speed following bilateral thalamic surgery in a patient with Parkinson's disease. *Journal of Communication Disorders, 18*, 329-349.

Chan, C. W. Y., Kearney, R. E., & Melvill-Jones, G. (1979). Tibialis anterior response to sudden ankle displacements in normal and Parkinsonian subjects. *Brain Research, 173*, 303-314.

Chase, W. G. (1978). Elementary information processes. In W. K. Estes (Ed.), *Handbook of learning and cognitive processes: Vol. 5. Human information processing* (pp. 19-90). Hillsdale, NJ: Erlbaum.

Cody, F. W. J., Macdermott, N., Matthews, P. B. C., & Richardson, H. C. (1986). Observations on the genesis of the stretch reflex in Parkinson's disease. *Brain, 109*, 229-249.

Cools, A. R., Van Den Bercken, J. H. L., Horstink, M. W. I., Van Spaendonck, K. P. M., & Berger, H. J. C. (1984). Cognitive and motor shifting aptitude disorder in Parkinson's disease. *Journal of Neurology, Neurosurgery, and Psychiatry, 47*, 443-453.

Dakof, G. A., & Mendelsohn, G. A. (1986). Parkinson's disease: The psychological aspects of a chronic illness. *Psychological Bulletin, 99*, 375-387.

Day, B. L., Dick, J. P. R., & Marsden, C. D. (1984). Patients with Parkinson's disease can employ a predictive strategy. *Journal of Neurology, Neurosurgery, and Psychiatry, 47*, 1299-1306.

De Ajuriaguerra, J. (1975). The concept of akinesia. *Psychological Medicine, 5*, 129-137.

Della Sala, S. D., Lorenzo, G. D., Giordano, A., & Spinnler, H. (1986). Is there a specific visuo-spatial impairment in Parkinsonians? *Journal of Neurology, Neurosurgery, and Psychiatry, 49*, 1258-1265.

DeLong, M. R., Alexander, G. E., Georgopoulos, A. P., Crutcher, M. D., Mitchell, S. J., & Richardson, R. T. (1984). Role of basal ganglia in limb movements. *Human Neurobiology, 2*, 235-244.

DeLong, M. R., Georgopoulos, A. P., & Crutcher, M. D. (1983). Cortico-basal ganglia relations and coding of motor performance. In J. Massion, J. Paillard, W. Schultz, & M. Wiesendanger (Eds.), *Experimental Brain Research: Supplementum: Vol. 7. Neural coding of motor performance* (pp. 30-40). Berlin: Springer.

Delwaide, P. J. (1985). Are there modifications in spinal cord functions of Parkinsonian patients? In P. J. Delwaide & A. Agnoli (Eds.), *Clinical neurophysiology in Parkinsonism: Contribution to assessment and pathophysiology* (pp. 19-32). Amsterdam: Elsevier.

Dichgans, J., & Diener, H. C. (1984). Clinical evidence of functional compartmentalization of the cerebellum. In J. R. Bloedel, J. Dichgans, & W. Precht (Eds.), *Cerebellar functions* (pp. 126-147). Berlin: Springer.

Dichgans, J., & Diener, H. C. (in press). Cerebellar disorders of stance and postural reflexes. In M. A. Gresty (Ed.), *Movement disorders of extra-pyramidal systems*. London: McMillan Press.

Dichgans, J., Diener, H. C., & Müller, A. (1985). Characteristics of increased postural sway and abnormal long loop responses in patients with cerebellar diseases and parkinsonism. In A. Struppler & A. Weindl (Eds.), *Electromyography and evoked potentials* (pp. 69-74). Berlin: Springer.

Dichgans, J., & Mauritz, K. H. (1983). Patterns and mechanisms of postural instability in patients with cerebellar lesions. In J. E. Desmedt (Ed.), *Advances in neurology: Vol. 39. Motor control mechanisms in health and disease* (pp. 633-643). New York: Raven Press.

Dick, J. P. R., Rothwell, J. C., Berardelli, A., Thompson, P. D., Gioux, M., Benecke, R., Day, B. L., & Marsden, C. D. (1986). Associated postural adjustments in Parkinson's disease. *Journal of Neurology, Neurosurgery, and Psychiatry, 49*, 1378-1385.

Diener, H. C., & Dichgans, J. (1986). Long loop reflexes and posture. In W. Bles & T. Brandt (Eds.), *Disorders of posture and gait* (pp. 41-51). Amsterdam: Elsevier.

Diener, H. C., Dichgans, J., Bacher, M., & Gompt, B. (1984). Quantification of postural sway in normals and patients with cerebellar diseases. *Electroencephalography and Clinical Neurophysiology, 57*, 134-142.

Diener, H. C., Müller, A., Thron, A., Poremba, M., Dichgans, J., & Rapp, H. (1986). Correlation of clinical signs with CT findings in patients with cerebellar disease. *Journal of Neurology, 233*, 5-12.

Draper, I. T., & Johns, R. S. (1964). The disorders of movement in Parkinsonism and the effect of drug treatment. *Bulletin of the Johns Hopkins Hospital, 115*, 465-480.

Eccles, J. C. (1973). The cerebellum as a computer: Patterns in space and time. *Journal of Physiology, 229*, 1-32.

Eccles, J. C. (1979). Introductory remarks. In J. Massion & K. Sasaki (Eds.), *Cerebro-cerebellar interactions* (pp. 1-18). Amsterdam: Elsevier.

Edelman, G. M., & Mountcastle, V. B. (1979). *The mindful brain*. Cambridge, MA: The MIT Press.

Evarts, E. V. (1985). Transcortical reflexes: Their properties and functional significance. In A. W. Goodwin & I. Darian-Smith (Eds.), *Experimental Brain Research: Supplementum: Vol. 10. Hand function and the neocortex* (pp. 130-154). Berlin: Springer.

Evarts, E. V., Terävåinen, H., & Calne, D. B. (1981). Reaction time in Parkinson's disease. *Brain, 104*, 167-186.

Fahn, S., Calne, D. B., & Shoulson, I. (1983). Preface. In S. Fahn, D. B. Calne, & I. Shoulson (Eds.), *Advances in neurology: Vol. 37. Experimental therapeutics of movement disorders* (pp. vii-viii). New York: Raven Press.

Flowers, K. A. (1976). Visual 'closed-loop' and 'open-loop' characteristics of voluntary movement in patients with Parkinsonism and intention tremor. *Brain, 99*, 269-310.

Flowers, K. (1978). Some frequency response characteristics of Parkinsonism on pursuit tracking. *Brain, 101*, 19-34.

Forno, L. S. (1982). Pathology of Parkinson's disease. In C. D. Marsden & S. Fahn (Eds.), *Movement disorders* (pp. 25-40). London: Butterworths.

Friedemann, H. H., Noth, J., Diener, J. C., & Bacher, M. (1987). Long latency EMG responses in hand and leg muscles: Cerebellar disorders. *Journal of Neurology, Neurosurgery, and Psychiatry, 50*, 71-77.

Gilman, S., Bloedel, J. R., & Lechtenberg, R. (1981). *Disorders of the cerebellum*. Philadelphia, PA: Davis.

Girotti, F., Grassi, M. P., Carella, F., Soliveri, P., Musicco, M., Lamperti, E., & Caraceni, T. (1986). Possible involvement of attention processes in Parkinson's disease. In M. D. Yahr & K. J. Bergmann (Eds.), *Advances in neurology: Vol. 45. Parkinson's disease* (pp. 425-429). New York: Raven Press.

Gopher, D., & Sanders, A. F. (1984). S-Oh-R: Oh stages! Oh resources! In W. Prinz & A. F. Sanders (Eds.), *Cognition and motor processes* (pp. 231-253). Berlin: Springer.

Gotham, A.-M., Brown, R. G., & Marsden, C. D. (1986). Depression in Parkinson's disease: A quantitative and qualitative analysis. *Journal of Neurology, Neurosurgery, and Psychiatry, 49*, 381-389.

Gresty, M. A., & Findley, L. J. (1981). Tremors in Parkinson's disease. In F. C. Rose & R. Capildeo (Eds.), *Research progress in Parkinson's disease* (pp. 75-97). Bath: Pitman Press.

Grimm, R. J. (1983). Program disorders of movement. In J. E. Desmedt (Ed.), *Motor control mechanisms in health and disease* (pp. 1-11). New York: Raven Press.

Grimm, R. J., & Nashner, L. M. (1978). Long loop dyscontrol. In J. E. Desmedt (Ed.), *Cerebral motor control in man: Long loop mechanisms* (pp. 70-84). Basel: Karger.

Gunilla, R., Oberg, E., & Divac, I. (1985). Levels of motor planning: Cognition and the control of movement. In E. V. Evarts, S. P. Wise, & D. Bousfield (Eds.), *The motor system in neurobiology* (pp. 284-287). Amsterdam: Elsevier.

Hallett, M. (1983). Analysis of abnormal voluntary and involuntary movements with surface electromyography. In J. E. Desmedt (Ed.), *Advances in neurology: Vol. 39. Motor control mechanisms in health and disease* (pp. 907-914). New York: Raven Press.

Hallett, M., & Khoshbin, S. (1980). A physiological mechanism of bradykinesia. *Brain, 103*, 301-314.

Hallett, M., & Marsden, C. D. (1981). Physiology and patho-physiology of the ballistic movement pattern. In J. E. Desmedt (Ed.), *Motor unit types, recruitment, and plasticity in health and disease* (pp. 331-346). Basel: Karger.

Hallett, M., Shahani, B. T., & Young, R. R. (1975a). EMG analysis of stereotyped voluntary movements in man. *Journal of Neurology, Neurosurgery, and Psychiatry, 38*, 1154-1162.

Hallett, M., Shahani, B. T., & Young, R. R. (1975b). EMG analysis of patients with cerebellar deficits. *Journal of Neurology, Neurosurgery, and Psychiatry, 38*, 1163-1169.

Hallett, M., Shahani, B. T., & Young, R. R. (1977). Analysis of stereotyped voluntary movements at the elbow in patients with

Parkinson's disease. *Journal of Neurology, Neurosurgery, and Psychiatry, 40,* 1129-1135.

Hoehn, M. M., & Yahr, M. D. (1967). Parkinsonism: Onset, progression, and mortality. *Neurology, 17,* 427-442.

Holmes, G. (1922). Clinical symptoms of cerebellar disease and their interpretation. The Croonian Lectures. II. *Lancet, 100,* 1231-1237.

Holmes, G. (1939). The cerebellum of man. *Brain, 62,* 1-30.

Hornykiewicz, O. (1982). Brain neurotransmitter changes in Parkinson's disease. In C. D. Marsden & S. Fahn (Eds.), *Movement disorders* (pp. 41-58). London: Butterworths.

Ingvarsson, P., Johnels, B., Lund, S., & Steg, G. (1986). Coordination of manual, postural, and locomotor movements during simple goal-directed motor tasks in Parkinsonian off and on states. In M. D. Yahr & K. J. Bergmann (Eds.), *Advances in neurology: Vol. 45. Parkinson's desease* (pp. 375-382). New York: Raven Press.

Ito, M. (1984). *The cerebellum and neural control.* New York: Raven Press.

Kelso, J. A. S. (1981). Contrasting perspectives on order and regulation in movement. In J. Long & A. Baddeley (Eds.), *Attention and performance IX* (pp. 437-457). Hillsdale, NJ: Erlbaum.

Knight, R. G., Godfrey, H. P. D., & Shelton, E. J. (1988). The psychological deficits associated with Parkinson's disease. *Clinical Psychology Review, 8,* 391-410.

Knutsson, E. (1972). An analysis of parkinsonian gait. *Brain, 95,* 475-486.

Knutsson, E., & Martensson, A. (1986). Posture and gait in parkinsonian patients. In W. Bles & T. Brandt (Eds.), *Disorders of posture and gait* (pp. 217-229). Amsterdam: Elsevier.

Korczyn, A. D., Inzelberg, R., Treves, T., Neufeld, M., Reider, I., & Rabey, P. M. (1986). Dementia of Parkinson's disease. In M. D.

Yahr & K. J. Bergmann (Eds.), *Advances in neurology: Vol. 45. Parkinson's disease* (pp. 399-403). New York: Raven Press.

Lamarre, Y., & Chapman, C. E. (1986). Comparative timing of neuronal discharge in cortical and cerebellar structures during a simple arm movement in the monkey. H. Heuer & C. Fromm (Eds.), *Experimental Brain Research: Supplementum: Vol. 15. Generation and modulation of action patterns* (pp. 14-27). Berlin: Springer.

Lamarre, Y., Spidalieri, G., & Chapman, C. E. (1983). A comparison of neuronal discharge recorded in the sensori-motor cortex, parietal cortex and dentate nucleus of the monkey during arm movements triggered by light, sound or somesthetic stimuli. In J. Massion, J. Paillard, W. Schultz, & M. Wiesendanger (Eds.), *Experimental Brain Research: Supplementum: Vol. 7. Neural coding of motor performance* (pp. 140-156). Berlin: Springer.

Lashley, K. S. (1951). The problem of serial order in behavior. In L. A. Jeffress (Ed.), *Cerebral mechanisms in behavior* (pp. 112-136). New York: Wiley.

Lee, R. G., Becker, W. J., Morrice, B. L., & White, D. G. (1987). Kinematics of cerebellar incoordination studied with a 3-dimensional movement analysis system. *Society for Neuroscience Abstracts*, Vol. 13, Part 1, 349.

Lee, R. G., Murphy, J. T., & Tatton, W. G. (1983). Long-latency myotatic reflexes in man: Mechanisms, functional significance, and changes in patients with Parkinson's disease or hemiplegia. In J. E. Desmedt (Ed.), *Advances in neurology: Vol. 39. Motor control mechanisms in health and disease* (pp. 489-508). New York: Raven Press.

Lee, R. G., & Tatton, W. G. (1975). Motor responses to sudden limb displacements in primates with specific CNS lesions and in human patients with motor system disorders. *The Canadian Journal of Neurologial Sciences*, 2, 285-293.

Lee, R. G., & Tatton, W. G. (1978). Long loop reflexes in man: Clinical applications. In J. E. Desmedt (Ed.), *Cerebral motor control in man: Long loop mechanisms* (pp. 320-333). Basel: Karger.

Margolin, D. I. (1984). The neuropsychology of writing and spelling: Semantic, phonological, motor, and perceptual processes. *Quarterly Journal of Experimental Psychology, 36A*, 459-489.

Marsden, C. D. (1982). The mysterious motor function of the basal ganglia: The Robert Wartenberg Lecture. *Neurology, 32*, 514-539.

Marsden, C. D. (1984). Function of the basal ganglia as revealed by cognitive and motor disorders in Parkinson's disease. *Canadian Journal of Neurological Science, 11*, 129-135.

Marsden, C. D. (1985). Defects of movement in Parkinson's disease. In P. J. Delwaide & A. Agnoli (Eds.), *Clinical neurophysiology in Parkinsonism* (pp. 107-115). Amsterdam: Elsevier.

Marsden, C. D., Merton, P. A., Morton, H. B., Hallett, M., Adam, J., & Rushton, D. M. (1977). Disorders of movement in cerebellar disease in man. In F. Rose (Ed.), *Physiological aspects of clinical neurology*. Oxford: Blackwell.

Marsden, C. D., Rothwell, J. C., & Day, B. L. (1983). Long-latency automatic responses to muscle stretch in man: Origin and function. In J. E. Desmedt (Ed.), *Advances in neurology: Vol. 39. Motor control mechanisms in health and disease* (pp. 509-539). New York: Raven Press.

Massion, J., & Sasaki, K. (1979). Cerebro-cerebellar interaction: Solved and unsolved problems. In J. Massion & K. Sasaki (Eds.), *Cerebro-cerebellar interactions* (pp. 261-287). Amsterdam: Elsevier.

McLellan, D. L. (1981). Rigidity. In F. C. Rose & R. Capildeo (Eds.), *Research progress in Parkinson's disease* (pp. 88-97). Bath: Pitman Press.

McLennan, J. E., Nakano, K., Tyler, H. R., & Schwab, R. S. (1972). Micrographia in Parkinson's disease. *Journal of Neurological Science, 15*, 141-152.

Megaw, E. D. (1974). Possible modification to a rapid on-going programmed manual response. *Brain Research, 71*, 425-441.

Meyer-Lohmann, J., Hore, J., & Brooks, V. B. (1977). Cerebellar participation in generation of prompt arm movements. *Journal of Neurophysiology, 40*, 1038-1050.

Moore, A. P. (1987). Impaired sensorimotor integration in Parkinsonism and dyskinesia: A role for corollary discharges? *Journal of Neurology, Neurosurgery, and Psychiatry, 50*, 544-552.

Moray, N. (1984). On the usefulness of experimental psychology. In K. M. J. Lagerspetz & P. Niemi (Eds.), *Psychology in the 1990's* (pp. 225-235). Amsterdam: North-Holland.

Morrice, B. L., Lee, R. G., Becker, W. J., White, D. G., & Hoffer, J. A. (1987). Viscous loading improves manual tracking accuracy in humans with cerebellar deficits. *Society for Neuroscience Abstracts*, Vol. 13, Part 3, 1700.

Mortimer, J. A., Pirozzolo, F. J., Hansch, E. C., & Webster, D. D. (1982). Relationship of motor symptoms to intellectual deficits in Parkinson's disease. *Neurology, 32*, 133-137.

Mortimer, J. A., & Webster, D. D. (1978). Relationships between quantitative measures of rigidity and tremor and the electromyographic responses to load perturbations in unselected normal subjects and Parkinson patients. In J. E. Desmedt (Ed.), *Cerebral motor control in man: Long loop mechanisms* (pp. 342-360). Basel: Karger.

Murray, M. P., Sepic, S. B., Gardner, G. M., & Downs, W. J. (1978). Walking patterns of men with Parkinsonism. *American Journal of Physical Medicine, 57*, 278-294.

Nagasaki, H., & Nakamura, R. (1982). Rhythm formation and its disturbances: A study based upon periodic response of a motor output system. *Journal of Human Ergology, 11*, 127-142.

Nakamura, R., Nagasaki, H., & Narabayashi, H. (1978). Disturbances of rhythm formation in patients with Parkinson's disease: Part I. Characteristics of tapping response to the periodic signals. *Perceptual and Motor Skills, 46*, 63-75.

Nashner, L. M. (1976). Adapting reflexes controlling human posture. *Experimental Brain Research, 26*, 59-72.

Nashner, L. M., & Grimm, R. J. (1978). Analysis of multiloop dyscontrols in standing cerebellar patients. In J. Desmedt (Ed.), *Cerebral motor control in man: Long loop mechanisms* (pp. 300-319). Basel: Karger.

Obesa, J. A., Quesada, P., Artieda, J., & Martinez-Lage, J. M. (1985). Reciprocal inhibition in rigidity and dystonia. In P. J. Delwaide & A. Agnoli (Eds.), *Clinical neurophysiology in Parkinsonism* (pp. 9-18). Amsterdam: Elsevier.

Parkinson, J. (1817). *An essay of the shaking palsy*. London: Sherwood, Neely, and Jones.

Pew, R. W. (1984). A distributed processing view of human motor control. In W. Prinz & A. F. Sanders (Eds.), *Cognition and motor processes* (pp. 19-27). Berlin: Springer.

Posner, M. I., Pea, R., & Volpe, B. (1982). Cognitive-neuroscience: Developments toward a science of synthesis. In J. Mehler, M. Garrett, & E. Walker (Eds.), *Perspectives on mental representation: Experimental and theoretical studies of cognitive processes and capacities* (pp. 251-276). Hillsdale, NJ: Erlbaum.

Pullman, S. L., Watts, R. L., Juncos, J. L., Chase, T. N., & Sanes, J. N. (1988). Dopaminergic effects on simple and choice reaction time performance in Parkinson's disease. *Neurology, 38*, 249-254.

Rafal, R. D., Inhoff, A. W., Friedman, J. H., & Bernstein, E. (1987). Programming and execution of sequential movements in Parkinson's disease. *Journal of Neurology, Neurosurgery, and Psychiatry, 50*, 1267-1273.

Rajput, A. H., Offord, K., Beard, C. M., & Kurland, L. T. (1984). Epidemiological survey of dementia in Parkinsonism and control population. In R. G. Hassler & J. F. Christ (Eds.), *Advances in neurology: Vol. 40. Parkinson-specific motor and mental disorders, role of pallidum: Pathophysiological, biochemical, and therapeutic aspects* (pp. 229-234). New York: Raven Press.

Rogers, D., Lees, A. J., Trimble, M., & Stern, G. M. (1986). Concept of bradyphrenia: A neuropsychiatric approach. In M. D. Yahr & K. J. Bergmann (Eds.), *Advances in neurology: Vol. 45. Parkinson's disease* (pp. 447-450). New York: Raven Press.

Rolls, E. T. (1983). The initiation of movements. In J. Massion, J. Paillard, W. Schultz, & M. Wiesendanger (Eds.), *Experimental Brain Research: Supplementum: Vol. 7. Neural coding of motor performance* (pp. 97-113). Berlin: Springer.

Rondot, P., Bathien, N., & Toma, S. (1979). Physiopathology of cerebellar movement. In J. Massion & K. Sasaki (Eds.), *Cerebrocerebellar interactions* (pp. 203-230). Amsterdam: Elsevier.

Rosenbaum, D. A. (1980). Human movement initiation: Specification of arm, direction, and extent. *Journal of Experimental Psychology: General, 109*, 444-474.

Rothwell, J. (1987). *Control of human voluntary movement.* Rockville, MD: Aspen.

Sanders, A. F. (1980). Some effects of instructed muscle tension on choice reaction time and movement time. In R. S. Nickerson (Ed.), *Attention and performance VIII* (pp. 59-74). Hillsdale, NJ: Erlbaum.

Sanes, J. N. (1985). Information processing deficits in Parkinson's disease during movement. *Neuropsychologia, 23*, 381-392.

Sanes, J. N., & Evarts, E. V. (1985). Psychomotor performance in Parkinson's disease. In P. J. Delwaide & A. Agnoli (Eds.), *Clinical neurophysiology in Parkinsonism* (pp. 117-132). Amsterdam: Elsevier.

Schmidt, R. A. (1975). A schema theory of discrete motor skill learning. *Psychological Review, 82*, 225-260.

Sheridan, M. R., Flowers, K. A., & Hurrell, J. (1987). Programming and execution of movement in Parkinson's disease. *Brain, 110*, 1247-1271.

Shimizu, N., Yoshida, M., & Nagatsuka, Y. (1986). Disturbance of two simultaneous motor acts in patients with Parkinsonism and cerebellar ataxia. In M. D. Yahr & K. J. Bergmann (Eds.), *Advances in neurology: Vol. 45. Parkinson's disease* (pp. 367-370). New York: Raven Press.

Stelmach, G. E., & Diggles, V. (1982). Control theories in motor behavior. *Acta Psychologica, 50*, 83-105.

Stelmach, G. E., Phillips, J. G., & Chau, A. W. (in press). Visuo-spatial processing in Parkinsonians. *Neuropsychologia.*

Stelmach, G. E., Teasdale, N., Phillips, J., & Worringham, C. J. (in press). Force production characteristics in Parkinson's disease. *Experimental Brain Research.*

Stelmach, G. E., & Worringham, C. J. (1988a). The preparation and production of isometric force in Parkinson's disease. *Neuropsychologia, 26*, 93-103.

Stelmach, G. E., & Worringham, C. J. (1988b). The control of bimanual aiming movements in Parkinson's disease. *Journal of Neurology, Neurosurgery, and Psychiatry, 51*, 223-231.

Stelmach, G. E., Worringham, C. J., & Strand, E. A. (1986). Movement preparation in Parkinson's disease: The use of advance information. *Brain, 109*, 1179-1194.

Stelmach, G. E., Worringham, C. J., & Strand, E. A. (1987). The programming and execution of movement sequences in Parkinson's disease. *International Journal of Neuroscience, 36*, 55-65.

Stern, Y. (1986). Patients with Parkinson's disease can employ a predictive motor strategy. *Journal of Neurology, Neurosurgery, and Psychiatry, 49*, 107-108.

Stern, Y., & Mayeux, R. (1986). Intellectual impairment in Parkinson's disease. In M. D. Yahr & K. J. Bergmann (Eds.), *Advances in neurology: Vol. 45. Parkinson's disease* (pp. 405-408). New York: Raven Press.

Tatton, W. G., Eastman, M. J., Bedingham, W., Verrier, M. C., & Bruce, I. C. (1984). Defective utilization of sensory input as the basis for bradykinesia, rigidity, and decreased movement repertoire in Parkinson's disease: A hypothesis. *The Canadian Journal of Neurological Sciences, 11*, 136-143.

Teasdale, N., Phillips, J., & Stelmach, G. E. (1989). *Temporal movement control in patients with Parkinson's disease.* Manuscript submitted for publication.

Teasdale, N., & Stelmach, G. E. (1988). Movement disorders: The importance of the movement context. *Journal of Motor Behavior, 20*, 186-191.

Terävälnen, H., & Calne, D. (1980). Quantitative assessment of Parkinsonian deficits. In U. K. Rinne, M. Klinger, & G. Stamm (Eds.), *Parkinson's disease: Current progress, problems and management* (pp. 145-164). Amsterdam: Elsevier.

Thach, W. T. (1978). Correlation of neural discharge with pattern and force of muscular activity. Joint position and direction of intended movement in motor cortex and cerebellum. *Journal of Neurophysiology, 41*, 654-676.

Traub, M. M., Rothwell, J. C., & Marsden, C. D. (1980). Anticipatory postural reflexes in Parkinson's disease and other akinetic-rigid syndromes and in cerebellar ataxia. *Brain, 103*, 393-412.

Wing, A. M. (1984). Disorders of movement. In M. M. Smyth & A. M. Wing (Eds.), *The psychology of human movement* (pp. 269-296). London: Academic Press.

Wing, A. M., Keele, S., & Margolin, D. I. (1984). Motor disorder and the timing of repetitive movements. *Annals of the New York Academy of Science, 423*, 183-192.

Wing, A. M., & Miller, E. (1984). Basal ganglia lesions and the psychological analyses of the control of voluntary movement. *Ciba Foundation Symposium 107: Functions of the basal ganglia* (pp. 242-257). London: Pitman.

Zetusky, W. J., Jankovic, J., & Pirozzolo, F. J. (1985). The heterogeneity of Parkinson's disease: Clinical and prognostic implications. *Neurology*, *35*, 522-526.

Perspectives on the Coordination of Movement
S.A. Wallace (Editor)

THE CONCEPT AND MEASUREMENT OF COORDINATION IN SPEECH DISORDERS§

Ray D. KENT* and Scott G. ADAMS

Department of Communicative Disorders
University of Wisconsin-Madison

ABSTRACT

Several speech disorders have been described as having impaired coordination of movement as at least one of their characteristics. Among these disorders are stuttering, some of the dysarthrias, apraxia of speech, Broca's aphasia, conduction aphasia, and deafness. This chapter considers the evidence for coordination impairment in these disorders. First, the concept of movement coordination in speech is discussed. Second, procedures for the identification and measurement of coordination impairment are described. Third, data on the speech disorders are reviewed with respect to the general issue of coordination. Finally, theories of coordinative control in speech are evaluated against these data.

For voluntary motor systems generally, coordination is constrained to some degree by the operations of the motor system relative to physical objects in the external world. For example, locomotion is a motor operation relative to a surface or medium; prehension is a motor operation relative to a manipulated object;

---

*Address correspondence to: Ray D. Kent, Department of Communicative Disorders, University of Wisconsin-Madison, 1975 Willow Drive, Madison, WI 53706, U. S. A.

§This work was supported in part by National Institutes of Health research grant NS22458.

and catching is a motor operation relative to a moving object. Speech, however, is not defined by operations on any physical structures other than the speech organs (the articulators) themselves. Stetson (1951) captured the essence of speech in saying that speech is movement made audible. Speech movements are in the main internal to the human organism and thus are accomplished in a protected and controlled environment, relatively free of unpredictable disturbances. The purpose of these movements is to create acoustic patterns that have communicative function. In this sense, the coordination of speech is constrained by the acoustic-phonetic structure of the speaker's language and by the auditory perception of that structure. As such, speech movements are played out against an eventual auditory interpretation by another listener or by an eventual auditory confirmation by the speaker.

Speech production often is described in terms of attributes such as rhythm, coordination, programming, quality, and intelligibility. As difficult as it has been to give these concepts operational definitions, it has been hard to avoid them in describing speech. Similarly, speech disorders have been impressionistically described as deviations in these and similar attributes. Perceptual studies of stuttering and dysarthria are instructive in this regard. Stuttering has been defined as a disturbance in the flow of speech; it traditionally has been described in terms of dysfluency categories such as repetitions (of sounds, syllables, words, or phrases), hesitations, prolongations, and interjections. Recent physiologic and acoustic studies have demonstrated a number of *subperceptual* abnormalities in stuttered speech, and indeed even in the apparently fluent episodes of stutterers' speech (Adams, 1981). Dysarthria is a speech impairment that results from muscular dysfunction associated with neurologic disease or damage. To the lay ear, dysarthric speech usually sounds slurred or indistinct. However, practiced clinicians frequently can judge site of lesion from the particular pattern of dysarthric speech. It appears that the motor behavior of speech is affected in fairly regular ways by damage to different parts of the nervous system.

Recent instrumental investigations are offering new perspectives on the nature of stuttering and disarthria. This chapter will review these two speech disorders as well as several others with respect to their classification as disorders of speech coordination. First, however, some general comments will be made on definitions and theories of coordination as they relate to speech and its disorders.

Coordination rarely is given a rigorous definition in the literature on speech disorders. More often, it seems to be assumed that everyone knows what coordination (and discoordination) is. Unfortunately, the description of a speech disorder as "sounding uncoordinated" does not really describe what is aberrant in the underlying movement pattern. As a working definition in this chapter, we consider coordination to be a *repeatable spatiotemporal pattern of movement in relation to a behavioral act or goal.* We feel that it is necessary to add the behavioral context to distinguish coordinated movements from movements that merely appear to be coordinated. For example, lacking any other information, an observer might decide that the Jacksonian march of movements in an epileptic seizure is a coordinated pattern. Similarly, one might identify some aspect of coordination in the involuntary movements of chorea or athetosis, given that these movements may appear to have a rhythmic character. However, all of these abnormal movements fail to qualify as coordinated movements in respect to the foregoing definition.

The criterion of repeatability is necessary in one sense just to insure reliability of observation, but the purpose goes beyond that. It is assumed that a coordinated movement is reliably performed in repeated trials. This is not to say that the finest details of movement must be stereotypically executed. Rather, the basic structure of movement is expected to be reliably performed, and some of the finer details may in fact vary from execution to execution. A simple example of the intended distinction is the repetition of a single syllable [ba] (the square brackets denote phonetic transcription). The closure of the lips for the consonant [b] can be made with various degrees of labial and mandibular participation. For some speakers, the degree of participation of these two articulators may vary over the syllable train. The variation is principled and preserves the essential coordination for the movement.

The literature on speech movements frequently conflates a number of terms and concepts. One example of this conflation is *coordination* and *coarticulation*, both key concepts in speech research. The latter term refers generally to an observation of simultaneous vocal tract adjustments for two or more phonetic segments. That is, speech movements are not sequenced in a way that allows unambiguous assignment of temporal intervals to a single phonetic segment (the presumed underlying control unit of speech). When most speakers produce the word *soon* (written [sun] in the symbols of the International Phonetic Alphabet, IPA), the lip rounding for the vowel [u] is apparent during the production of the (normally) unrounded fricative [s]. That is, an articulatory feature of the vowel is anticipated during the production of the preceding consonant. The fricative [s] is not rounded when it occurs before an unrounded vowel, like that in the word *seen* (IPA [sin]). Patterns of coarticulation are phonetically principled insofar as these patterns preserve the essential phonetic contrasts of a language. The actual movement pattern that gives rise to the observation of coarticulation can be defined in terms of the coordination of its elements. In the example given, the lip rounding movement is coordinated with the tongue movement for the fricative [s].

To a significant degree, speech movements have defied detailed description, partly because the movements are multiarticulate and partly because many of the structures are hidden from direct observation. X-ray techniques have generated the largest sets of data on simultaneous movements of the tongue, velum, jaw, and lips. Even with restriction of movement data to two dimensions in the X-ray planar projection (Figure 13.1), speech is a complex process. Figure 13.2 shows movements of the velum, tongue, lips, and jaw during production of the sentence *Soon the snow began to melt.* A phonetic transcription of the utterance appears on the illustration as a guide to segmentation. But segmentation is in fact often arbitrary because the movements overlap one another in a shingled pattern. Sometimes, however, two or more structures move simultaneously. For example, at Time Point 7 in Figure 13.2, synchrony can be observed for lowering of the velum (related phonetically to the upcoming nasal consonant *n* in *began*), depression of the tongue body (in preparation for the second vowel in *began*), and lowering of the jaw (also related to the second vowel in *began*). Data such as these have shown that coarticulation is the rule and not the exception in speech. The data also have shown that speech movements are precisely timed, with some movement events being reliably separated by about 10 ms (Kent & Moll, 1975). Lackner and Levine (1975) suggested that speech is the most precisely timed motor behavior in humans.

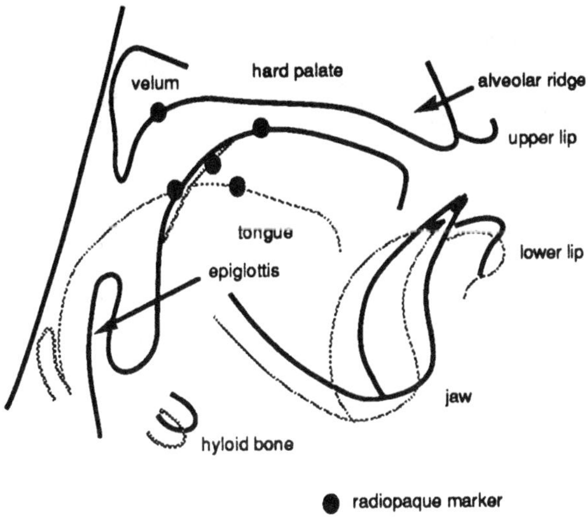

Figure 13.1. Superimposed tracings of two frames from a cinefluorographic film of the vocal tract. One frame applies to the production of vowel [i] (as in *he*), and the other, to vowel [a] (as in *ha*). The lateral projection provides an approximate midsaggital section. Also shown are the positions of radiopaque markers attached to the tongue dorsum and the knee of the velum (only one marker position is shown for the velum because the positions were essentially the same for the two vowel productions). Figure 13.2 presents movement data for structures shown here.

Several recent studies of speech disorders have examined anticipatory coarticulation (coarticulation in which an articulatory property of an upcoming phonetic segment is produced before the other defining properties of that segment are apparent). If the speech-disordered individual fails to exhibit the same coarticulatory effect determined for control speakers, the conclusion has been that the speech-disordered subjects have a delay (Ziegler & von Cramon, 1985, 1986a, 1986b) or deficiency (Tuller & Story, 1986) in anticipatory coarticulation. However, Katz (1987) reported that (a) the disordered speakers he studied began their coarticulatory gestures at least as early in the movement sequence as control subjects did, and (b) those coarticulatory abnormalities that did appear could be explained by "stimulus-specific motor and timing information" (p. 237). Abnormal coarticulation was most likely for phonetic sequences that require the *coordination of two or more independent articulators* (e.g., tongue and velopharynx, lips and

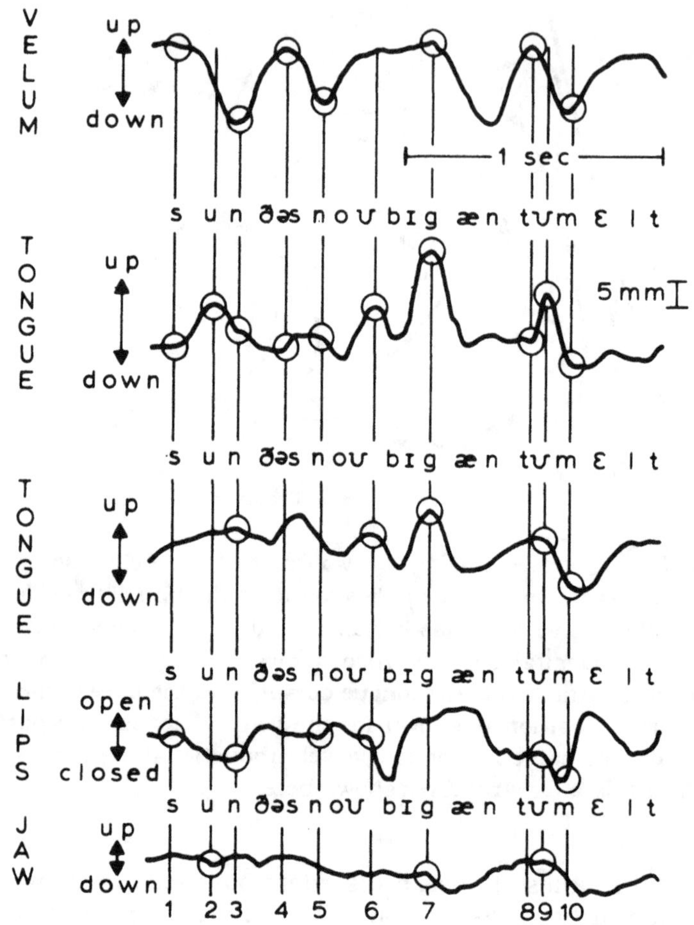

Figure 13.2. Movements recorded by cinefluorography (100 frames/s) during production of the sentence, *Soon the snow began to melt,* by a normal speaker. The traces, from top to bottom, are: vertical movements of a radiopaque marker attached to the velum; vertical movements of a radiopaque marker attached to the tongue dorsum; vertical movements of another marker, placed posterior to the preceding one; separation of the lips (degree of labial opening); and vertical position of the lower central incisor. See Figure 13.1 for an illustration of the structures. A phonetic transcription (International Phonetic Alphabet) of the utterance appears on the illustration to provide a rough guide to segmentation.

vocal folds). The question, then, becomes whether the putative coarticulatory differences between speech-disordered and control subjects should really be interpreted at a more fundamental level of coordination.

A further complication is that coarticulation (and perhaps coordination) varies with speaking rate. Coarticulatory interaction can be reduced at slow speaking rates, which are typical of many speech disorders. Wieneke, Janssen, and Belderbos (1987) explained the beneficial effect of slow rate in the treatment of many speech disorders as a matter of allowing greater preparation time for movements and reducing kinematic interaction and coarticulation. By this reasoning, some reduction of coarticulatory effects would be expected solely because of the slow rate that is typical of many speech disorders.

Several speech disorders have been described as being wholly or in part disorders of coordination. An extensive review of the relevant literature would be voluminous, but it is possible to highlight the characteristics of these disorders that relate to descriptions of discoordination. In the following sections, we review briefly the nature of the movement abnormality in the following speech disorders: stuttering, motor speech disorders, speech produced under disrupted sensory feedback, and speech of the hearing impaired. The narrow goal of this review is to identify what has been said about coordination or the lack of it. A broader goal is to consider these disorders with respect to the definition of coordination given earlier in this chapter.

STUTTERING

Some kind of failure of timing control has been a frequently proposed explanation of stuttering. But the nature of this failure has been difficult to isolate, and the problem is further compounded by the lack of a generally accepted and sufficiently complete account of speech timing per se. In a trivial sense, any disruption in the flow of speech might be related to a failure of timing control. Obviously, something more is intended by those who speculate that stutterers have a specific liability in the temporal planning or execution of speech. It is difficult to specify the impairment of timing control outside a model of speech timing or, at least, a fairly precise description of the timing requirements of normal speech. The problem, then, is twofold: (a) Describe the timing of normal speech, and (b) describe the way in which stuttered speech deviates from the timing of normal speech. Unfortunately, the regulation of speech timing is far from resolved. It therefore may be preferable to take the conservative route and simply compare the timing data of stutterers' speech with data on normal controls.

It is commonly reported across a number of speech or nonspeech tasks that stutterers, compared to nonstuttering controls, have increased latencies of movement onset and appear to be less precise in the temporal patterning of a sequence of movements (Borden, 1983; McMillan & Pindzola, 1986; Webster, 1986b). But the picture seems to become more clouded rather than clearer when one searches for a detailed description of the timing disruption in speech. For example, some studies indicate that stutterers spend a longer time in static articulatory configurations, or the relatively unchanging postures associated with vowel or consonant steady states (Pindzola, 1987; Zimmerman, 1980), whereas other investigators have noted differences in transitional segments (movements between steady states) but not in steady states (Starkweather & Myers, 1979).

Perhaps the empirical impasse is more apparent than real. Both kinds of temporal differences could be accommodated by an explanation that posits a specific disturbance in the regulation of timing. The manifestation of this disturbance at the articulatory and acoustic levels depends on several factors, including speaker differences, context effects, and accompanying behaviors. It is not uncommon for both steady states and transitional segments to be altered in some motor speech disorders (Rosenbek, Kent, & LaPointe, 1984). Many of these speakers with neurologic impairments exhibit both kinds of abnormality even though one may prevail for a given speaker and a given task. But both forms presumably reflect the same impairment at the motor control level. Similarly, stuttering may be characterized by several different manifestations of abnormal timing. The question is not so much which manifestation is dominant but rather what concept of disordered timing can account for the different manifestations.

If in fact a timing disorder is at the heart of stuttering, then a full description of it may depend not only on a theory of speech timing control but also on individual differences in timing. Keele, Ivry, and

Porkorny (1987) recently proposed three general factors of coordination that differentiate subjects across a variety of movement tasks—maximum rate of repetitive movements, precision of motor and perceptual timing, and force control. Although it is premature to make any rigid classification of stutterers with respect to these factors, it appears from the work of Borden (1983), Webster (1986b), Caruso, Abbs, and Gracco (1988), and other research summarized by Kent (1983a, 1984) that stutterers are likely to differ from nonstuttering controls, especially with respect to the second factor, precision of motor and perceptual timing. Caruso et al. (1988) reported that in their study, the 6 stutterers had abnormal sequencing of movement onsets and peak velocities in the face of apparently normal performance with respect to dynamic movement composition and intermovement motor equivalence. These researchers concluded that stutterers do not have a general dysfunction of movement coordination but rather a specific one.

The evidence that stutterers evince rather similar difficulties with speech and nonspeech movement sequences helps to identify an underlying coordination dysfunction. However, the puzzle is by no means solved until this dysfunction can be suitably related to a model of speech timing control to explain stuttering as a *disorder of speech*. Some researchers, such as Stromsta (1986), regard stuttering as primarily a problem of coarticulation. One difficulty with this conception is that coarticulation is an observed phenomenon relating to the sequencing of movements. As mentioned earlier, coarticulation is a general term that refers to simultaneous adjustments of the vocal tract for two or more sounds. For instance, during the production of the bilabial stop [b] in *bee*, the tongue is already in position for the following vowel [i]. Coarticulation is not in itself a theory or even a well-defined behavior. It is simply a term given to the observed overlapping of movements in speech.

An adequate theory of speech production will explain coarticulation as the consequence of some underlying movement regulation process (Kent & Minifie, 1977). This is not to dispute that stutterers may differ from nonstuttering controls in patterns of coarticulation but rather to argue that any such differences are reflections of differences in an underlying, as yet undisclosed, process. Coarticulation is not a unitary phenomenon. Some overlapping of movements is inevitable given that the biomechanical properties of the speaking apparatus do not permit infinite accelerations or truly independent structural movements. One pattern of coarticulation arises as the consequence of what Shaffer (1982) called *compound trajectories*, or the blending of two discrete movements into a single movement that accomplishes the goals of both (Kent, 1986a, 1986b). Another pattern of coarticulation arises as the components in a sequence are timed to yield a rapid but reliable succession of acoustically significant movements (Kent & Moll, 1975).

It appears from studies of both stuttered and normal speech that the first segment in a phonetic sequence carries special importance and special risk. As Bonnot, Chevrie-Muller, Arabia-Guidet, Maton, and Greiner (1986) characterize the matter,

> The initial position functions as a reference with a local reappraisal working at the syllabic and intrasegmental levels allowing the preservation of the essential characteristics of the phonemic units and of their correct seriation, and finally 'mechanical' adjustments occurring at the segmental borders. (p. 94)

Stutterers not only show difficulties beginning a sequence but also are most inclined to error on the first one to three components of a sequence. Stuttering on the final words of a sentence is so rare as to be implausible.

In some sense, the first position is an important factor in the regulation of the remainder of the sequence. One reason for its vulnerability in stuttering may be that it is at this point that the paralinguistic and segmental components are first reconciled. Paralinguistic features include emotion, loudness, and speaking rate. The segmental components are the speech sounds (phonetic elements). If these two components are not reconciled, fluent speech simply does not come about. If they are reconciled, then the tempo and rhythm of the remainder of the utterance have a satisfactory reference.

Stuttering may involve coordination difficulties at more than one level. One level is the relative timing of two or more articulations, such as those required at the larynx and vocal tract to produce the voiceless consonant in a word like *time*. This level involves the temporal regulation of individual movements. Another, higher, level is the coordination of prosodic and segmental information in which the articulatory movements are registered against a syllable pattern of a certain rate and rhythm. This level involves the integration of complementary frames of information. The interlocking of these two aspects of speech organization apparently is most uncertain at the beginning of an utterance, and it is at this initiation stage that dysfluencies are most likely.

The coordination of two reference frames, one prosodic and one segmental, may translate to the coordination of activity in the two cerebral hemispheres (Kent, 1983a; Webster, 1986a). Webster (1986a) concluded that one element of stuttering is a vulnerability of left-hemisphere processing to interference by concurrent right-hemisphere processing. If prosodic information is processed primarily in the right hemisphere and segmental information in the left, then coordination of the two types of information could be impaired by anything that interferes with the integration of their concurrent, and largely complementary, activity.

MOTOR SPEECH DISORDERS

In this chapter, motor speech disorders refer to the dysarthrias, apraxia of speech, and, possibly, some of the aphasias. These disorders have in common a neurologic origin, that is, damage or disease affecting the nervous system. Dysarthrias are speech disorders associated with the clinical signs of weakness, slowness, incoordination, or other muscular abnormalities. Apraxia of speech is a speech disorder that affects only movements for speech; the same musculature appears to be essentially normal for the performance of nonspeech motor acts. Aphasia is a language disorder, but recent evidence indicates that some individuals classified as aphasic have a speech disorder that partners the language impairment. The term *discoordination* or its equivalents frequently have been applied to the motor speech patterns in dysarthria, apraxia of speech, and some aphasias (particularly Broca's aphasia).

For many writers on the subject, coordination appears to represent some kind of spatiotemporal interlocking of movements such that one movement is reliably related to some other movement or movements in a sequence. Observationally, then, coordination can be defined as the correlation among movements. When coordination is impaired, as is suspected for some or all of the speech disorders considered in this chapter, then the correlation among movements is weakened. This weakening may take the form of a greater variability in the patterning of movements.

To take one specific example, voice onset time (VOT) is defined as the time between the release of a consonantal oral articulation and the onset of voicing for a following vowel. For the stop + vowel sequence in *two* (IPA [tu]), VOT is measured as the time between the release of the tongue tip constriction for [t] and the onset of vocal fold vibration for [u]. That is, a lingual movement is coordinated with the voicing adjustment at the larynx. In normal speakers, VOT values for repeated productions of a word like *two* typically fall in a narrow range between 40 and 80 ms. In contrast, the VOT values for the voiced cognate [d] as in *do* (IPA [du]) normally fall in a narrow range between 0 and 20 ms. That is, the voiced and voiceless stop consonants [d] and [t] in syllable-initial position can be distinguished on a linear scale representing the relative timing of oral and laryngeal events (Figure 13.3). However, for speakers with certain neurologic disorders, particularly apraxia of speech, the VOT values for several tokens of a syllable like *two* are widely distributed and may overlap the distribution of VOT values for tokens of a syllable like *do* (Blumstein, Cooper, Goodglass, Statlender, & Gottlieb, 1980; Freeman, Sands, & Harris, 1978; Hoit-Dalgaard, Murry, & Kopp, 1983; Itoh et al., 1982; Ziegler, 1987; Ziegler & von Cramon, 1986b).

The basic idea is that normal coordination is characterized by a small variability of movement intervals, whereas in abnormal coordination, the variability of these intervals is larger. In this way, coordination has come to be regarded as a continuous variable, an orthogonal dimension of speech motor control that varies from normal to abnormal as a function of the variance of interval measurements. According to this criterion, it has been established that subjects in the following groups are more poorly coordinated (more variable) than young adult normal speakers: children (Chermak & Schneiderman, 1986; Hawkins, 1984; Kent & Forner, 1980), elderly speakers (Sweeting & Baken, 1982), apraxic speakers (Ziegler, 1987), conduction aphasics (Kent & McNeil, 1987), and Broca's aphasics (Blumstein et al., 1980). Variability, then, appears to be one criterion by which the maturation and integrity of the speech motor control system can be assessed. (A more rigorous account of variability in speech movements would include factors such as those

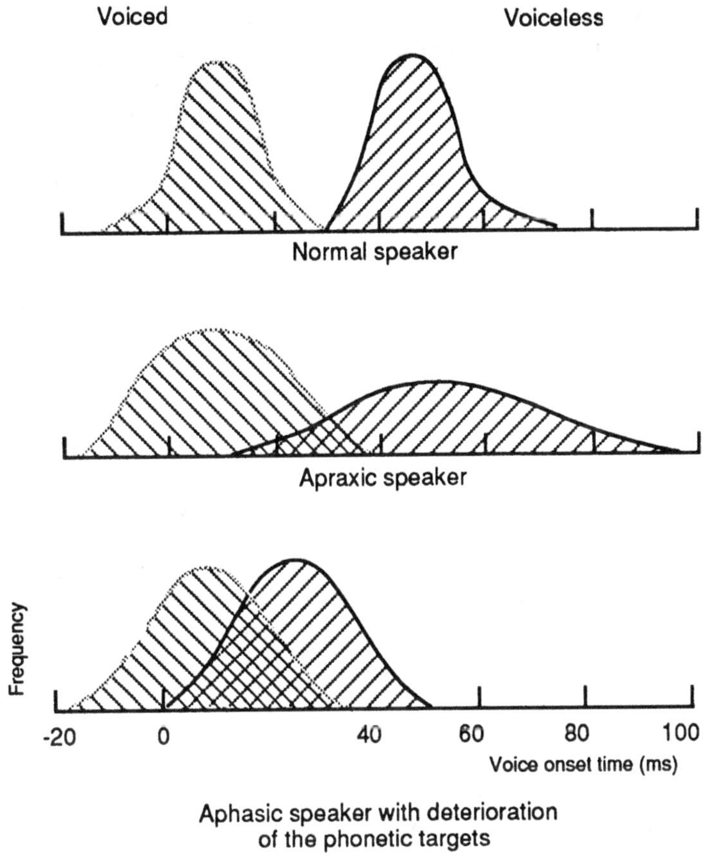

Figure 13.3. Hypothetical distributions of voice onset time (VOT) data for voiced and voiceless stops (e.g., [d] in *do* and [t] in *two*). The normal speaker maintains two nonoverlapping distributions with means at about 10 and 45 ms. The apraxic speaker has distributions with about the same means but with increased variance, so that overlapping occurs. The aphasic speaker has a reduced distance between the means, and the distributions overlap considerably.

described by Hatze, 1986, but an extended discussion is outside the scope of this chapter.)

This conception of coordination as temporal precision may not be sufficient, however. Some researchers have proposed that some patterns of abnormal coordination are qualitatively, not just quantitatively, different from normal coordination. For example, Kent (1983b) described a coordination pattern for the speech of children with athetoid cerebral palsy in which "everything moves at once." The distinction between this kind of abnormal coordination and the normal coordination of speech is illustrated in Figures 13.4 and 13.5. When a normal speaker (Figure 13.4) produces the alveolar stop [d] in word-initial position or following a nonnasal sound like a vowel, the velum elevates in advance of other articulatory movements to ensure velopharyngeal closure at the onset of the word. The closure of the velopharyngeal port allows air pressure containment for the plosive [d]. After velopharyngeal closure is achieved, movements of the tongue and jaw are executed for the initial stop [d]. In contrast, the cerebral-palsied child (Figure 13.5) makes essentially simultaneous movements of velar elevation, jaw elevation, and approach of the tongue tip toward the alveolar ridge. The normal speaker sequences the movements in time; the cerebral-palsied speaker bunches the movements as co-occurring gestures. Possibly, the latter pattern of coordination represents the child's compensation for a motor system that is unreliable owing to involuntary movements. If the child elevates the velum in advance of the oral consonantal articulation, he or she risks the premature loss of velopharyngeal closure before the consonant is formed (a very likely result, given that velopharyngeal closure for this child frequently was lost before the end of the word). But if the child makes essentially simultaneous movements of several articulators, they are appropriately positioned for at least a short interval.

Another illustration of the same phenomenon is shown in Figure 13.6. Articulatory movements have been traced from cinefluorographic films to depict the articulatory movements for the stop consonants in nonsense utterances formed of a vowel + stop + vowel sequence. The three stops differ in place of articulation: [b] is made by closure of the lips, [d] involves contact of the tongue tip against the alveolar ridge, and [g] is produced by elevation of the tongue dorsum against the roof of the mouth. Figure 13.6 shows that for each of these stops, the same pattern of coordination applies—essentially simultaneous movements of the jaw, velum, and the lip or tongue. Normal speakers reliably elevate the velum before production of the utterance-initial vowel, not with production of the consonant.

The abnormal coordination shown by the cerebral-palsied subject may be a pattern that copes with a disturbed and unreliable motor

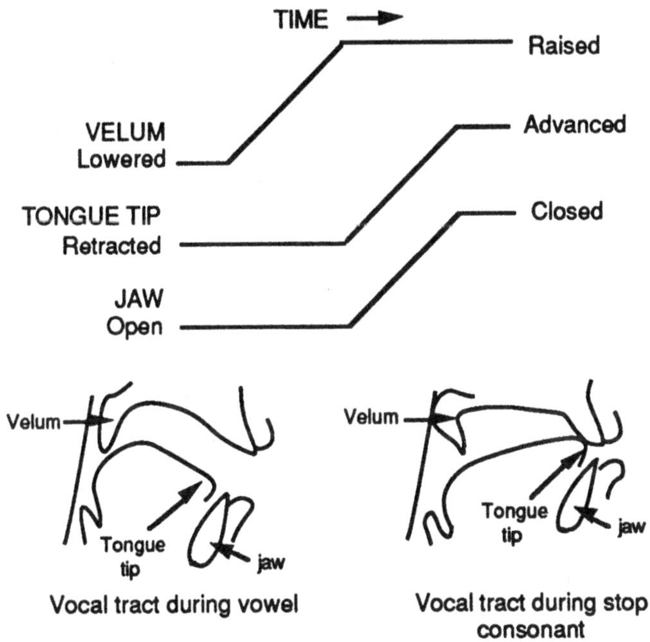

Figure 13.4. Articulatory coordination of a normal speaker's production of an alveolar stop [d] after a vowel element. The velum elevates well in advance of the tongue tip and jaw movements to insure velopharyngeal closure for the stop production.

system. The pattern perhaps should not be described as discoordinated, but rather alternatively coordinated in a way that contends with a disordered motor system. The simultaneity of movements in the athetoid child could also reflect the preservation of an immature movement pattern in which synchrony is a major feature. This kind of pattern is seen in many infantile behaviors such as walking. Thelen and Cooke (1987) noted that newborn stepping has highly synchronous flexions and extensions of the hip, knee, and ankle joint. They proposed that an essential process in the maturation of the motor patterns in walking is "the individuation of joint action from the obligatory synergy of the newborn period" (p. 392). The neurologic impairment in cerebral palsy may retard or prevent such individuation of movements.

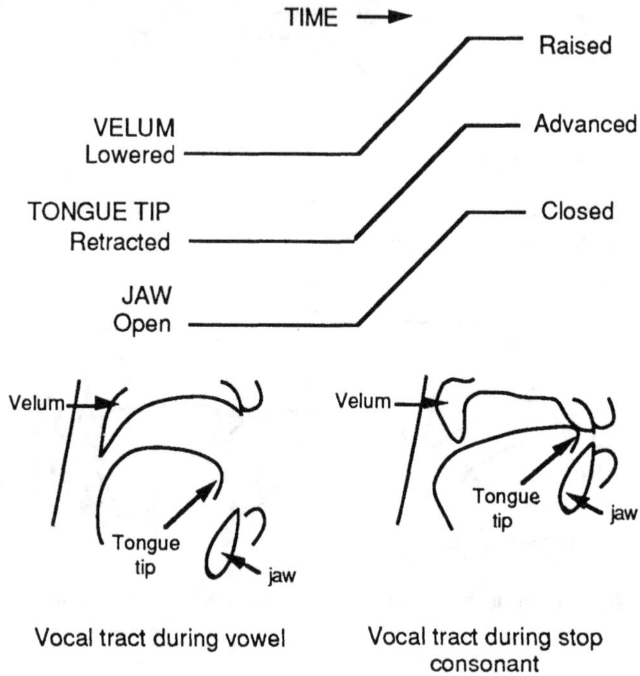

Figure 13.5. Articulatory coordination of a cerebral-palsied individual's production of an alveolar stop [d] after a vowel element. Notice the simultaneity of the movements of the velum, tongue tip, and jaw.

Disturbed patterns of coarticulation also have been presented as evidence for abnormalities in speech motor control. Ziegler and von Cramon's (1986b) recent study of 8 patients with apraxia of speech is a case in point. Data were collected on three movement patterns: (a) lingual-laryngeal phasing, measured as VOT for voiced versus voiceless stop consonants, (b) lingual-velar phasing, measured by the smoothness of the sound pressure level for utterances containing nasal consonants, and (c) lingual-labial phasing, measured by spectral evidence of anticipatory lip protrusion during prevocalic [t]. For each of these three patterns, the apraxic speakers showed evidence of abnormal sequencing of movements. Ziegler and von Cramon (1986b) interpreted the motor deficit as one of disturbed interarticulatory phasing and, in agreement with earlier studies, commented that this kind of motor deficit could lead to the impression of phonemic errors, such as substitutions of one phoneme for

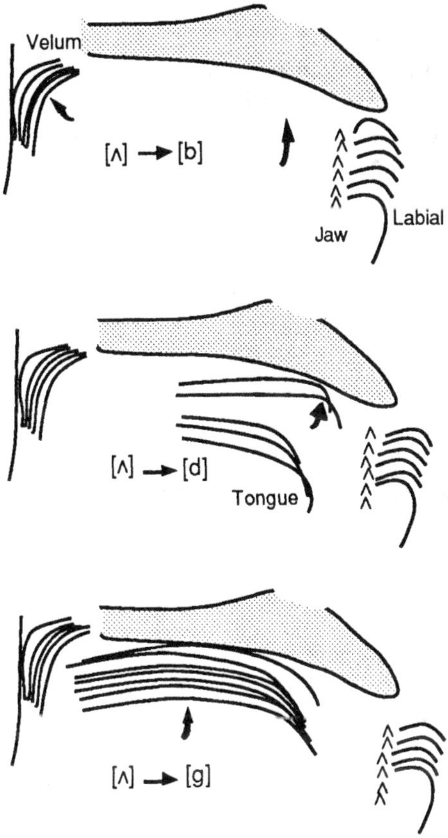

Figure 13.6. Superimposed tracings from a cinefluorographic film (24 frames/s) during the vowel-to-consonant movement for three stops differing in place of articulation. The speaker has athethoid cerebral palsy. The movement patterns show the synchrony of velar, jaw, and labial or tongue movements. Such synchrony is an abnormal coordinative pattern for these utterances. Normal speakers elevate the velum in advance of the consonantal movements.

another (cf. Itoh, Sasanuma, Hirose, Yoshioka, & Ushijima, 1980; Rosenbek et al., 1984).

Diminished organizational coherence in the sequencing of movements for speech conceivably could arise for several reasons. First, such a pattern could result from a deliberate (compensatory) or pathological slowing of speech rate (Wieneke et al., 1987). A slowing of the rate of speech generally reduces the overlap or coarticulation in the movement pattern. Second, diminished coherence could result from ineffectiveness in the routines of prearticulatory coding, that is, the processes that determine the articulatory requirements for the production of segments in particular contexts. Third, the deviant patterns observed by Ziegler and von Cramon (1986b) could be the consequence of a coordination dysfunction per se. The first alternative could be tested by observation of the effects on coarticulation as apraxic subjects produce speech materials at different speaking rates. If coarticulation is not affected by rate, then a simple rate explanation can be rejected. The second and third explanations are more difficult to test and in fact may be cooperative in the disorder of apraxia of speech (see evidence considered by Rosenbek et al., 1984). However, the third explanation is consistent with the frequently observed timing variability in apraxia of speech and is sufficient to account for many of the documented errors in this disorder.

It is unclear to what extent acquired dysarthrias in adults disrupt the temporal organization of movements in a phonetic sequence. The dysarthrias stand in contrast to apraxia of speech in that dysarthria results from a demonstrable muscle impairment for speech and nonspeech functions whereas apraxia of speech specifically impairs speech but not other functions of the same musculature. Some data indicate that the temporal organization of movement sequences in dysarthric individuals is not disrupted to the degree that individual movements are (Kent & Netsell, 1975; Ziegler & von Cramon, 1983; 1986c). In many, if not most, forms of dysarthria, movements tend to be reduced in magnitude or velocity, or both (Kent & Netsell, 1978; Kent, Netsell, & Abbs, 1979; Hirose, 1986; Hunker, Abbs, & Barlow, 1982; Weismer, Kimelman, & Gorman, 1985). What is not known is whether temporal abnormalities go beyond the aberrant kinematic properties of individual movements. If they do not, an important lesson is at hand: Coordination of movements in a pattern can be essentially normal in the face of impaired discrete movements.

SPEECH PRODUCED WITH DIMINISHED ORAL SENSATION

Most people who have been anesthetized for dental work have experienced the effects of reduced oral sensation on speech. In fact, the effects probably seem much more significant to the partially anesthetized speaker than to listeners. Most studies of speech produced with anesthetization of the oral tissues indicate that the effects on speech production are relatively minor (Borden, Harris, & Catena, 1973; Borden, Harris, & Oliver, 1973; Gammon, Smith, Daniloff, & Kim, 1971; Horii, House, Li, & Ringel, 1973; Scott & Ringel, 1971). The rather slight effects of reduced afference may be surprising given that the oral structures, especially the tongue tip and lips, are generously supplied with afferent units and are capable of some of the most refined somatosensory discriminations in the human body (Bosma, 1967, 1970). Perhaps in recognition of this potential, Dubner, Sessle, and Storey (1978) believed that "the precise motor function associated with speech requires sensory feedback from intraoral structures as well as from auditory cues" (p. 74). How is it, then, that oral anesthetization does not disrupt speech completely? In particular, why does the reduction of oral sensory information not result in dramatic disturbances of coordination?

First, it is open to question whether the attempts to disrupt oral sensory feedback have accomplished their purpose. As Abbs (1981) noted, elimination of afferent information from the speech production system is difficult. Several types of sensory information are available, and the use of the information may be highly adaptive, such that disruption of one type of afference is readily compensated by reliance on other types. Data from animals indicate that hypoglossal afferents could play a role in tongue proprioception (Lowe, 1982). This is not to say that investigators have not been diligent in their efforts to reduce oral sensation. For example, Kelso and Tuller (1983) studied the abilities of subjects to produce vowels while (a) their jaw positions were fixed at various degrees of opening with a bite block, (b) their temporomandibular joints were anesthetized bilaterally, (c) a topical anesthetic was applied to the subjects' oral mucosa, and (d) their speech output was masked by white noise. Despite these disruptions, the subjects compensated for the bite blocks and were able to produce vowels that were not acoustically different from vowels produced in a control condition. Either the subjects were able to use alternative sources of afference or they could dispense with sensory feedback for the vowel production task.

A second factor to be considered is that speech is a highly practiced skill, perhaps practiced to such a degree that reliance on moment-to-moment afference for the guidance of movements is unnecessary and inefficient. Temporary disruption of feedback therefore does not significantly affect performance. Long-term disruptions seem to be another story. As will be discussed later in this chapter, a hearing loss acquired after speech is developed usually results in a deterioration of speech articulation. Similarly, even though experimental attempts to disrupt trigeminal afference have relatively slight affects on speech, patients with trigeminal sensory neuropathy complain of difficulties with oral motor functions generally—chewing, swallowing, and speech (Lecky, Hughes, & Murray, 1987). It appears that speech motor control does not depend critically on continuous sensory feedback. However, with prolonged sensory disruption, speech loses its precision and ease.

The role of sensory feedback in mature speech may be primarily to ensure that expected movements and their acoustic consequences are produced, and, if not, to amend the regulatory processes that govern movement. This is the core of an adaptive model theory proposed by Neilson and Neilson (1987). This model proposes adaptive control based on feedback that establishes, verifies, and—when required—modifies the relationship between motor commands and their multimodal sensory consequences. The model relies fundamentally on an adaptable neuronal substrate of information based on sensory feedback. This substrate is used to transform a planned sensory trajectory into a corresponding set of motor commands. Experimental disruption of trigeminal afference may have relatively minor effects on speech because the internal standard of speech is essentially intact and motor commands related to that standard can be executed with satisfactory results. Those compensations that do occur seem to be general in scope, extending to muscles whose innervations are independent of the anesthetized nerves (Borden, Harris, & Catena, 1973).

Standing in apparent conflict with this conclusion are the reports that autogenic and nonautogenic responses in either the lips or jaw compensate for perturbations to speech movements of these structures (Shaiman, 1988). For example, if the jaw is perturbed during a closing movement, the lower lip movement increases (a nonautogenic response) to achieve the bilabial closure for a consonant like [p]. Such results indicate that afferent mechanisms may play a continuous or nearly continuous role in speech regulation.

Perhaps there are two complementary control processes at work. Le Magnen (1975, 1987) proposed that cephalic sensory systems are dually organized into neocortical and subcortical systems. Evidence for such a dual organization has been described for vision, audition, gustation, and olfaction (Le Magnen, 1987). The subcortical system is thought to participate in "unlearned prewired sensory processing and responses" (Le Magnen, 1987, p. 97). The neocortical system is involved in the learning and relearning of sensory processing and responses. In speech, the subcortical system may be involved in such functions as perioral reflexes. Although the role of these reflexes in speech is arguable, Smith, McFarland, Weber, and Moore (1987) proposed that the perioral reflex could serve to maintain an anterior oral seal during chewing. Similarly, during speech, perioral reflexes might ensure that when the lips are directed to achieve bilabial closure, they accomplish their objective despite unexpected perturbations.

The neocortical system for speech regulation could operate in the sense described in Neilson and Neilson's (1987) adaptive model theory. The monitoring and revisory capabilities of this system would be used primarily for major, continuing loads or perturbations to the speech articulators and possibly to provide feedforward information to be used by the subcortical system. Recent evidence shows that the perioral reflex is task-dependent, participating in some movements but not others (Abbs & Gracco, 1984; Shaiman, 1988). This conception is fundamentally compatible with the three-level system of motor regulation proposed by Feldman (1986). This simplified model has at its lowest level the loads or forces that counteract muscle activity, at its intermediate level the stretch reflex (servosystem) and comparable mechanisms, and at its highest level the neuronal (descending) influence on motoneurons.

SPEECH OF THE HEARING IMPAIRED

A variety of types of damage to the auditory system can result in hearing impairments ranging from mild impairment to deafness. Most studies of the speech of hearing-impaired individuals have focused on those persons with a severe to profound hearing loss. This loss prevents the auditory reception of most speech sounds produced below about 75-90 dB hearing level (normal conversational speech is produced at about 40-60 dB hearing level). Generally, individuals with this loss have had the impairment from birth or early infancy. The damage is sensorineural, affecting the cochlea or peripheral auditory neural pathways. In most persons with severe to profound hearing loss since birth or infancy, speech production deficits are dramatic. Unless otherwise indicated, the discussion in this chapter is restricted to this group.

As mentioned earlier, we believe that speech coordination is constrained by the auditory perception of the acoustic-phonetic structure of a speaker's language, and the speech of the hearing impaired therefore provides an opportunity to examine the long-term effects of reduced or distorted auditory perception on the development of coordinated speech. For the hearing-impaired child, there is a disruption of the normal course of development in which the child acquires experience with speech sounds and, in so doing, establishes mappings between vocal tract configurations and movements with acoustic events. Deprivation of this auditory experience produces speech patterns that have been described as deviant in both articulatory posturing and articulatory timing.

Articulatory posture refers to the overall configuration of the speech production system over time. Postural characteristics are not easily ascribed to either segmental (phonetic) or suprasegmental (prosodic effects such as rate, stress pattern, or intonation) aspects of speech because these characteristics span entire utterances and have both local and far-ranging consequences. Examples of postural errors in the speech of the hearing impaired include deviant voice quality and pitch, abnormal nasalization, and distorted vowels. These speech errors are believed to be the result of an abnormal or ineffective posturing of the vocal folds, velopharynx, and tongue body, respectively (Stevens, Nickerson, & Rollins, 1983).

Osberger and McGarr (1982) also suggested that the tendency of hearing-impaired speakers to initiate speech at abnormally low lung volumes and speak over restricted lung volume ranges is evidence of postural errors in control of the respiratory system. They proposed a parallel between the postural aspects of speech breathing and the postural aspects of nonspeech activities such as locomotion. Just as locomotion can be described as a series of rapid, rhythmic, and highly coordinated movements superimposed on a broad postural base, speech can be considered as a complex and rapidly changing articulatory-phonatory process overlaid on a more slowly changing respiratory substrate.

It might be concluded from the examples of postural abnormalities just cited that posture is associated with a relatively stable or continuous and relatively slowly evolving position of one body part relative to a more rapidly changing and discrete movement in another body part. Given this interpretation, it seems possible that the postural role of a particular structure may change from movement to movement and that these postural aspects must be continuously coordinated in the overall production of a motor act. Gahéry and Massion (1981) described a central motor command for action that triggers two interacting subprograms, one for the voluntary movement itself and the other for the associated postural adjustments. In this view, posture can be considered to be intricately involved in a complex network or nesting of coordinated movements that include coordination of the postural movement itself; coordination of the postural with the nonpostural, discrete aspects of the movement; and coordination of the discrete, voluntary movement itself. It can be imagined that the failure of hearing-impaired speakers to adopt an appropriate posture in one subsystem of speech could lead to difficulties in the coordination of movements in other subsystems. A challenge to future research in this area is to determine the degree to which the speech posturing errors of the hearing impaired influence, or directly result in, abnormal patterns of coordination of speech.

It is also possible that the movements and positions for some speech sounds provide the postural base for others. Öhman (1966) modeled the articulatory movements in vowel + consonant + vowel sequences as a diphthongal vowel-to-vowel substrate on which the consonant articulation was superimposed as a perturbation. This model gives to the vowel articulation a dual aspect: first, a discrete set of movements related to the vowel element itself, and second, a postural configuration on which consonant movements are predicated.

The term *posture* may be misleading if one looks for close analogies with the skeletal musculature. The primary distinction is one of the relative independence of the vowel and consonant movements. Although vowel articulation is influenced to some degree by consonant articulation, the reverse influence, vowel upon consonant, usually is much more striking. It is perhaps appropriate to think of the coordination of vocal tract muscles in speech as governed in large measure by the coproduction of vowel and consonant. This coordinative linkage is one factor in explaining the primacy of the consonant + vowel syllable in children's speech development (Kent & Bauer, 1985; Lindblom & Zetterstrom, 1986). The linkage appears to be matched on the sensory side by a particular sensitivity to the acoustic structure of consonant + vowel sequences (Kent, 1987). Significantly, hearing-impaired infants lag their normal-hearing peers in the age of acquisition of babbling formed from consonant-vowel sequences (Kent, Osberger, Netsell, & Hustedde, 1987; Oller, 1986).

The timing errors in the speech of the hearing impaired cross several levels including the timing of suprasegmental, segmental, and microsegmental (i.e., microphonetic) units. Examples of suprasegmental and segmental timing errors include excessively prolonged segments, frequent and lengthy pauses, and abnormalities in the relative timing of stressed versus unstressed syllables (Osberger & Levitt, 1979). These timing errors can be interpreted as reflecting a difficulty in the coordination of the overall temporal pattern or rhythm of speech. This conclusion is significant in its implication that in an intact speech system, musculature does not in itself generate suitable rhythms to support speech. Auditory information either facilitates, or is essential to, the establishment of the rhythm of speech.

At the microsegmental level, interarticulatory timing errors are observed. Many of these take the form of what was referred to earlier as disruptions in the spatiotemporal interlocking of movements. These interarticulator timing errors have been reported for the coordination of laryngeal and tongue or lip movements (Mashie & Conture, 1983; McGarr & Lofquist, 1982), lip and tongue movements (McGarr & Harris, 1983), and respiratory and laryngeal movements (Whitehead, 1983). They take the form of both abnormal and variable timing in the coordinated onset of two movements.

Another possible explanation of some of the abnormalities in the speech of the hearing impaired is an abnormal regulation of phasic and tonic inputs to the speech muscles. Ostry, Feltham, and Munhall (1984) concluded that speech development in children can be described in part by the somewhat independent control of phasic and tonic activity. They interpreted their data to mean that tonic activity (related to stiffness regulation in the muscles and, at the phonological level, to the production of stress contrasts) was not complete until about 5-6 years of age. Hearing-impaired children may fail to develop proper coordination of the phasic and tonic elements, and this failure may result in abnormal timing patterns (including stress relationships) in their speech. The speech difficulties of the hearing impaired would be compounded if the phasic control also were abnormal, as appears to be the case.

Osberger and McGarr (1982) suggested that reduced experience with speech auditory cues leads to the development of an abnormal linguistic-phonological system. Support for this notion comes from the numerous studies showing that the hearing impaired have deficits across a range of linguistic levels and operations—for example, lexical, morphological, syntactic, and semantic levels, and reading (Moeller, Osberger, & Eccarius, 1986; Osberger, Moeller, Eccarius, Robbins, & Johnson, 1986). Mashie and Conture (1983) further proposed that the interarticulatory coordination difficulties associated with deaf speakers' laryngeal regulation may be attributed to both an aberrant linguistic system and inadequate speech motor control skills.

To understand the processes underlying coordination in speech, it seems important to determine the degree to which abnormalities in the linguistic-phonological system may contribute to the abnormal coordination observed in the speech of the hearing impaired. One reasonable approach is to examine the speech of the adventitiously deaf, or individuals who have lost their hearing subsequent to the development of a normal linguistic-phonological system. Zimmerman and Rettaliata (1981) reported on a kinematic analysis of the speech of a 34-year-old man who had a history of progressive hearing loss beginning at age 12. He exhibited the following abnormalities: a relatively immobile tongue posture, abnormalities in the timing of tongue dorsum and laryngeal movements relative to movements of other articulators, and longer movement times and utterance durations. These results for the adventitiously deaf individual are in essential agreement with many of the postural and timing errors that have been described for speakers who never had normal hearing. It appears, then, that a significant portion of the coordination abnormalities in the speech of hearing-impaired persons is the result of a disruption of speech motor regulation rather than the expression of a linguistic deficit through an essentially normal speech motor system.

It has been stated that the speech of adventitiously deafened individuals deteriorates gradually. One might infer, then, that the maintenance of coordinated speech does not require auditory information on a movement-to-movement basis. Instead, auditory information is seen as important for the long-term monitoring and adjustment of coordinated speech (Zimmerman & Rettaliata, 1981). In an attempt to determine the short-term participation of auditory feedback in speech, researchers have conducted a number of studies with normal-hearing subjects who are temporarily deprived of auditory information through the use of intense masking noise. In one study, Forrest, Abbas, and Zimmerman (1986) obtained detailed kinematic data on speech during masking and reported a variety of spatiotemporal changes relative to the no-noise control condition. These changes included alterations in vocal tract shape, interarticulator timing, and segment durations. Some of these changes, such as interarticulator timing, are similar to the patterns that have been reported for hearing-impaired speakers. Overall, however, the changes observed across the normal speakers contending with noise masking were inconsistent and appeared to reflect individual differences in the strategies adopted to compensate for loss of auditory feedback. In addition, these researchers reported that the kinematic changes did not appear to effect the overall quality of the speech produced. These results indicate that auditory information does play a role in preserving some of the movement-to-movement precision and coordination of speech but that a longer term deprivation of auditory information is required to produce severe problems with coordination of speech movement.

CONCLUSION

Particularly with the application of acoustic and physiologic research methods to the study of speech, a better understanding of the coordinative properties of this complex motor skill is rapidly developing. Certainly, much remains to be learned. Because speech is a uniquely human behavior and because neurophysiologic experimentation on humans is severely limited by safety considerations, the study of speech disorders plays a critical role in the continuing effort to understand speech as a motor behavior and to describe the operations of the nervous system underlying speech production. Although speech is unique in some respects, the sensorimotor processes that govern speech production may bear important similarities to those in other motor systems. The commonalities are important, but so are the differences, and it is through a consideration of both that speech motor control will be understood in the broader field of sensorimotor investigation.

Although speech is remarkable for its compensatory capabilities (even some aglossic individuals are intelligible), it is susceptible to a number of disorders. Many of these purportedly involve a disturbance of coordination. This review has considered some of the major types of evidence used to identify coordinative breakdowns. The sources of difficulty are numerous, as might be expected from a motor behavior that is (a) intimately linked with the human capability for symbolic expression, (b) ultimately directed toward generation of an acoustic signal that can be interpreted linguistically, and (c) accomplished with a complex system with about 100 muscles, with potentially the finest somatosensory discrimination in the body, and with a design not principally for speech but for respiratory and digestive functions.

Lessons about the coordination of speech movements can be learned both from successful compensations for a disorder process and from disorder processes in which the compensation is incomplete or apparently nonexistent. Over the short term, speech is remarkably resistant to disruptions of sensory feedback. However, continued sensory disruption leads to a gradual deterioration in refined spatiotemporal properties. The interlocking of speech movements, precisely maintained in the normal speaker, is weakened by certain types of disease or damage. Full understanding of these breakdowns probably will require recognition of several levels of coordination or integration, including a segmental-suprasegmental (phonetic-prosodic) convergence in the final motor act, an auditory-motor linkage, a somatosensory-motor connection, and coordinative patterns among individual muscles or muscle systems.

In addition, a significant part of the puzzle of speech coordination may be solved by further research on two dualistic hypotheses of speech motor control. One of these, described by Öhman (1966) and subsequently presented in modified form by others, is the idea that speech has a quasi-postural component, consisting of the vowel-to-vowel substrate, and a more discrete component, consisting of the superimposed consonantal movements. The other dualistic hypothesis is the phasic-tonic distinction in muscle regulation. Ostry et al. (1984) reported that children develop the phasic input to the speech muscles earlier than they do the tonic input. Thus, the two inputs are relatively independent in development. They may be similarly independent in their vulnerability to disruption. Moreover, the timing of the phasic and tonic inputs may be modified by disease or damage. If either of these is true, then a dual regulation model may explain some aspects of disordered coordination.

REFERENCES

Abbs, J. H. (1981). Neuromotor mechanisms of speech production. In J. K. Darby, Jr. (Ed.), *Speech evaluation in medicine* (pp. 181-198). New York: Grune and Stratton.

Abbs, J. H., & Gracco, V. L. (1984). Motor control of multi-movement behaviors: Orofacial muscle responses to load perturbations of the lips during speech. *Journal of Neurophysiology, 51*, 705-723.

Adams, M. R. (1981). The speech production abilities of stutterers: Recent, ongoing, and future research. *Journal of Fluency Disorders, 6*, 311-326.

Blumstein, S. E., Cooper, W. E., Goodglass, H., Statlender, S., & Gottlieb, J. (1980). Production deficits in aphasia: A voice-onset time analysis. *Brain and Language, 9*, 153-170.

Bonnot, J. F. P., Chevrie-Muller, C., Arabia-Guidet, C., Maton, B., & Greiner, G. F. (1986). Coarticulation and motor encoding of labiality and nasality in CVCVCV nonsense words. *Speech Communication, 5*, 83-95.

Borden, G. J. (1983). Initiation versus execution time during manual and oral counting by stutterers. *Journal of Speech and Hearing Research, 26*, 389-396.

Borden, G. J., Harris, K. S., & Catena, L. (1973). Oral feedback: II. An electromyographic study of speech under nerve-block anesthesia. *Journal of Phonetics, 1*, 297-308.

Borden, G. J., Harris, K. S., & Oliver, W. (1973). Oral feedback: I. Variability of the effect of nerve-block anesthesia upon speech. *Journal of Phonetics, 1*, 289-295.

Bosma, J. F. (Ed.). (1967). *Symposium on oral sensation and perception*. Springfield, IL: Thomas.

Bosma, J. F. (Ed.). (1970). *Second symposium on oral sensation and perception*. Springfield, IL: Thomas.

Caruso, A. J., Abbs, J. H., & Gracco, V. L. (1988). Kinematic analysis of multiple movement coordination during speech in stutterers. *Brain, 111*, 439-455.

Chermak, G., & Schneiderman, C. (1986). Speech timing variability of children and adults. *Journal of Phonetics, 13*, 477-480.

Dubner, R., Sessle, B. J., & Storey, A. T. (1978). *The neural basis of oral and facial function*. New York: Plenum.

Feldman, A. G. (1986). Once more on the equilibrium-point hypothesis (l model) for motor control. *Journal of Motor Behavior, 18*, 17-54.

Freeman, F. J., Sands, E. S., & Harris, K. S. (1978). Temporal coordination of phonation and articulation in a case of verbal apraxia: A voice onset time study. *Brain and Language, 6*, 106-111.

Forrest, K., Abbas, P. J., & Zimmerman, G. N. (1986). Effects of white noise masking and low pass filtering on speech kinematics. *Journal of Speech and Hearing Research, 29*, 549-562.

Gahéry, Y., & Massion, J. (1981). Coordination between posture and movement. *Trends in Neurosciences, 4*, 199-202.

Gammon, A., Smith, P. J., Daniloff, R. G., & Kim, C. W. (1971). Articulation and stress/juncture production under oral anesthetization and masking. *Journal of Speech and Hearing Research, 14*, 271-282.

Hatze, H. (1986). Motion variability—its definition, quantification, and origin. *Journal of Motor Behavior, 18*, 5-16.

Hawkins, S. (1984). On the development of motor control of speech: Evidence from studies of temporal coordination. In N. J. Lass (Ed.), *Speech and language: Advances in basic research and practice* (Vol. 10, pp. 317-374). New York: Academic Press.

Hirose, H. (1986). Pathophysiology of motor speech disorders (dysarthria). *Folia Phoniatrica, 38*, 61-88.

Hoit-Dalgaard, J., Murry, T., & Kopp, H. G. (1983). Voice onset time production and perception in apraxic subjects. *Brain and Language, 20*, 329-339.

Horii, Y., House, A. S., Li, K.-P., & Ringel, R. L. (1973). Acoustic characteristics of speech produced without oral sensation. *Journal of Speech and Hearing Research, 16*, 67-77.

Hunker, C. J., Abbs, J. H., & Barlow, S. M. (1982). The relationship between parkinsonian rigidity and hypokinesia in the orofacial system: A quantitative analysis. *Neurology, 32*, 749-754.

Itoh, M., Sasanuma, S., Hirose, H., Yoshioka, H., & Ushijima, T. (1980). Abnormal articulatory dynamics in a patient with apraxia of speech: X-ray microbeam observation. *Brain and Language, 11*, 66-75.

Itoh, M., Sasanuma, S., Tatsumi, I. F., Murakami, S., Fukusako, Y., & Suzuki, T. (1982). Voice onset time characteristics in apraxia of speech. *Brain and Language, 17*, 193-210.

Katz, W. F. (1987). Anticipatory labial and lingual coarticulation in aphasia. In J. H. Ryalls (Ed.), *Phonetic approaches to speech production in aphasia and related disorders* (pp. 221-242). San Diego, CA: College-Hill.

Keele, S., Ivry, R., & Porkorny, R. (1987). Force control and its relation to timing. *Journal of Motor Behavior, 19*, 96-114.

Kelso, J. A. S., & Tuller, B. (1983). "Compensatory articulation" under conditions of reduced afferent information: A dynamic formulation. *Journal of Speech and Hearing Research, 26*, 217-224.

Kent, R. D. (1983a). Facts about stuttering: Neuropsychologic perspectives. *Journal of Speech and Hearing Disorders, 48*, 249-255.

Kent, R. D. (1983b). The segmental organization of speech. In P. F. MacNeilage (Ed.), *The production of speech* (pp. 57-89). New York: Springer.

Kent, R. D. (1984). Stuttering as a deficiency in temporal programming. In W. H. Perkins & R. Curlee (Eds.), *Nature and treatment of stuttering: New directions* (pp. 283-301). San Diego, CA: College Hill.

Kent, R. D. (1986a). The iceberg hypothesis: The temporal assembly of speech movements. In J. S. Perkell & D. H. Klatt (Eds.), *Invariance and variability in speech processes* (pp. 234-242). Hillsdale, NJ: Erlbaum.

Kent, R. D. (1986b). Is a paradigm change needed? *Journal of Phonetics, 14*, 111-115.

Kent, R. D. (1987). The power of babble [Review of *Precursors of early speech*]. *Contemporary Psychology, 32*, 712-713.

Kent, R. D., & Bauer, H. R. (1985). Vocalizations of one-year-olds. *Journal of Child Language, 12*, 491-526.

Kent, R. D., & Forner, L. L. (1980). Speech segment durations in sentence recitations by children and adults. *Journal of Phonetics, 8*, 157-168.

Kent, R. D., & McNeil, M. R. (1987). Relative timing of sentence repetition in apraxia of speech and conduction aphasia. In J. H. Ryalls (Ed.), *Phonetic approaches to speech production in aphasia and related disorders* (pp. 181-220). San Diego, CA: College-Hill.

Kent, R. D., & Minifie, F. D. (1977). Coarticulation in recent speech production models. *Journal of Phonetics*, *5*, 115-133.

Kent, R. D., & Moll, K. L. (1975). Articulatory timing in selected consonant sequences. *Brain and Language*, *2*, 304-323.

Kent, R. D., & Netsell, R. (1975). A case study of an ataxic dysarthric: Cineradiographic and spectrographic observations. *Journal of Speech and Hearing Disorders*, *40*, 115-134.

Kent, R. D., & Netsell, R. (1978). Articulatory abnormalities in athetoid cerebral palsy. *Journal of Speech and Hearing Disorders*, *43*, 353-373.

Kent, R. D., Netsell, R., & Abbs, J. H. (1979). Acoustic characteristics of dysarthria associated with cerebellar lesion. *Journal of Speech and Hearing Research*, *22*, 627-648.

Kent, R. D., Osberger, M. J., Netsell, R., & Hustedde, C. G. (1987). Phonetic development in identical twins differing in auditory function. *Journal of Speech and Hearing Disorders*, *52*, 64-75.

Lackner, J. R., & Levine, K. B. (1975). Speech production: Evidence for syntactically and phonologically determined units. *Perception and Psychophysics*, *17*, 107-113.

Lecky, B. R. F., Hughes, R. A. C., & Murray, N. M. F. (1987). Trigeminal sensory neuropathy: A study of 22 cases. *Brain*, *110*, 1463-1485.

Le Magnen, J. (1975). Olfactory endocrine interactions and regulatory behavior. In D. A. Denton & J. P. Coghlan (Eds.), *Olfaction and taste* (Vol. 5, pp. 381-388). New York: Academic Press.

Le Magnen, J. (1987). Central processing of sensory information in the control of feeding. In H. Autrum, D. Ottoson, E. R. Perl, R. F. Schmidt, H. Shimazu, & W. D. Willis (Eds.), *Progress in sensory physiology* (Vol. 8, pp. 95-128). Berlin: Springer.

Lindblom, B., & Zetterstrom, R. (Eds.). (1986). *Precursors of early speech*. New York: Stockton Press.

Lowe, A. A. (1982). The neural regulation of tongue movements. *Progress in Neurobiology, 15*, 295-344.

Mashie, J. J., & Conture, E. G. (1983). Deaf speakers' laryngeal behavior. *Journal of Speech and Hearing Research, 26*, 550-559.

McGarr, N., & Harris, K. (1983). Articulatory control in a deaf speaker. In I. Hochberg, H. Levitt, & M. Osberger (Eds.), *Speech of the hearing-impaired: Research, training and personnel preparation* (pp. 75-96). Baltimore, MD: University Park Press.

McGarr, N. S., & Lofquist, A. (1982). Obstruent production by hearing-impaired speakers: Interarticulatory timing and acoustics. *Journal of the Acoustical Society of America, 72*, 34-42.

McMillan, M. O., & Pindzola, R. H. (1986). Temporal disruptions in the "accurate" speech of articulatory defective speakers and stutterers. *Journal of Motor Behavior, 18*, 279-286.

Moeller, M. P., Osberger, M. J., & Eccarius, M. (1986). Receptive language skills. In M. J. Osberger (Ed.), *Language and learning skills of hearing-impaired students* (ASHA Monographs No. 23, pp. 41-53). Rockville, MD: American Speech-Language-Hearing Association.

Neilson, M. D., & Neilson, P. D. (1987). Speech motor control and stuttering: A computational model of adaptive sensory-motor processing. *Speech Communication, 6*, 325-333.

Öhman, S. E. G. (1966). Coarticulation of VCV utterances: Spectrographic measurements. *Journal of the Acoustical Society of America, 39*, 151-168.

Oller, D. K. (1986). Metaphonology and infant vocalizations. In B. Lindblom & R. Zetterstrom (Eds.), *Precursors of early speech* (pp. 21-35). New York: Stockton Press.

Osberger, M. J., & Levitt, H. (1979). The effect of timing errors on the intelligibility of deaf children's speech. *Journal of the Acoustical Society of America, 66*, 1316-1324.

Osberger, M. J., & McGarr, N. S. (1982). Speech production characteristics of the hearing impaired. In N. J. Lass (Ed.), *Speech and language: Advances in basic research and practice* (pp. 221-283). New York: Academic Press.

Osberger, M. J., Moeller, M. P., Eccarius, M., Robbins, A. M., & Johnson, D. (1986). Expressive language skills. In M. J. Osberger (Ed.), *Language and learning skills of hearing-impaired students* (ASHA Monographs No. 23, pp. 54-65). Rockville, MD: American Speech-Language-Hearing Association.

Ostry, D., Feltham, R., & Munhall, K. (1984). Characteristics of speech motor development in children. *Developmental Psychology, 20*, 859-871.

Pindzola, R. H. (1987). Durational characteristics of the fluent speech of stutterers and nonstutterers. *Folia Phoniatrica, 39*, 90-97.

Rosenbek, J. C., Kent, R. D., & LaPointe, L. L. (1984). Apraxia of speech: An overview and some perspectives. In J. C. Rosenbek, M. R. McNeil, & A. Aronson (Eds.), *Apraxia of speech: Physiology-acoustics-linguistics-management* (pp. 1-72). San Diego, CA: College-Hill.

Scott, C. M., & Ringel, R. L. (1971). Articulation without oral sensory control. *Journal of Speech and Hearing Research, 14*, 804-818.

Shaffer, L. H. (1982). Rhythm and timing in skill. *Psychological Review, 89*, 109-122.

Shaiman, S. (1988). *Task-dependent sensorimotor mechanisms during speech.* Doctoral dissertation, University of Wisconsin-Madison.

Smith, A., McFarland, D. H., Weber, C. M., & Moore, C. A. (1987). Spatial organization of human perioral reflexes. *Experimental Neurology, 98*, 233-248.

Starkweather, C. W., & Myers, M. (1979). Duration of subsegments within the intervocalic interval in stutterers and nonstutterers. *Journal of Fluency Disorders, 4*, 205-214.

Stetson, R. H. (1951). *Motor phonetics* (2nd ed.). Amsterdam: North-Holland.

Stevens, K. N., Nickerson, R. S., & Rollins, A. M. (1983). Suprasegmental and postural aspects of speech production and their effect on articulatory skills and intelligibility. In I. Hochberg, H. Levitt, & M. Osberger (Eds.), *Speech of the hearing impaired: Research, training and personnel preparation* (pp. 35-51). Baltimore, MD: University Park Press.

Stromsta, C. (1986). *Elements of stuttering*. Oshtemo, MI: Atsmorts.

Sweeting, P. M., & Baken, R. J. (1982). Voice onset time in a normal-aged population. *Journal of Speech and Hearing Research, 25*, 129-134.

Thelen, E., & Cooke, D. W. (1987). Relationships between newborn stepping and later walking: A new interpretation. *Developmental Medicine and Child Neurology, 29*, 380-393.

Tuller, B., & Story, R. (1986). Co-articulation in aphasic speech. *Journal of the Acoustical Society of America, 80*, (Suppl. 1), Abstract No. MM17.

Webster, W. G. (1986a). Neuropsychological models of stuttering: II. Interhemispheric interference. *Neuropsychologia, 24*, 737-741.

Webster, W. G. (1986b). Response sequence organization and reproduction by stutterers. *Neuropsychologia, 24*, 813-821.

Weismer, G., Kimelman, M. D. Z., & Gorman, S. (1985). More on speech production deficit associated with Parkinson's disease. *Journal of the Acoustical Society of America, 78*, S55.

Whitehead, R. (1983). Some respiratory and aerodynamic patterns in the speech of the hearing impaired. In I. Hochberg, H. Levitt, & M. Osberger (Eds.), *Speech of the hearing impaired: Research, training and personnel preparation* (pp. 97-116). Baltimore, MD: University Park Press.

Wieneke, G., Janssen, P., & Belderbos, H. (1987). The influence of speaking rate on the duration of jaw movements. *Journal of Phonetics, 15*, 111-126.

Ziegler, W. (1987). Phonetic realization of phonological contrast in aphasic patients. In J. H. Ryalls (Ed.), *Phonetic approaches to speech production in aphasia and related disorders* (pp. 163-179). San Diego, CA: College-Hill.

Ziegler, W., & von Cramon, D. (1983). Vowel distortion in traumatic dysarthria: A formant study. *Phonetica, 40*, 63-78.

Ziegler, W., & von Cramon, D. (1985). Anticipatory coarticulation in a patient with apraxia of speech. *Brain and Language, 26*, 117-130.

Ziegler, W., & von Cramon, D. (1986a). Disturbed coarticulation in apraxia of speech: Acoustic evidence. *Brain and Language, 29*, 34-47.

Ziegler, W., & von Cramon, D. (1986b). Timing deficits in apraxia of speech. *European Archives of Psychiatric and Neurological Science, 236*, 44-49.

Ziegler, W., & von Cramon, D. (1986c). Spastic dysarthria after acquired brain injury: An acoustic study. *British Journal of Disorders of Communication, 21*, 173-187.

Zimmerman, G. N. (1980). Stuttering: A disorder of movement. *Journal of Speech and Hearing Research, 23*, 122-136.

Zimmerman, G. N., & Rettaliata, P. (1981). Articulatory patterns of an adventitiously deaf speaker: Implications for the role of auditory information in speech production. *Journal of Speech and Hearing Research, 24*, 169-178.

ACKNOWLEDGMENT

The epigraph by H. H. Pattee was quoted from page 176 of *Boston Studies in the Philosophy of Science: Vol. 27, Topics in the Philosophy of Biology*, edited by M. Grene and E. Mendelsohn, 1976, Boston: Reidel. Copyright 1976 by D. Reidel Publishing Co. Reprinted by permission.

INDEX